Mathematical Models for Speech Technology

Mathematical Models for Speech Technology

Stephen E. Levinson
University of Illinois at Urbana-Champaign, USA

John Wiley & Sons, Ltd

Other Wiley Editorial Offices

John Wiley & Sons Inc., 111 River Street, Hoboken, NJ 07030, USA

Jossey-Bass, 989 Market Street, San Francisco, CA 94103-1741, USA

Wiley-VCH Verlag GmbH, Boschstr. 12, D-69469 Weinheim, Germany

John Wiley & Sons Australia Ltd, 33 Park Road, Milton, Queensland 4064, Australia

John Wiley & Sons (Asia) Pte Ltd, 2 Clementi Loop #02-01, Jin Xing Distripark, Singapore 129809

John Wiley & Sons Canada Ltd, 22 Worcester Road, Etobicoke, Ontario, Canada M9W 1L1

Wiley also publishes its books in a variety of electronic formats. Some content that appears in print may not be available in electronic books.

Library of Congress Cataloging-in-Publication Data

Levinson, Stephen C.
Mathematical models for speech technology / Stephen Levinson.
p. cm.
Includes bibliographical references and index.
ISBN 0-470-84407-8 (cloth)
1. Speech processing systems. 2. Computational linguistics. 3. Applied linguistics–Mathematics. 4. Stochastic processes. 5. Knowledge, Theory of. I. Title.
TK7882.S65L48 2005
006.4′54′015118–dc22

2004026215

British Library Cataloguing in Publication Data

A catalogue record for this book is available from the British Library

ISBN 0-470-84407-8

Typeset in 10/12 Times by Laserwords Private Limited, Chennai, India
Printed and bound in Great Britain by Antony Rowe Ltd, Chippenham, Wiltshire
This book is printed on acid-free paper responsibly manufactured from sustainable forestry in which at least two trees are planted for each one used for paper production.

To my parents
Doris R. Levinson
and
Benjamin A. Levinson

Contents

Preface **xi**

1 Introduction **1**

1.1 Milestones in the history of speech technology 1
1.2 Prospects for the future 3
1.3 Technical synopsis 4

2 Preliminaries **9**

2.1 The physics of speech production 9
2.1.1 The human vocal apparatus 9
2.1.2 Boundary conditions 14
2.1.3 Non-stationarity 16
2.1.4 Fluid dynamical effects 16
2.2 The source–filter model 17
2.3 Information-bearing features of the speech signal 17
2.3.1 Fourier methods 19
2.3.2 Linear prediction and the Webster equation 21
2.4 Time–frequency representations 23
2.5 Classification of acoustic patterns in speech 27
2.5.1 Statistical decision theory 28
2.5.2 Estimation of class-conditional probability density functions 30
2.5.3 Information-preserving transformations 39
2.5.4 Unsupervised density estimation – quantization 42
2.5.5 A note on connectionism 43
2.6 Temporal invariance and stationarity 44
2.6.1 A variational problem 45
2.6.2 A solution by dynamic programming 47
2.7 Taxonomy of linguistic structure 51
2.7.1 Acoustic phonetics, phonology, and phonotactics 52
2.7.2 Morphology and lexical structure 55
2.7.3 Prosody, syntax, and semantics 55
2.7.4 Pragmatics and dialog 56

3 Mathematical models of linguistic structure **57**

3.1 Probabilistic functions of a discrete Markov process 57
3.1.1 The discrete observation hidden Markov model 57
3.1.2 The continuous observation case 80
3.1.3 The autoregressive observation case 87
3.1.4 The semi-Markov process and correlated observations 88
3.1.5 The non-stationary observation case 99
3.1.6 Parameter estimation via the EM algorithm 107
3.1.7 The Cave–Neuwirth and Poritz results 107
3.2 Formal grammars and abstract automata 109
3.2.1 The Chomsky hierarchy 110
3.2.2 Stochastic grammars 113
3.2.3 Equivalence of regular stochastic grammars and discrete HMMs 114
3.2.4 Recognition of well-formed strings 115
3.2.5 Representation of phonology and syntax 116

4 Syntactic analysis **119**

4.1 Deterministic parsing algorithms 119
4.1.1 The Dijkstra algorithm for regular languages 119
4.1.2 The Cocke–Kasami–Younger algorithm for context-free languages 121
4.2 Probabilistic parsing algorithms 122
4.2.1 Using the Baum algorithm to parse regular languages 122
4.2.2 Dynamic programming methods 123
4.2.3 Probabilistic Cocke–Kasami–Younger methods 130
4.2.4 Asynchronous methods 130
4.3 Parsing natural language 131
4.3.1 The right-linear case 132
4.3.2 The Markovian case 133
4.3.3 The context-free case 133

5 Grammatical Inference **137**

5.1 Exact inference and Gold's theorem 137
5.2 Baum's algorithm for regular grammars 137
5.3 Event counting in parse trees 139
5.4 Baker's algorithm for context-free grammars 140

6 Information-theoretic analysis of speech communication **143**

6.1 The Miller *et al.* experiments 143
6.2 Entropy of an information source 143
6.2.1 Entropy of deterministic formal languages 144
6.2.2 Entropy of languages generated by stochastic grammars 150
6.2.3 Epsilon representations of deterministic languages 153
6.3 Recognition error rates and entropy 153
6.3.1 Analytic results derived from the Fano bound 154
6.3.2 Experimental results 156

7 Automatic speech recognition and constructive theories of language 157

7.1 Integrated architectures 157
7.2 Modular architectures 161
7.2.1 Acoustic-phonetic transcription 161
7.2.2 Lexical access 162
7.2.3 Syntax analysis 165
7.3 Parameter estimation from fluent speech 166
7.3.1 Use of the Baum algorithm 166
7.3.2 The role of text analysis 167
7.4 System performance 168
7.5 Other speech technologies 169
7.5.1 Articulatory speech synthesis 169
7.5.2 Very low-bandwidth speech coding 170
7.5.3 Automatic language identification 170
7.5.4 Automatic language translation 171

8 Automatic speech understanding and semantics 173

8.1 Transcription and comprehension 173
8.2 Limited domain semantics 174
8.2.1 A semantic interpreter 175
8.2.2 Error recovery 182
8.3 The semantics of natural language 189
8.3.1 Shallow semantics and mutual information 189
8.3.2 Graphical methods 190
8.3.3 Formal logical models of semantics 190
8.3.4 Relationship between syntax and semantics 194
8.4 System architectures 195
8.5 Human and machine performance 197

9 Theories of mind and language 199

9.1 The challenge of automatic natural language understanding 199
9.2 Metaphors for mind 199
9.2.1 Wiener's cybernetics and the diachronic history 201
9.2.2 The crisis in the foundations of mathematics 205
9.2.3 Turing's universal machine 210
9.2.4 The Church–Turing hypothesis 212
9.3 The artificial intelligence program 213
9.3.1 Functional equivalence and the strong theory of AI 213
9.3.2 The broken promise 214
9.3.3 Schorske's causes of cultural decline 214
9.3.4 The ahistorical blind alley 215
9.3.5 Observation, introspection and divine inspiration 215
9.3.6 Resurrecting the program by unifying the synchronic and diachronic 216

10 A Speculation on the prospects for a science of mind 219

10.1 The parable of the thermos bottle: measurements and symbols 219
10.2 The four questions of science 220
10.2.1 Reductionism and emergence 220
10.2.2 From early intuition to quantitative reasoning 221
10.2.3 Objections to mathematical realism 223
10.2.4 The objection from the diversity of the sciences 224
10.2.5 The objection from Cartesian duality 225
10.2.6 The objection from either free will or determinism 225
10.2.7 The postmodern objection 226
10.2.8 Beginning the new science 227
10.3 A constructive theory of mind 228
10.3.1 Reinterpreting the strong theory of AI 228
10.3.2 Generalizing the Turing test 228
10.4 The problem of consciousness 229
10.5 The role of sensorimotor function, associative memory and reinforcement learning in automatic acquisition of spoken language by an autonomous robot 230
10.5.1 Embodied mind from integrated sensorimotor function 231
10.5.2 Associative memory as the basis for thought 231
10.5.3 Reinforcement learning via interaction with physical reality 232
10.5.4 Semantics as sensorimotor memory 234
10.5.5 The primacy of semantics in linguistic structure 234
10.5.6 Thought as linguistic manipulation of mental representations of reality 235
10.5.7 Illy the autonomous robot 235
10.5.8 Software 237
10.5.9 Associative memory architecture 238
10.5.10 Performance 238
10.5.11 Obstacles to the program 239
10.6 Final thoughts: predicting the course of discovery 241

Bibliography 243

Index 257

Preface

Evolution

This monograph was written over the past four years to serve as a text for advanced graduate students in electrical engineering interested in the techniques of automatic speech recognition and text to speech synthesis. However, the book evolved over a considerably longer period for a significantly broader purpose. Since 1972, I have sought to demonstrate how mathematical analysis captures and illuminates the phenomena of language and mind.

The first draft was written in 1975 during my tenure as a J. Willard Gibbs instructor at Yale University. The manuscript grew out of my lecture notes for a graduate course in pattern recognition, the main component of which was a statistical approach to the recognition of acoustic patterns in speech. The connection to language and mind was the result of both incorporating syntactic and semantic information into the statistical decision-theoretic process and observing that the detection and identification of patterns is fundamental to perception and intelligence.

The incomplete manuscript was set aside until 1983, at which time an opportunity to resurrect it appeared in the guise of a visiting fellowship in the Engineering Department of Cambridge University. A revised draft was written from lecture notes prepared for another course in pattern recognition for third-year engineering students. This time, topics of syntax and semantics were augmented with several other aspects of linguistic structure and were encompassed by the notion of composite pattern recognition as the classification of complicated patterns composed of a multi-leveled hierarchy of smaller and simpler ones. This second draft also included a brief intellectual history of the collection of ideas designated by what I will later argue is an unfortunate name, artificial intelligence (AI), and a recognition of its role in speech.

Once again the manuscript was set aside until the occasion of my appointment to the Department of Electrical and Computer Engineering at the University of Illinois. In 1997 I began organizing a program of graduate study in speech signal processing that would include both instruction in the existing technology and research to advance it. In its present form, the program comprises three tightly integrated parts: a course devoted to speech as an acoustic signal, another course on the linguistic structure of the acoustic signal, and research directed at automatic language acquisition. The first course required little innovation as there are several texts that provide standard and comprehensive coverage of the material. This book is a modification of my long-dormant manuscript and is now the basis for both the second course covering mathematical models of linguistic structure and the research project studying automatic language acquisition.

Goals and Methods

Linguists, electrical engineers, and psychologists have collectively contributed to our knowledge of speech communication. In recognition of the interdisciplinary nature of the subject, this book is written so that it may be construed as either a mathematical theory of language or an introduction to the technologies of speech recognition and synthesis. This is appropriate since the speech technologies rest on psycholinguistic concepts of the modularity of the human language engine. On the other hand, the models and techniques developed by electrical engineers can quite properly be regarded as the single most comprehensive collection of linguistic knowledge ever assembled. Moreover, linguistic theories can only be applied and tested by embedding them in a mathematically rational and computationally tractable framework. However, mathematical and computational models are useful only to the extent that they capture the essential structure and function of the language engine.

To the best of my knowledge, no single text previously existed that both covers all of the relevant material in a coherent framework and presents it in such a multidisciplinary spirit. A course of this nature could, heretofore, have been taught only by using several texts and a collection of old scholarly papers published in a variety of journals. Moreover, when significant portions of the material have been included in books on speech processing, they have been, without exception, presented as immutable canon of the subject. The unpleasant fact is that while modern speech technology is a triumph of engineering, it falls far short of constructing a machine that is able to use natural spoken language in a manner even approaching normal human facility. There is, at present, only an incomplete science of speech communication supporting a correspondingly limited technology. Based on the assumption that the shortcomings of our technology are the consequence of gaps in our knowledge rather than pervasive error, it does not seem unreasonable to examine our current knowledge with an eye toward extracting some general principles, thereby providing students with the background required to read the existing literature critically and to forge a strategy for research in the field that includes both incremental improvements and revolutionary ideas. Sadly, the recent literature is almost exclusively about technical refinements.

There are several specific pedagogic techniques I have adopted to foster this perspective. Discussions of all of the mathematical models of linguistic structure include their historical contexts, their underlying early intuitions and the mechanisms by which they capture the essential features of the phenomena they are intended to represent. Wherever possible, it is shown how these models draw upon results from related disciplines. Since topics as diverse as acoustics and semantics are included, careful attention has been paid to reconciling the perspectives of the different disciplines, to unifying the formalisms, and to using coherent nomenclature.

Another guiding principle of this presentation is to emphasize the meaningful similarities and relationships among the mathematical models in preference to their obvious but superficially common features. For example, not all models that have state spaces or use dynamic programming to explore them serve identical purposes, even if they admit of identical formal descriptions. Conversely, there are some obscure but significant similarities amongst seemingly disparate models. For example, hidden Markov models and stochastic formal grammars are quite different formally yet are similar in the important sense that they both have an observable process to account for measurements and an underlying but hidden process to account for structure.

Finally, students should know what the important open questions in the field are. The orientation of this book makes it possible to discuss explicitly some of the current theoretical debates. In particular, most current research is aimed at transcribing speech into text without any regard for comprehension of the message. At the very least, this distorts the process by placing undue emphasis on word recognition accuracy and ignoring the more fundamental roles of syntax and semantics in message comprehension. At worst, it may not even be possible to obtain an accurate transcription without understanding the message. Another mystery concerns the relative importance of perceptual and cognitive processes. Informed opinion has vacillated from one extreme to the other and back again. There is still no agreement, as different arguments are often organized along disciplinary boundaries.

When this book is used as a text for a graduate course on speech technology, Chapters 1 and 2 should be considered a review of a prerequisite course on speech signal processing. Chapters 3 through 8 contain the technical core of the course and Chapters 9 and 10 place the material in its scientific and philosophical context. These last two chapters are also intended as guidance and motivation for independent study by advanced students.

Whereas a technical synopsis of the contents of this book is given in Chapter 1, here I shall analyze it in a more didactic manner. The prerequisite material covered in Chapter 2 comprises succinct if standard presentations of the physics of speech generation by the vocal apparatus, methods of spectral analysis, methods of statistical pattern recognition for acoustic/phonetic perception, and a traditional taxonomy of linguistic structure. From these discussions we extract a few themes that will appear frequently in the succeeding chapters. First, the speech signal is a non-stationary time–frequency distribution of energy. This both motivates the importance of the short-duration amplitude spectrum for encoding the intelligence carried by the signal and justifies the use of the spectrogram which is shown to be an optimal representation in a well-defined sense. Linear prediction is seen as a particularly useful spectral parameterization because of its close relationship to the geometry and physics of the vocal tract.

Second, speech is literate. Thus, the spectral information must encode a small finite alphabet of symbols, the sequencing of which is governed by a hierarchy of linguistic rules. It follows, then, that any useful analysis of the speech signal must account for the representation of structured sequences of discrete symbols by continuous, noisy measurements of a multivariate, non-stationary function of time. This is best accomplished using non-parametric methods of statistical pattern recognition that employ a topological metric as a measure of perceptual dissimilarity. These techniques not only are optimal in the sense of minimum error, but also provide a justification for the direct normalization of time scales to define a metric that is invariant with respect to changes of time scale in signal.

The next six chapters are devoted to a detailed examination of techniques that address precisely these unique properties of the speech signal and, in so doing, capture linguistic structure. We begin with a study of probabilistic functions of a Markov process. Often referred to in the literature as hidden Markov models (HMMs), they have become a ubiquitous yet often seriously misunderstood mathematical object. The HMM owes its widespread application to the existence of a class of techniques for robust estimation of its parameters from large collections of data. The true value of the HMM, however, lies not in its computational simplicity but rather in its representational power. Not only does it intrinsically capture non-stationarity and the transformation of continuous measurements into discrete symbols, it also provides a natural way to represent acoustic phonetics, phonology, phonotactics, and even prosody.

In this book we develop the mathematical theory incrementally, beginning with the simple quantized observation case. We include a standard proof of Baum's algorithm for this case. The proof rests on the convexity of the log-likelihood function and is somewhat opaque, providing little insight into the reestimation formulas. However, by relating the parameter estimation problem for HMMs to the classical theory of constrained optimization, we are able to give a novel, short, and intuitively appealing geometric proof showing that the reestimation formulas work by computing a finite step in a direction that has a positive projection on the gradient of the likelihood function. We then progress to models of increasing complexity, including the little-known cases of non-stationary observation distributions and semi-Markov processes with continuous probability density functions for state duration.

We end the presentation of Chapter 3 with an account of two seminal but often overlooked experiments demonstrating the remarkable power of the HMM to discover and represent linguistic structure in both text and speech. The Cave–Neuwirth and Poritz experiments are then contrasted with the common formulation based on the special case of the non-ergodic HMM as a means of treating piecewise stationarity.

As powerful and versatile as it is, the HMM is not the only nor necessarily the best way to capture linguistic structure. We continue, therefore, with a treatment of formal grammars in the Chomsky hierarchy and their stochastic counterparts. The latter are seen to be probabilistic functions of an unobservable stochastic process with some similarities to the HMM. For example, we observe that the right linear grammar is equivalent to the discrete symbol HMM. However, the more complex grammars provide greater parsimony for fixed representational power. In particular, they provide a natural way to model the phonology and syntax of natural language.

Based on these formalisms, Chapter 4 approaches the problem of parsing, that is, determining the syntactic structure of a sentence with respect to a given grammar. Despite its central role in linguistics, this problem is usually ignored in the speech processing literature because it is usually assumed that word order constraints are sufficient for transcription of an utterance and the underlying grammatical structure is superfluous. We prefer the position that transcription is only an intermediate goal along the way to extracting the meaning of the message, of which syntactic structure is a prerequisite. Later we advance the idea that, in fact, transcription without meaning is a highly error-prone process. Parsing a spoken utterance is beset by two sources of uncertainty, variability of the acoustic signal and ambiguity in the production rules of the grammar. Here we show that these uncertainties can be accounted for probabilistically in two complementary ways, assigning likelihoods to the words conditioned on the acoustic signal and placing fixed probabilities on the rules of the grammar. Both of these ideas can be efficiently utilized at the first two levels of the Chomsky hierarchy and, in fact, they may be combined. We develop probabilistic parsing algorithms based on the Dijkstra and Cocke–Kasami–Younger algorithms for the right linear and context-free cases, respectively.

In Chapter 5, we address the inverse of the parsing problem, that of grammatical inference. This is the problem of determining a grammar from a set of possibly well-formed sentences, the syntactic structure of which is not provided. This is a classical problem and is usually ignored by linguists as too difficult. In fact, the difficulty of this problem is regarded by strict Chomskians as proof that the human language engine is innate. We, however, treat the problem of grammatical inference as one simply of parameter estimation. We show that the reestimation formulas for the discrete symbol HMM and the little-known Baker algorithm for stochastic context-free grammars are actually grammatical inference algorithms. Once the stochastic grammars are estimated, their deterministic counterparts are easily constructed. Finally, we show how parsing algorithms can be used to provide the sufficient statistics required by the EM algorithm so that it may be applied to the inference problem.

Chapter 6 is a divertimento in which we reflect on some of the implications of our mathematical models of phonology, phonotactics, and syntax. We begin by recalling an instructive experiment of Miller *et al.* demonstrating quantitatively that human listeners use linguistic structure to disambiguate corrupted utterances. This phenomenon is widely interpreted in the speech literature to mean that the purpose of grammar is to impose constraints on word order and thereby reduce recognition error rates in the presence of noise or other naturally occurring variability in the speech signal. Moreover, this analysis of Miller is the unstated justification for ignoring the grammatical structure itself and using only word order for transcription.

The information-theoretic concept of entropy is correctly used in the literature on speech recognition as a measure of the uncertainty inherent in word order, leading to the intuition that recognition error rate rises with increasing entropy. Entropy is typically estimated by playing the Shannon game of sequential prediction of words from a statistical analysis of large corpora of text or phonetic transcriptions thereof. Here we take a unique approach showing how the entropy of a language can be directly calculated from a formal specification of its grammar. Of course, entropy is a statistical property most easily obtained if the grammar is stochastic. However, we show that entropy can be obtained from a deterministic grammar simply by making some weak assumptions about the distributions of sentences in the language. Taking this surprising result one step further, we derive from the Fano bound a quantitative relationship among the entropy of a language, the variability intrinsic to speech, and the recognition error rate. This result may be used to explain how grammar serves as the error-correcting code of natural language.

All of the foregoing material is unified in Chapter 7 into a constructive theory of language or, from the engineer's perspective, the design of a speech recognition machine. We discuss two basic architectures, one integrated, the other modular. The latter approach is inspired by psycholinguistic models of human language processing and depends crucially on the Cave–Neuwirth and Poritz experiments featured in Chapter 3. We note the use of the semi-Markov model to represent aspects of prosody, phonotactics, and phonology. We also demonstrate the ability of the modular system to cope with words not contained in its lexicon.

In evaluating the performance of these systems, we observe that their ability to transcribe speech into text without regard for the meaning of the message arguably exceeds human performance on similar tasks such as recognizing fluent speech in an unknown language. And yet, this remarkable achievement does not provide speech recognition machines with anything remotely like human linguistic competence.

It seems quite natural, then, to try to improve the performance of our machines by providing them with some method for extracting the meaning of an utterance. On the rare occasions when this idea is discussed in the literature, it is often inverted so that the purpose of semantic analysis becomes simply that of improving word recognition accuracy. Of course, this is a very narrow view of human linguistic behavior. Humans use language to convey meaningful messages to each other. Linguistic competence consists in the ability to express meaning reliably, not to simply obtain faithful lexical transcriptions. It is in this ability to communicate that our machines fail. Chapter 8, therefore, is devoted to augmenting the grammatical model with a semantic one and linking them in a cooperative way.

We begin with a description of a laboratory prototype for a speech understanding system. Such a system should not simply be a transcription engine followed by a text processing semantic module. We note that such a system would require two separate syntax analyzers. Whereas, if the parsing algorithms described in Chapter 4 are used, the requisite syntactic structure is derived at the same time that the word order constraints are applied to reduce the lexical transcription error rate.

The most straightforward approach is to base the understanding system on the simplified semantics of a carefully circumscribed subset of natural language. Such formal artificial languages bear a strong resemblance to programming languages and can be analyzed using compiler techniques. Such systems may be made to carry out dialogs of limited scope with humans. However, the communication process is quite restricted and brittle. Extension of the technique to another domain of discourse is time-consuming because little if any data can be reused.

What is required to enable the machine to converse in colloquial discourse is a generalized model of unrestricted semantics. There are many such models, but they all reduce to mathematical logic or searching labeled, directed graphs. The former rests on the intuition that the extraction of meaning is equivalent to the derivation of logically true statements about reality, said statements being expressed formally in first-order logic. The latter model rests on the intuition that meaning emerges out of the properties of and relationships among objects and actions and can be extracted by finding suitable paths in an abstract graph. Such ideas have yet to be applied to speech processing. Thus, Chapter 8 concludes in an unsatisfying manner in that it provides neither theoretical nor empirical validation of a model of semantic analysis.

Up to this juncture, the exposition is presented in the customary, turgid scientific style. The mathematics and its application are objective and factual. No personal opinions regarding their significance are advanced. For Chapters 9 and 10, that conservatism is largely discarded as the unfinished work of the first eight chapters deposits us directly on the threshold of some of the very deepest and most vociferously debated ideas in the Western philosophical tradition. We are forced to confront the question of what sort of theory would support the construction of a machine with a human language faculty and we are obliged to assess the role of our present knowledge in such a theory. This profound shift of purpose must be emphasized. In the two concluding chapters, then, the mathematics nearly vanishes and, to preserve some semblance of intellectual responsibility, I employ the first person singular verb form.

It is my strongly held belief that a simulation of the human language engine requires nothing less than the construction of a complete human mind. Although this goal has proved to be utterly elusive, I insist that there is no inherent reason why it cannot be accomplished. There is, however, a cogent reason for our quandary revealed by a critical review of the intellectual history of AI.

In a remarkable work entitled *Fin-de-Siècle Vienna: Politics and Culture*, Carl E. Schorske gives a highly instructive explanation for the unkept promises of AI. He convincingly argues that cultural endeavors stagnate and fail when they become ahistorical by losing contact with both their diachronic history (i.e. their intellectual antecedents) and their synchronic history (i.e. their connections to independently developed but related ideas), and become fixated in the technical details of contemporary thought. Although Schorske did not include science in his analysis, his thesis seems highly appropriate there, too, with AI as a striking instance. Specifically, the loss of history in rapid response to an overwhelming but narrow discovery is made manifest by comparing the work of Norbert Wiener and Alan Turing.

The first edition of Wiener's *Cybernetics* was published in 1948, the very year that ENIAC, the first electronic, stored program, digital computer became operational. From this very early vantage point, Wiener has a fully diachronic perspective and recognizes that from ancient times to the present, metaphors for mind have always been expressed in the high technology of the day. Yet he clearly sees that the emerging computer offers a powerful tool with which to study and simulate, information and control in machines and organisms alike.

By 1950, Turing, on the other hand, had developed a deep understanding of the implications of his prior work in the foundations of mathematics for theories of mind. Since the Universal Turing Machine, and, hence, its reification in the form of the digital computer, is capable of performing almost any symbolic manipulation process, it is assumed sufficient for creating a mental model of the real world of our everyday experience. This intuition has evolved into what we today refer to as the "strong theory of AI". It is an almost exclusively contemporary view and was, in fact, Turing's preferred interpretation of thought as a purely abstract symbolic process. There is, however, a historical aspect to the remarkable 1950 paper. This is not surprising since the ideas it expresses date from the mid-1930s, at which time the metaphors for mind derived from classical electromechanical devices. In the penultimate paragraph of the paper, Turing offers an astounding and often overlooked alternative to the technical model of thought as symbolic logic. He suggests that the symbols and the relations among them could be inferred from real-world sensory data, a cybernetic and hence, historical view.

Unfortunately, the next generation of thinkers following Wiener and Turing fully endorsed the mind–software identity and en route lost all semblance of the historical trajectory. Based on my interpretation of Schorske, I submit that there have been no conceptual advances since then in the AI tradition. There has been some technical progress but no enlightenment. This is a rather frustrating conclusion in light of the elegance of Turing's theory which seemed to promise the immediate construction of an indisputably mechanical mind.

The key to revitalizing research on the theory of mind lies in synthesizing the synchronic and diachronic histories in what I call the cybernetic paradigm. This presently unfashionable mode of interdisciplinary thought unifies Turing's and Wiener's work and comprises much of the material of the first eight chapters of this volume. My synthesis leads to the following constructive theory of brain, mind, and language. The disembodied mind is a fantasy. A well integrated sensorimotor periphery is required. Thought is almost exclusively the product of associative memory rather than symbolic logic. The memory is highly sensitive to spatiotemporal order and its episodic structure integrates all sensorimotor stimuli. Thus, there are no isolated perceptual or cognitive functions. Memory is built up from instincts by the reinforcement of successful behavior in the real world at large. As a cognitive model of reality is acquired, a linguistic image of it is formed using specialized brain structures. This "language engine" is primarily responsive to semantic information while other levels of linguistic structure exist to make semantics robust to ambiguity. I note in passing that this theory is in direct opposition to the well-known Chomskian view that language is grammar. That is, the difficulty in language acquisition is precisely the difficulty of learning the acoustic/phonetic, phonological, morphological, prosodic, and syntactic rules that define language. Whereas, according to the theory described above, Chomskian grammar is both an error-correcting code that makes communication reliable and a framework upon which semantics is built. When the language is fully acquired, most mental processes are mediated linguistically and we appear to think in our native language, which we hear as our mind's voice.

Finally, I describe a means of testing this theory of cognition by building an autonomous intelligent robot. For the purposes of this experiment, sensorimotor function includes binaural audio, stereo video, tactile sense, and proprioceptive control of motion and manipulation of objects. Thus, I am able to exploit the synergy intrinsic in the combined sensorimotor signals. This sensory fusion is essential for the development of a mental representation of reality. The contents of the associative memory must be acquired by the interaction of the machine with the physical world in a reinforcement training regime. The reinforcement signal is a direct, real-time, on-line evaluation of only the success or failure of the robot's behavior in response to some stimulus. This signal comes from three sources: autonomous experimentation by the robot including imitation, instruction of the robot by a benevolent teacher as to the success or failure of its behavior, and instruction of the robot by the teacher in the form of direct physical demonstration of the desired behavior (e.g. overhauling the robot's actuators). Such instruction makes no use of any supervised training based on preclassified data. Nor does the robot use any predetermined representation of concepts or algorithms. There is no research known to me which is based on quite the combination of ideas I have described or quite the spirit in which I invoke them. A unique feature of the approach I advocate is the central role of language in the formation of the human mind.

As of this writing, my experiments have produced a robot, trained as described above, of sufficient complexity to be able to carry out simple navigation and object manipulation tasks in response to naturally spoken commands. The linguistic competence of the robot is acquired along with its other cognitive abilities in the course of its training. This result is due to the synergistic effect that the behavior of a complex combination of simple parts can be much richer than would be predicted by analyzing the components in isolation. Of course, I make no claim to have built a sentient being and I recognize that my hypotheses are controversial. However, in my best scientific and technical judgment, when a mechanical mind is eventually constructed, it will much more closely resemble the ideas expressed in the final two chapters than it does those of the previous six which are so vigorously pursued at present.

I am, of course, fully aware that those readers who find the technical aspects of this book worthwhile may well regard the final two chapters as a wholly inappropriate flight of fancy. Conversely, those who are intrigued by my metaphysics may judge the plethora of technical detail in the first eight chapters to be hopelessly boring. After 35 years of research on this subject, my fondest hope is that a few will find the presentation, as a whole, a provocative albeit controversial reflection on some significant scientific ideas and, at the same time, an exciting approach to an advanced technology of the future.

Acknowledgments

Since this book evolved over a period of years, it is hardly surprising that many people influenced the thought processes that determined its ultimate form. My interest in automatic speech understanding originated in my doctoral research at the University of Rhode Island under the patient direction of D. W. Tufts. Foremost among the many invaluable contributions he made to this book was that he allowed me the freedom to explore the very ideas that lead to this book despite his ambivalence toward them. He also recognized the relevance of and acquainted me with the early pattern recognition literature, much of which was applied to the acoustic patterns of speech.

At Yale University, where this book was first conceived, I greatly benefited from my interaction with F. Tuteur and H. Stark who encouraged me in many ways, not the least of which was their faithful attendance of and participation in my first offering of a course on pattern recognition. At the same time I was introduced to the crucial relationship of dynamic programming to parsing by R. J. Lipton and L. Snyder.

My 22 years at the Bell Telephone Laboratories, mostly under the supervision of J. L. Flanagan, was an extraordinary learning experience. I still marvel at my good fortune to have been a member of his Acoustics Research Department and Information Principles Research Laboratory. During that entire period, Dr. Flanagan was my most steadfast advocate. I must also single out three colleagues, L. R. Rabiner, A. E. Rosenberg, and M. M. Sondhi, without whose continuous collaboration I would never have understood numerous fundamental concepts nor had access to software and experimental data with which to study them.

My first real understanding of the theory of hidden Markov models resulted from my very stimulating association with the mathematicians, A. B. Poritz, L. R. Liporace and J. Ferguson of the Institute for Defense Analyses.

The second partial draft of this book was written as the result of an invitation from F. Fallside to come to the Engineering Department of Cambridge University. Through his efforts and with the generous support of Bell Laboratories, a visiting fellowship sponsored by the Science and Engineering Research Council of the UK was arranged. It is a source of deep regret to me that Prof. Fallside did not live to see this final draft.

I finally completed this manifesto as the result of a joint appointment to the Beckman Institute and the Department of Electrical and Computer Engineering at the University of Illinois. To rescue me from a pervasive climate of indifference to my ideas, S. M. Kang, J. Jonas, and T. S. Huang encouraged me to come to the Prairie and supported me in pursuit of the ideas to which I have devoted my entire professional life.

I am especially indebted to my loving wife of 25 years, Dr. Diana Sheet's, whose perspectives as a historian sensitized me to the tidal forces that shaped the intellectual and cultural waterfront of the early twentieth century. Her suggested reading list and our many discussions of the effect of generational rebellion in the development of a discipline and the tension between synchronic and diachronic analyses are reflected in the final two chapters.

I offer a final note of recognition and thanks to Sharon Collins who expertly typed the manuscript, often from barely legible handwritten drafts, and illustrations drawn in India ink on mylar, and rewarded my frequent complaints with only an occasional barbed reply. Thanks are also due to my colleague Mark Hasegawa-Johnson for his critical reading of Chapters 9 and 10 and to the graduate students at Yale, Cambridge, and Illinois who suffered through the early versions of the course, identified numerous opacities and errors in the text, and made important suggestions to improve the presentation.

Unfortunately, one can never truly chronicle the evolution of even his own ideas. Many of mine derive from frequent interactions with colleagues throughout my career. I have been particularly fortunate in the exchange of ideas afforded me by my membership in several IEEE Technical Societies and the Acoustical Society of America. More recently, I have expanded my horizons greatly as a result of discussions with several participants in the 1998 semester-long faculty seminar entitled "Mind, Brain and Language", sponsored by the Center for Advanced Study at the University of Illinois.

Of course, any mathematical blunders or poor judgment in the selection of subject matter I may have made or any intellectual improprieties in which the reader feels I have indulged are attributable not to my aforementioned benefactors but to me alone. I am gratified by the opportunity to proffer my theories, however incisive or specious they eventually prove to be.

S. E. Levinson
Urbana-Champaign

1

Introduction

1.1 Milestones in the History of Speech Technology

From antiquity, the phenomenon of speech has been an object of both general curiosity and scientific inquiry. Over the centuries, much effort has been devoted to the study of this remarkable process whereby our eating and breathing apparatus is used to transform thoughts in the mind of a speaker into vibrations in the air and back into congruent thoughts in the mind of a listener. Although we still do not have satisfactory answers to many of the questions about speech and language that the ancients pondered, we do have substantial scientific knowledge of the subject and an evolving technology based on it.

It is difficult to select a particular event or discovery as the origin of speech technology. Perhaps the speaking machine of W. von Kempelen [153] in the mid-eighteenth century qualifies. We can, however, safely say that the great body of classical mathematics and physics enabled the invention of the telephone, radio, audio recording, and the digital computer. These technologies gradually became the primary components of the growing global telecommunications network in which the conflicting criteria of high-fidelity and low-bandwidth transmission demanded that attention be focused on the nature of the speech signal. In the 1940s, basic research was conducted at Bell Telephone Laboratories and the Research Laboratory of Electronics at the Massachusetts Institute of Technology in auditory physiology, the psychophysics of acoustic perception, the physiology of the vocal apparatus, and its physical acoustics. Out of this effort a coherent picture of speech communication emerged. New instruments such as the sound spectrograph and the vocoder were devised for analyzing and generating speech signals. Much of this knowledge was encapsulated in the source–filter model of speech production which admitted of both a mathematical formulation and a real electrical implementation. Building on this foundation, analog circuitry was invented for both narrowband voice transmission and recognition of spoken numbers by classification of acoustic patterns extracted from the speech signal.

By the early 1950s, it had been recognized that the digital computer would become the tool of choice for analyzing signals in general and speech in particular. As a result, speech research spent the next two decades or so converting the analog circuitry to its digital equivalent. The relationship between digital signal processing (DSP) and speech analysis was mutually beneficial. Because the bandwidth of the speech signal was well matched to the processing speeds of the early computers, the new DSP techniques proved

Mathematical Models for Speech Technology. Stephen Levinson
 ISBN: 0-470-84407-8

to be easy to use, efficient, and effective. Many DSP algorithms, such as those for linear prediction and Fourier analysis, were particularly appropriate for speech and were quickly adopted. This, in turn, resulted in the development of new and more general theories and methods of DSP.

The mathematical theories of information and communication, random processes, detection and estimation, and spectral analysis went through similar transformations as they were adapted for digital implementations. One important outcome of this metamorphosis was the development of statistical pattern analysis. Such techniques were precisely what was needed for automatic speech recognition and they were quickly applied. As in the case of DSP, the success of pattern recognition for speech processing led to the development of new general methods of pattern recognition. During this period, another basic new mathematical theory appeared, that of probabilistic functions of a Markov process, commonly known as hidden Markov models (HMMs) [27]. This theory was destined to become the core of most modern speech recognition systems.

Concurrently, microelectronic technologies were rapidly developing. In particular, new devices for fast arithmetic and special addressing schemes appeared, making small, low-power speech processors readily available. These devices were responsible for the debut in the early 1970s of the first of several generations of inexpensive speech recognition systems for industrial applications.

The availability of all of these new digital techniques brought about spectacular advances in speech recognition and naturally encouraged research on ever more difficult problems such as recognition of fluent utterances independent of the speaker. The unbounded enthusiasm for these endeavors prompted John Pierce to write his infamous letter entitled "Whither Speech Recognition". Published in 1969, it was a scathing criticism of speech recognition research warning that until cognitive processes were understood and included in speech recognition machines, no progress would be made.

Perhaps Pierce was unaware that his concerns were being addressed elsewhere independent of the work in speech recognition. Studies of language were under way. In particular, Zelig Harris [119] and Noam Chomsky [45] had proposed formal specifications of grammar and theories of their role in natural language. Marvin Minsky [222] and his students at the MIT AI Laboratory proposed computational methods for representing the semantics of natural language. Finally, in 1970 Allen Newell and several colleagues [233] drafted a report to the Advanced Research Project Authority (ARPA) suggesting that formal models of syntax and semantics be incorporated into acoustic pattern recognition algorithms to enable the construction of more sophisticated systems that could understand spoken messages in the context of a simple, well-specified task.

The first attempt to realize the goals set forth in the Newell report was the ARPA speech understanding initiative. Under this program several efforts were undertaken to construct a speech recognition system based on the standard, modular model of the human language engine. Naive implementations of this model failed. This was both disappointing and surprising in light of the success of this model in speech synthesis. Although several components and partial systems were built by teams at Carnegie Mellon University [206] and Bolt Beranek and Newman, Inc. [332], they were never effective in speech recognition.

While the ARPA project was collapsing, Frederick Jelinek and his colleagues at IBM and, independently, James Baker, then a student at Carnegie Mellon University, introduced the hidden Markov model to speech recognition [20, 147]. The theoretical papers on the

HMM had been written by Leonard Baum and his colleagues at the Institute for Defense Analyses in the late 1950s [25, 26, 27, 28, 29]. Sadly, the applications of their work did not appear in the open literature at that time, which may account for the delay of nearly a decade before the method was used for speech recognition. In the HMM-based systems, all aspects of linguistic structure are integrated into a monolithic stochastic model the parameters of which can be determined directly from a corpus of speech. The architecture also supports an optimal statistical decision-theoretic algorithm for automatic speech recognition. Due to these important properties, the HMM methodology succeeded where the naive linguistic model failed and an important lesson was learned. Chaotic, rule-based implementations of the otherwise useful modular model cannot be optimized since they lack the mathematical rationality and computational tractability of the HMM-based systems. At present, all speech recognition systems use the integrated HMM-based approach. Some versions of it are now commercially available for use on personal computers; however, their performance is not as reliable as one might wish.

The success of the HMM-based system focused attention on the transcription of speech into text for use in a voice-operated typewriter or dictation machine. One important aspect of the modular approach that the integrated HMM-based system does not address is that of message comprehension. This is because only word order constraints have computationally tractable implementations that can be naturally fit into the HMM framework. Although the need for semantics and underlying syntactic structure is obvious, the lack of a compatible mathematical formulation makes it less attractive. At the present time, the use of syntactic structure and semantic analysis is still an open question. Some early speech understanding systems were actually constructed by Raj Reddy [272, 178] and this author [179, 180] based on straightforward application of compiler technology to carefully circumscribed data retrieval tasks. Unlike the HMM-based recognition systems, these experiments remained in the laboratory for considerable time, ultimately appearing in greatly simplified form in some telephone-based applications.

On the other hand, there are some simple, commercially successful uses of speech understanding. These limited applications substitute automatic recognition of isolated words and phrases from a limited vocabulary for a small number of single keystrokes on a telephone touch pad. This straightforward exchange allows a speaker to perform some simple functions selected from a carefully constructed menu. Such systems are used by travel agencies and financial institutions over the public telephone network. They are quite robust and well tolerated by the general population.

This brief account brings us to the present state of the art. In the sequel, we examine in detail the theories and techniques that brought us to this juncture and we consider how we might advance beyond it.

1.2 Prospects for the Future

The ultimate goal of speech technology is the construction of machines that are indistinguishable from humans in their ability to communicate in natural spoken language. As noted, the performance of even the best existing machines falls far short of the desired level of proficiency. Yet, a variety of human–machine communication tasks have been demonstrated as research prototypes and some of that technology is now available commercially.

Solving the ultimate puzzle is valuable both as an intellectual achievement and for the practical benefits it would confer on society. Eventually, telecommunications will be provided by a vast digital packet switched network the terminal devices of which, whether they be fixed or portable, will be more like computers than telephones and will be on-line continuously. The present-day Internet has provided us enough of a glimpse of this future to know that its value lies in its ability to connect every terminal to every known source of information on the planet. If everyone is to take full advantage of this remarkable resource, it must appear to every network subscriber as if he has his own personal librarian to help him acquire whatever information or service he requires. Since there are not enough trained librarians to go around, the service must be automated. The point-and-click interface of today is inadequate for that purpose, especially for hand-held devices. By contrast, a perfected speech technology would provide universal access to most of the information available on the Internet by means of ordinary conversation. This would greatly improve the ease and efficiency with which a mass society purchases goods and services, maintains financial, medical, and other personal records, and obtains information. An advanced technology could also be a component of prosthetic aids for people afflicted with speech, hearing, and even cognitive disorders.

Most practitioners of speech technology believe that this futuristic vision is close at hand. It is commonly supposed that the performance of today's best experimental systems is only an order-of-magnitude in error rate away from human performance and that even existing technology is commercially viable. It is also a widely held view that the order of magnitude improvement required for human-like performance will be achieved by incremental improvement rather than revolutionary new insights and techniques [185]. Regardless of how the technology advances – and there is no reason to suppose it will not – it is reasonable to expect that when the ultimate goal has been achieved, some of the existing technology, imperfect though it may be, will have survived in familiar form. It is prudent, therefore, to study the present state of the art while looking for radical new methods to advance it.

1.3 Technical Synopsis

Modern speech processing technology is usually considered to comprise the three related subfields of speech coding, speech synthesis, and automatic speech recognition. The latter two topics refer to techniques for transforming text into speech and speech into text, respectively. Speech coding is the process of faithful and efficient reproduction of speech usually for communication between humans. We shall not address speech coding here except to note, in passing, that a system composed of a speech recognition device and a speech synthesizer could be made into the ultimate coder in which speech is transcribed into text, transmitted at 50 bits per second and then converted back to speech. Methods for speech recognition and synthesis are more naturally applied to the construction of systems for human–machine communication by voice. It is the theory of such systems to which these pages are largely devoted. We begin with the acoustic signal and proceed level by level through the hierarchy of linguistic structure up to and including the determination of meaning in the context of a conversation.

Chapter 2 is a review of the material considered to be prerequisite for our mathematical analysis of the structure of language. This material is presented in highly condensed form as there are definitive texts for each of the four topics covered. First we review the physics of speech production in the human vocal apparatus. Readers wishing a thorough treatment of this subject are urged to consult Flanagan [86].

The physics of speech generation leads to the source–filter model of Dudley [69] and to the importance of the short-duration amplitude spectrum. Representation of the spectrum using Fourier analysis leads to the optimal formulation of the spectrogram while linear prediction analysis yields a particular parameterization of the filter closely related to the governing physics and geometry. Comprehensive studies of these topics may be found in Riley [274] and Markel and Gray [211], respectively. Fletcher [89] provides a thorough treatment of categorical perception, the process by means of which humans classify acoustic patterns. This function is well described by the theory of statistical pattern recognition. Here, we follow Patrick [241], emphasizing the non-parametric, Bayesian approach.

Finally, we review the types of linguistic structure for which we will later develop detailed, faithful, mathematical models. We adopt the broad taxonomy of C. S. Peirce [244] and then refine and augment it with the classical presentation found in Chomsky and Halle [47].

Chapters 3 through 8 provide mathematical models of several aspects of linguistic structure. We begin with two powerful analytical tools, the probabilistic function of a Markov process, otherwise known as the hidden Markov model, and the formal grammar. First the HMM is developed in full mathematical detail, beginning with the basic discrete symbol case. We then proceed to generalize the elementary case to that of elliptically symmetric distributions, of which the Gaussian is a special instance. Then we advance to the universal case of Gaussian mixtures and two special cases, the autoregressive process, related to linear prediction, and the non-stationary autoregressive case. Next, turning our attention to the hidden process, we relax the constraint of exponentially decreasing state durations and consider semi-Markov processes and the problem of correlated observations.

In a similar manner, we develop the formal grammar by considering the members of the Chomsky hierarchy in order of increasing parsimony of expression. For reasons of computational complexity, the detailed analyses are confined to the right-linear and context-free cases.

Finally, we recount two classical experiments based on the HMM demonstrating how these models discover and represent linguistic structure. We show how both models can be used to capture acoustic phonetics, phonology, phonotactics, syntax, and even some aspects of prosody.

These mathematical models have desirable properties. They reflect the natural constraints on the order in which words and sounds are allowed to appear and they specify the permissible phrase structures of the well-formed sequences. Phrase structure will later be seen to be important to representation of meaning. In Chapter 4 we develop parsing algorithms that enable both the optimal use of ordering constraints in speech recognition and the determination of the underlying structure for subsequent use as an outline of semantics. The parsing algorithms are seen to be applicable to both deterministic and stochastic specifications of linguistic structure for either right-linear or context-free grammars. The simple right-linear case has an equivalent HMM. Finally, we show how these models may be used to express the syntax of natural language.

Chapter 5 addresses the problem of inference of linguistic structure from data. First we cast this as a generic problem of parameter estimation. The computational requirements of the estimation problem can be reduced by using parsing algorithms to count the occurrences of particular types of structures. This allows us to transform the problem into one of statistical estimation for which the well-known EM algorithm is ideally suited. The classical experiments described in Chapter 3 may now be considered as instances of grammatical inference.

Chapter 6 provides an information-theoretic characterization of and explanation for the classical results of a set of experiments carried out by Miller *et al.* [220]. Their results confirm the intuitively appealing notion that grammar is an error-correcting code. We show how the classical Fano bound can be used to relate the entropy of a formal language, the equivocation of an acoustic pattern recognizer, and the error probability of a speech recognition system.

Chapter 7 combines the results of all of the foregoing chapters for the purpose of designing speech recognition systems. We contrast two architectures, the integrated system and the modular system. In the former, which is based on the non-ergodic HMM, all levels of linguistic structure are assimilated into a single stochastic model and recognition is based on an algorithm for evaluating its likelihood function. The latter is based on the ergodic HMM as used in the Poritz [250] experiment but requires different models for each specific aspect of linguistic structure. The individual models operate sequentially according to the traditional conception of the human language engine. Finally, we demonstrate how speech synthesis algorithms can be used to aid in the construction of both of these systems.

Whereas Chapter 7 is concerned with systems that recognize speech by transcribing it into ordinary text, Chapter 8 addresses the problem of understanding a spoken message in the sense of executing a command as intended by the speaker. This requires not only the incorporation of semantics – which, for this purpose, is defined as an internal, symbolic representation of reality – but also a mapping from lexical and syntactic structure to meaning. The simplest means to accomplish this is to adapt the semantics of programming languages to building a compiler for a useful subset of natural language. We describe a particularly instructive example of a system of this kind that is capable of performing some of the functions of a travel agent.

Unfortunately, this method cannot be extrapolated to encompass unrestricted conversation. We must, then, consider more general models that might be capable of representing the semantics of natural language. There are two such models available, mathematical logic and labeled, directed graph searching. These more general models of semantics have yet to be incorporated into a speech understanding system. Thus, our theory of language and our experiments on human–machine communication by voice are incomplete.

In the final two chapters, we take up the challenge of advancing our theories and technologies. Obviously there is a significant component of speculation in so doing. Chapter 9 begins with the premise that communication with machines in natural spoken language requires nothing less than a complete, constructive theory of mind. We carefully examine the two existing theories, the information- and control-theoretic (i.e. cybernetic) perspective of Wiener [330], and the symbolic computation view of Turing [319]. We then offer an explanation why such cogent theories have, thus far, failed to yield the expected results.

The reason is simply that the theories are complementary but have, to date, always been studied independently.

In Chapter 10 we propose a new theory of mind and outline an experimental program to test its validity. The theory is a version of the notion of embodied mind which is a synthesis of the cybernetic and computational perspectives. The experimental platform is an autonomous robot that acquires cognitive and linguistic abilities by interacting with the real world. When this approach was first suggested by Turing, it was technologically infeasible. Today it is plausible. In fact, we give a detailed description of our experiments with an autonomous robot that has acquired some limited abilities to navigate visually, manipulate objects and respond to spoken commands. Of course, we have not yet succeeded in building a sentient being. In fact, there are some daunting obstacles to extending our methods to that point. However, at the time of this writing, a community of researchers [1] around the world is organizing itself to pursue this ambitious goal by a variety of approaches all in the same spirit as the one described here. This kind of interdisciplinary research has an impeccable scientific pedigree and it offers the prospect of new insights and corresponding technological advances.

2

Preliminaries

2.1 The Physics of Speech Production

Speech is the unique signal generated by the human vocal apparatus. Air from the lungs is forced through the vocal tract, generating acoustic waves that are radiated at the lips as a pressure field. The physics of this process is well understood, giving us important insights into speech communication. The rudiments of speech generation are given in Sections 2.1.1 and 2.1.2. Thorough treatments of this important subject may be found in Flanagan [86] and Rabiner and Schafer [265].

2.1.1 The Human Vocal Apparatus

Figure 2.1 shows a representation of the midsagittal section of the human vocal tract due to Coker [51]. In this model, the cross-sectional area of the oral cavity $A(x)$, from the glottis, $x = 0$, to the lips, $x = L$, is determined by five parameters: a_1, tongue body height; a_2, anterior/posterior position of the tongue body; a_3, tongue tip height; a_4, mouth opening; and a_5, pharyngeal opening. In addition, a sixth parameter, a_6, is used to additively alter the nominal 17-cm vocal tract length. The articulatory vector **a** is $(a_1, a_2, \dots, a_6)$.

The vocal tract model has three components: an oral cavity, a glottal source, and an acoustic impedance at the lips. We shall consider them singly first and then in combination.

As is commonly done, we assume that the behavior of the oral cavity is that of a lossless acoustic tube of slowly varying (in time and space) cross-sectional area, $A(x)$, in which plane waves propagate in one dimension (see Fig. 2.2). Sondhi [303] and Portnoff [252] have shown that under these assumptions, the pressure, $p(x, t)$, and volume velocity, $u(x, t)$, satisfy

$$-\frac{\partial p}{\partial x} = \frac{\rho}{A(x,t)} \frac{\partial u}{\partial t} \tag{2.1a}$$

and

$$-\frac{\partial u}{\partial x} = \frac{A(x,t)}{\rho c^2} \frac{\partial p}{\partial t}, \tag{2.1b}$$

Mathematical Models for Speech Technology. Stephen Levinson
 ISBN: 0-470-84407-8

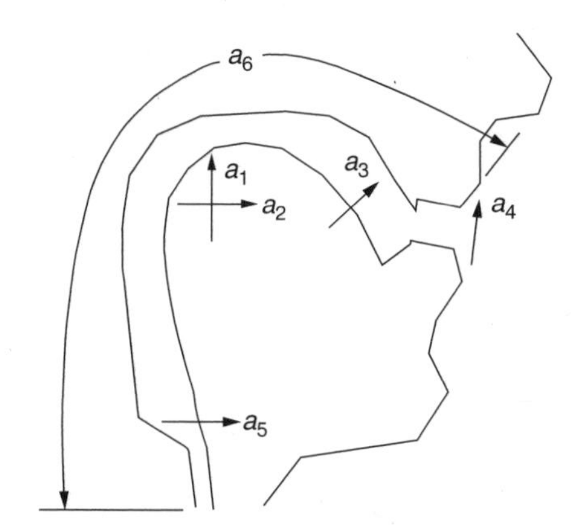

Figure 2.1 Coker's articulatory model

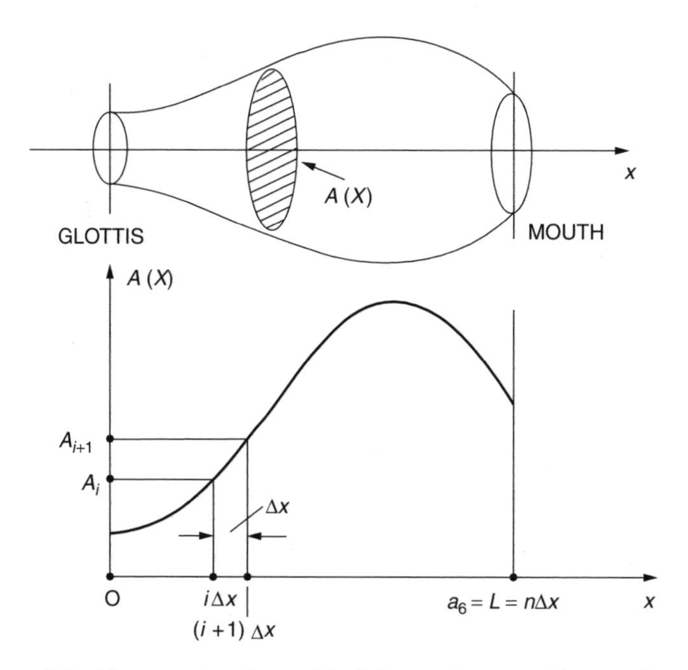

Figure 2.2 The acoustic tube model of the vocal tract and its area function

which express Newton's law and conservation of mass, respectively. In (2.1) ρ is the equilibrium density of the air in the tube and c is the corresponding velocity of sound.

Differentiating (2.1a) and (2.1b) with respect to time and space, respectively, and then eliminating the mixed partials, we get the well-known Webster equation [327] for pressure,

$$\frac{\partial^2 p}{\partial x^2} + \frac{1}{A(x,t)} \frac{\partial p}{\partial x} \frac{\partial A}{\partial x} = \frac{1}{c^2} \frac{\partial^2 p}{\partial t^2}. \tag{2.2}$$

The eigenvalues of (2.2) are taken as formant frequencies. We elect to use the Webster equation (in volume velocity) to compute a sinusoidal steady-state transfer function for the acoustic tube including the effects of thermal, viscous, and wall losses.

To do so we let $p(x,t) = P(x,\omega)e^{j\omega t}$ and $u(x,t) = U(x,\omega)e^{j\omega t}$, where ω is angular frequency and j is the imaginary unit. When p and u have this form, (2.1a) and (2.1b) become (cf. [252])

$$-\frac{dP}{dx} = Z(x,\omega)U(x,\omega) \tag{2.3a}$$

and

$$-\frac{dU}{dx} = Y(x,\omega)P(x,\omega), \tag{2.3b}$$

respectively. In order to account for the losses we define $Z(x,\omega)$ and $Y(x,\omega)$ to be the generalized acoustic impedance and admittance per unit length, respectively. Differentiating (2.3b) with respect to x and substituting for $-dP/dx$ and P from (2.3a) and (2.3b), respectively, we obtain

$$\frac{d^2U}{dx^2} = \frac{1}{Y(x,\omega)}\frac{dU}{dx}\frac{dY}{dx} - Y(x,\omega)Z(x,\omega)U(x,\omega), \tag{2.4}$$

which is recognized as the "lossy" Webster equation for the volume velocity.

The sinusoidal steady-state transfer function of the vocal tract can be computed by discretizing (2.4) in space and obtaining approximate solutions to the resulting difference equation for a sequence of frequencies. Let us write U_i^k to signify $U(i\Delta x, k\Delta\omega)$ where the spatial discretization assumes $\Delta x = L/n$ with $i = 0$ at the glottis and $i = n$ at the lips, as is shown in Fig. 2.3. Similarly, we choose $\Delta\omega = \Omega/N$ and let $0 \le k \le N$. We shall define A_i, Y_i^k, and Z_i^k in an analogous manner.

Approximating second derivatives by second central differences and first derivatives by first backward differences, the finite difference representation of (2.4) is just

$$\frac{U_{i+1}^k - 2U_i^k + U_{i-1}^k}{(\Delta x)^2} = \left(\frac{1}{Y_i^k}\right)\left(\frac{U_i^k - U_{i-1}^k}{\Delta x}\right)\left(\frac{Y_i^k - Y_{i-1}^k}{\Delta x}\right) + Z_i^k Y_i^k U_i^k, \tag{2.5}$$

which is easily simplified to the three-point recursion formula

$$U_{i+1}^k = U_i^k\left(3 + (\Delta x)^2 Z_i^k Y_i^k - \frac{Y_{i-1}^k}{Y_i^k}\right) + U_{i-1}^k\left(\frac{Y_{i-1}^k}{Y_i^k} - 2\right). \tag{2.6}$$

Given suitable values for U_0^k and U_1^k for $0 \le k \le N$, we can obtain the desired transfer function from (2.6). We shall return to consider the numerical properties of this formula later.

First, however, we must find appropriate expressions for Y and Z to account for the losses. Losses arise from thermal effects and viscosity but are primarily due to wall vibrations. A detailed treatment of the wall losses is found in Portnoff [252] and is neatly

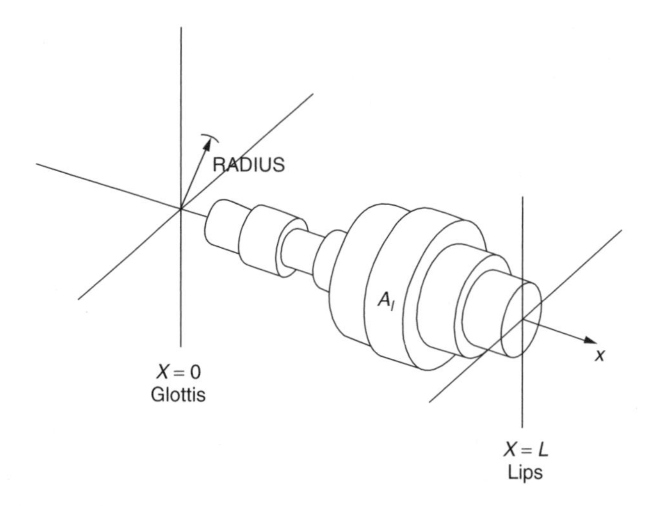

Figure 2.3 The discretized acoustic tube model of the vocal tract

summarized by Rabiner and Schafer [265]. Portnoff assumes that the walls are displaced $\xi(x, t)$ in a direction normal to the flow due to the pressure at x only. The vocal tract walls are modeled by a damped spring-mass system for which the relationship between pressure and displacement is

$$p(x, t) = M\frac{\partial^2\xi}{\partial t^2} + b\frac{\partial\xi}{\partial t} + k(x)\xi(x, t), \tag{2.7}$$

where M, b, and $k(x)$ are the unit length wall mass, damping coefficient, and spring constant, respectively.

The displacement of the walls is assumed to perturb the area function about a neutral position according to

$$A(x, t) = A(x) + S(x)\xi(x, t), \tag{2.8}$$

where $A(x)$ and $S(x)$ are the neutral area and circumference, respectively. By substituting (2.1a) into (2.1b) and, ignoring higher-order terms, transforming into the frequency domain, Portnoff goes on to observe that the effect of vibrating walls is to add a term Y_W to the acoustic admittance in (2.3b), where

$$Y_W(x, \omega) = jwS(x, \omega)\left(\frac{[k(x) - \omega^2 M] - j\omega b}{[k(x) - \omega^2 M]^2 + \omega^2 b^2}\right). \tag{2.9}$$

The other losses that we wish to consider are those arising from viscous friction and thermal conduction. The former can be accounted for by adding a real quantity Z_v to the acoustic impedance in (2.3a), where

$$Z_v(x, \omega) = \frac{S(x)}{A^2(x)}\left(\frac{\omega\rho\mu}{2}\right)^{1/2}, \tag{2.10}$$

where μ is the viscosity of air.

The thermal losses have an effect which is described by adding a real quantity Y_T to the acoustic admittance in (2.3b), where

$$Y_T(x, \omega) = \frac{S(x)(\eta - 1)}{\rho c^2} \left(\frac{\lambda \omega}{2 C_p \rho} \right)^{1/2}, \tag{2.11}$$

in which λ is the coefficient of heat conduction, η is the adiabatic constant, and C_p is the heat capacity. All the constants are, of course, for the air at the conditions of temperature, pressure, and humidity found in the vocal tract.

In view of (2.1), (2.9), (2.10), and (2.11) it is possible to set

$$Z(x, \omega) = j\omega\rho / A(x) + Z_v(x, \omega) \tag{2.12}$$

and

$$Y(x, \omega) = j\omega A(x)/\rho c^2 + Y_W(x, \omega) + Y_T(x, \omega). \tag{2.13}$$

There are two disadvantages to this approach. First, (2.12) and (2.13) are computationally expensive to evaluate. Second, (2.9) requires values for some physical constants of the tissue forming the vocal tract walls. Estimates of these constants are available in [139] and [86].

A computationally simpler empirical model of the losses which agrees with the measurements has been proposed by Sondhi [303] in which

$$Z(x, \omega) = j\omega\rho / A(x) \tag{2.14}$$

and

$$Y(x, \omega) = \frac{A(x)}{\rho c^2} \left(j\omega + \frac{\omega_0^2}{\alpha + j\omega} + (\beta j \omega)^{1/2} \right). \tag{2.15}$$

Sondhi has chosen values for the constants, $\omega_0 = 406\pi$, $\alpha = 130\pi$, $\beta = 4$, which he then shows give good agreement with measured formant bandwidths. Moreover, the form of the model agrees with the results of Portnoff, as becomes clear when we observe that $Y_W(x, \omega)$ in (2.9) will have the same form as the second term on the right-hand side of (2.15) if $k(x) \equiv 0$ and the ratio of circumference to area is constant. In fact, Portnoff used $k(x) = 0$ and the second assumption is not unreasonable. The third term on the right-hand side of (2.15) may be seen to be of the same form as (2.10) and (2.11) (under the assumption that the ratio of S to A is constant) by noting that

$$(j\omega)^{1/2} = (1 + j)(\omega/2)^{1/2}. \tag{2.16}$$

2.1.2 Boundary Conditions

With a description of the vocal tract in hand, we can turn our attention to the boundary conditions. Following Flanagan [86], we have assumed the glottal excitation to be a constant volume source with an asymmetric triangular waveform of amplitude V. Dunn *et al.* [71] have analyzed such a source in detail. What is relevant is that the spectral envelope decreases with the square of frequency. We have therefore taken the glottal source $U_g(\omega)$ to be

$$U_g(\omega) = V/\omega^2. \tag{2.17}$$

For the boundary condition at the mouth we use the well-known Portnoff [252] and Rabiner and Schafer [265] relationship between sinusoidal steady-state pressure and volume velocity,

$$P(L, \omega) = Z_r(\omega)U(L, \omega), \tag{2.18}$$

where the radiation impedance Z_r is taken as that of a piston in an infinite plane baffle, the behavior of which is well approximated by

$$Z_r(\omega) = j\omega L_r/(1 + j\omega L_r/R). \tag{2.19}$$

Values of the constants which are appropriate for the vocal tract model are given by Flanagan [86] as

$$R = 128/9\pi^2 \tag{2.20}$$

and

$$L_r = 8[A(L)/\pi]^{1/2}/3\pi c. \tag{2.21}$$

It is convenient to solve (2.4) with its boundary conditions (2.17) and (2.18) by solving a related initial-value problem for the transfer function

$$H(\omega) = U(L, \omega)/U(0, \omega). \tag{2.22}$$

At $x = L$,

$$-\left.\frac{dU}{dx}\right|_{x=L} = \frac{A(L)}{\rho c^2}(j\omega)P(L, \omega) \tag{2.23}$$

from which the frequency domain difference equation

$$-\frac{U_n^k - U_{n-1}^k}{\Delta x} = jk\Delta\omega\frac{A_n}{\rho c^2}P_n^k \tag{2.24}$$

can be derived. Let

$$U_n^k = 1 \tag{2.25}$$

and note that, from (2.18),

$$P_n^k = Z_r(k\Delta\omega)U_n^k. \tag{2.26}$$

Substituting (2.25) and (2.26) into (2.24), we see that

$$U_{n-1}^k = 1 + jk\Delta\omega\frac{A_n\Delta x}{\rho c^2}Z_r(k\Delta\omega). \tag{2.27}$$

Now solving (2.6) for U_{i-1}^k, we get the reversed three-point recursion relation

$$U_{i-1}^k = [1/(Y_{i-1}^k/Y_i^k - 2)]\{U_{i+1}^k - U_i^k[3 + (\Delta x)^2 Z_i^k Y_i^k - Y_{i-1}^k/Y_i^k]\}, \tag{2.28}$$

where

$$Z_i^k = jk\Delta\omega\rho/A_i \tag{2.29}$$

and

$$Y_i^k = (A_i/\rho c^2)[jk\Delta\omega + \omega_0^2/(\alpha + jk\Delta\omega) + (\beta jk\Delta\omega)^{1/2}]. \tag{2.30}$$

It is worthy of note that the ratio of admittances can be simplified and the $Z_i^k Y_i^k$ product is independent of i, so that (2.28) becomes

$$U_{i-1}^k = (A_{i-1}/A_i - 2)^{-1}\{U_{i+1}^k - U_i^k[3 + (\Delta x)^2 Z(k)Y(k) - A_{i-1}/A_i]\}. \tag{2.31}$$

Given the initial conditions of (2.25) and (2.27), we can compute U_0^k by evaluating (2.31) for $i = n-1, n-2, \ldots, 1$. Then from (2.22),

$$H(k\Delta\omega) = U_n^k/U_0^k = 1/U_0^k. \tag{2.32}$$

Finally, we compute the vocal tract output by multiplying the transfer function by the excitation from (2.17) and the radiation load from (2.19),

$$\hat{P}_k = \hat{P}(k\Delta\omega) = H(k\Delta\omega)U_g(k\Delta\omega)Z_r(k\Delta\omega), \tag{2.33}$$

for $1 \le k \le N$.

Figure 2.4 shows the power spectrum, $10\log_{10}(|\hat{P}(\omega)|^2)$, plotted in dB and some parameters for the phoneme $/ah/$. The area function used was obtained from X-ray measurements and appears in Flanagan [86].

Figure 2.4 illustrates the single most important aspect of the speech. The intelligence in speech is encoded in the power spectrum of the acoustic pressure wave. Different articulatory configurations result in signals with different spectra, especially different resonance frequencies called formants, which are perceived as different sounds. We shall return to consider how these sounds form the basis of the linguistic code of speech in Section 2.7.

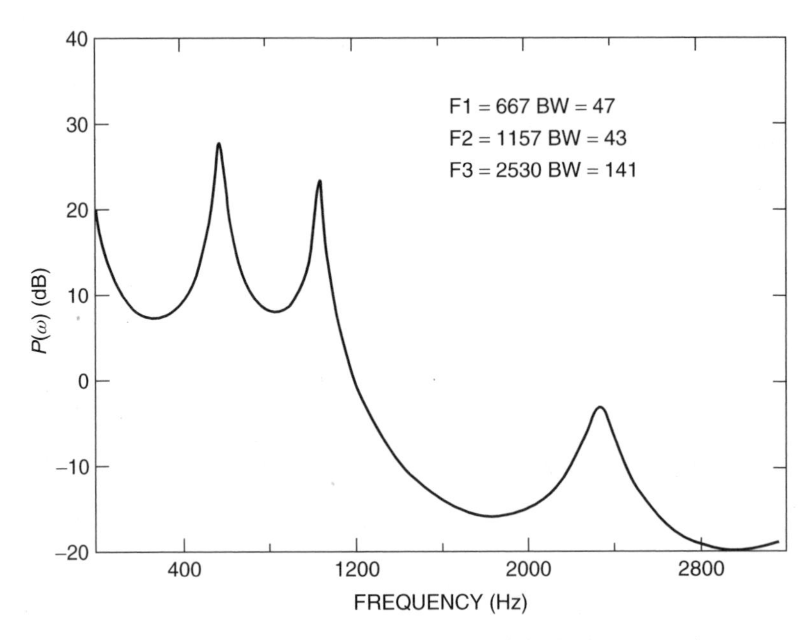

Figure 2.4 Frequency domain solution of the Webster equation

2.1.3 *Non-Stationarity*

The speech signal, $p(t)$, is the solution to (2.2). Since the function $A(x, t)$ is continuously varying in time, the solution, $p(t)$, is a non-stationary random change in time. Fortunately, $A(x, t)$ is slowly time-varying with respect to $p(t)$. That is,

$$\left|\frac{\partial A}{\partial t}\right| \ll \left|\frac{\partial p}{\partial t}\right|. \tag{2.34}$$

Equation (2.34) may be taken to mean that $p(t)$ is quasi-stationary or piecewise stationary. As such, $p(t)$ can be considered to be a sequence of intervals within each one of which $p(t)$ is stationary. It is true that there are rapid articulatory gestures that violate (2.34), but in general the quasi-stationary assumption is useful. However, as we shall discuss in Sections 2.3, 2.4, and 2.6, special techniques are required to treat the non-stationarity of $p(t)$.

2.1.4 *Fluid Dynamical Effects*

Equation (2.2) predicts the formation of planar acoustic waves as a result of air flowing into the vocal tract according to the boundary condition of (2.17). However, the Webster equation ignores any effects that the convertive air flow may have on the function $p(t)$. If, instead of (2.1a) and (2.1b), we consider two-dimensional wave propagation, we can write the conservation of mass as

$$\frac{\partial u}{\partial x} = \frac{\partial u}{\partial y} = -M^2\frac{\partial p}{\partial t}, \tag{2.35}$$

where M is the Mach number. We can also include the viscous and convective effects by observing

$$\frac{\partial u_x}{\partial t} = -\frac{\partial p}{\partial x} - \frac{\partial}{\partial x}\left(u_x u_y\right) + \frac{\partial}{\partial x}\left[\frac{1}{N_R}\left(\frac{\partial u_x}{\partial x} + \frac{\partial u_y}{\partial x}\right) - \overline{\mu_x \mu_y}\right], \tag{2.36a}$$

$$\frac{\partial u_y}{\partial t} = -\frac{\partial p}{\partial y} - \frac{\partial}{\partial y}\left(u_x u_y\right) + \frac{\partial}{\partial y}\left[\frac{1}{N_R}\left(\frac{\partial u_x}{\partial y} + \frac{\partial u_y}{\partial y}\right) - \overline{\mu_x \mu_y}\right]. \tag{2.36b}$$

In (2.36a) and (2.36b) the first term on the right-hand side is recognized as Newton's law expressed in (2.1a) and (2.1b). The second term is the convective flow. The third term accounts for viscous shear and drag at Reynolds number, N_R, and the last term for turbulence.

Equations (2.35) and (2.36) are known as the normalized, two-dimensional, Reynolds averaged, Navier–Stokes equations for slightly compressible flow. These equations can be solved numerically for $p(t)$. The solutions are slightly different from those obtained from (2.2) due to the formation of vortices and transfer of energy between the convective and wave propagation components of the fluid flow. Typical solutions for the articulatory configuration of Fig. 2.2 are shown in Figs. 2.4 and 2.5. There is reason to believe that (2.35) and (2.36) provide a more faithful model of the acoustics of the vocal apparatus than the Webster equation does [327].

2.2 The Source–Filter Model

The electrical analog of the physics discussed in Section 2.1 is the source–filter model of Dudley [69] shown in Fig. 2.6. In this model, the acoustic tube with time-varying area function is characterized by a filter with time-varying coefficients. The input to the filter is a mixture of a quasi-periodic signal and a noise source. When the filter is excited by the input signal the output is a voltage analog of the sound pressure wave $p(t)$. The source–filter model is easily implemented in either analog or digital hardware and is the basis for all speech processing technology.

2.3 Information-Bearing Features of the Speech Signal

The conclusion to be drawn from the previous two sections is that information is encoded in the speech signal in its short-duration amplitude spectrum [86]. This implies that by estimating the power spectrum of the speech signal as a function of time, we can identify the corresponding sequence of sounds. Because the speech signal $x(t)$ is non-stationary it has a time-varying spectrum that can be obtained from the time-varying Fourier transform, $X_n(\omega)$. Note that $x(t)$ is the voltage analog of the sound pressure wave, $p(t)$, obtained by solving (2.2).

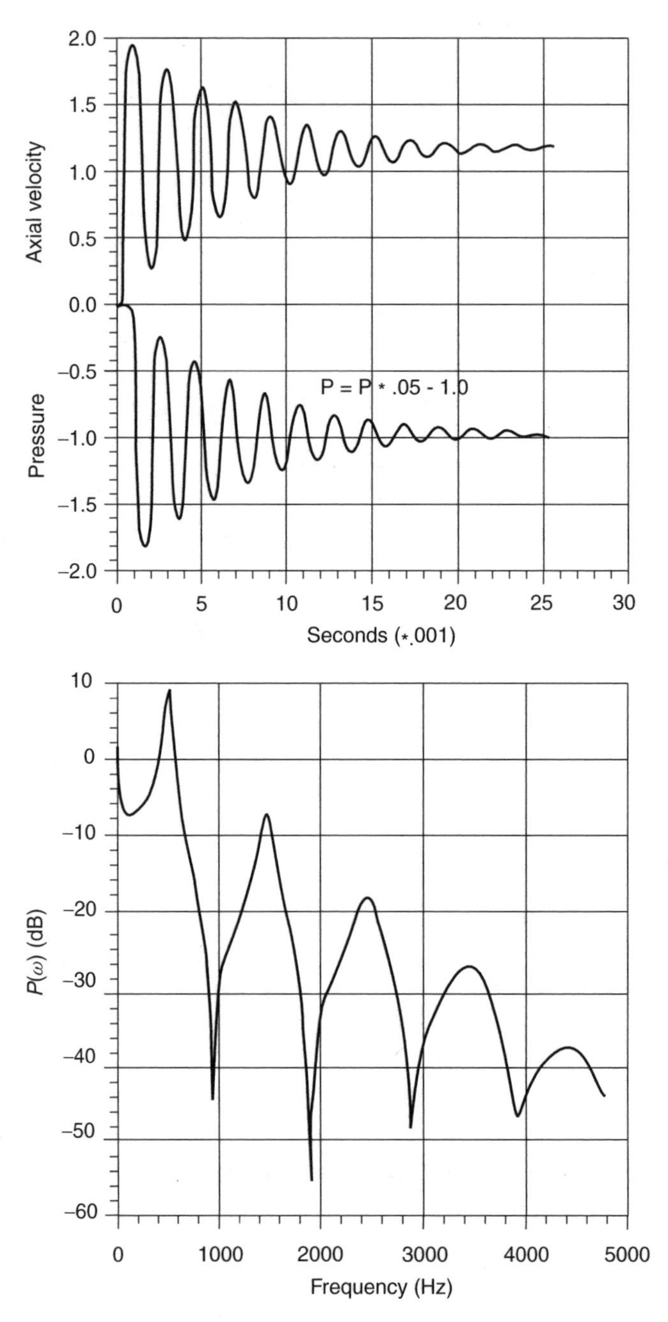

Figure 2.5 Speech signals and their spectrum obtained by solving the Navier–Stokes equations

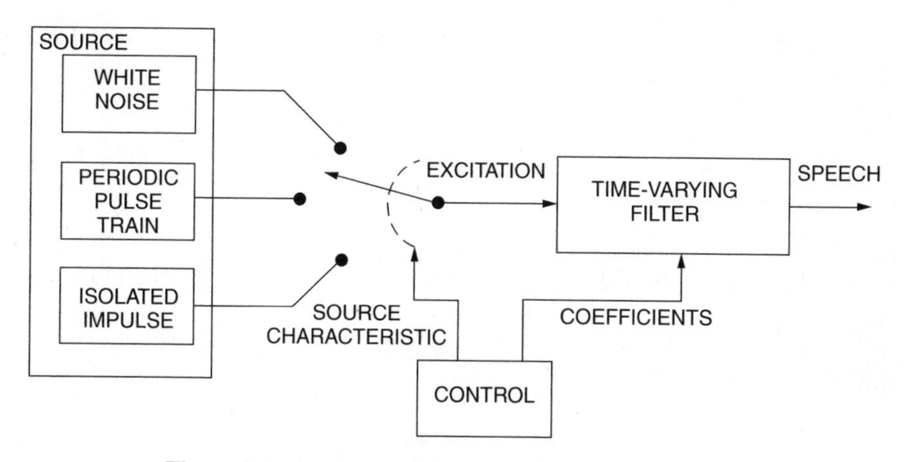

Figure 2.6 The source–filter model of speech production

2.3.1 Fourier Methods

The short-time Fourier transform is computed by observing the speech signal $x(t)$ through a finite window, $\{w_n\}_{n=0}^{N-1}$, where $w_n = .54 - .46\cos(\frac{2\pi n}{N})$. For computation $x(t)$ is sampled yielding the time series $\{x_n\}_{n=0}^{\infty}$, where $x_n = x(nT)$ and T is the sampling interval measured in seconds.

The short-time Fourier transform, $X_n(\omega)$, of the signal x_n is given by

$$X_n(\omega) = \sum_{m=-\infty}^{\infty} w_{n-m} x_m e^{-j\omega m}. \tag{2.37}$$

Recall that the index, n, refers to time nT, indicating that $X_n(\omega)$ is a function of both time and angular frequency, ω.

The signal, $x(t)$, can be recovered from its short-time Fourier transform according to

$$x_n = \frac{1}{w_0 2\pi} \int_{-\pi}^{\pi} X_n(\omega) e^{j\omega n} d\omega \tag{2.38}$$

The power spectrum, $S_n(\omega)$, of the signal at time nT is just

$$S_n(\omega) = |X_n(\omega)|^2, \tag{2.39}$$

and, for $0 \le \omega \le B_x$ and $n = 0, 1, \ldots,$ this is called the spectrogram. $S_n(\omega)$ is characteristic of the sound $\{x_n\}_{n=0}^{N-1}$.

Equation (2.37) can be conveniently evaluated at frequencies $\omega_k = \frac{2\pi k}{N}$ by means of the discrete Fourier transform (DFT)

$$X_n(\omega_k) = \sum_{m=0}^{N-1} w_m x_m e^{\frac{-j2\pi km}{N}}. \tag{2.40}$$

The computation of (2.40) is performed using the well-known fast Fourier transform (FFT).

Then the spectrogram, S_{kn}, at frequency ω_k and time nT is just

$$S_{kn} = |X_n(\omega_k)|^2, \tag{2.41}$$

and S_{kn} is an information-bearing feature of $x(t)$.

Because $X_n(\omega)$ changes with time, it must be sampled at a rate sufficient to permit the reconstruction of $x(t)$. The bandwidth, B_x, of the speech signal $x(t)$, is approximately 5 kHz, so the sampling rate, F_s, is 10 kHz. For the Hamming window, of length $N = 100$, $\{w_n\}$, the bandwidth, B, is

$$B = \frac{2F_s}{N} = \frac{20\,000}{100} = 200\,\text{Hz}. \tag{2.42}$$

Thus the Nyquist rate for the short-time Fourier transform is $2B = 400\,\text{Hz}$. which at $F_s = 10\,000$ requires a value of $X_n(\omega_k)$ every 25 samples. Since $N = 100$, the windows should overlap by 75%.

A typical spectrogram computed from (2.41) is shown in Fig. 2.7. Time is on the horizontal axis, frequency is on the vertical, and the power is the level of black at a given time and frequency.

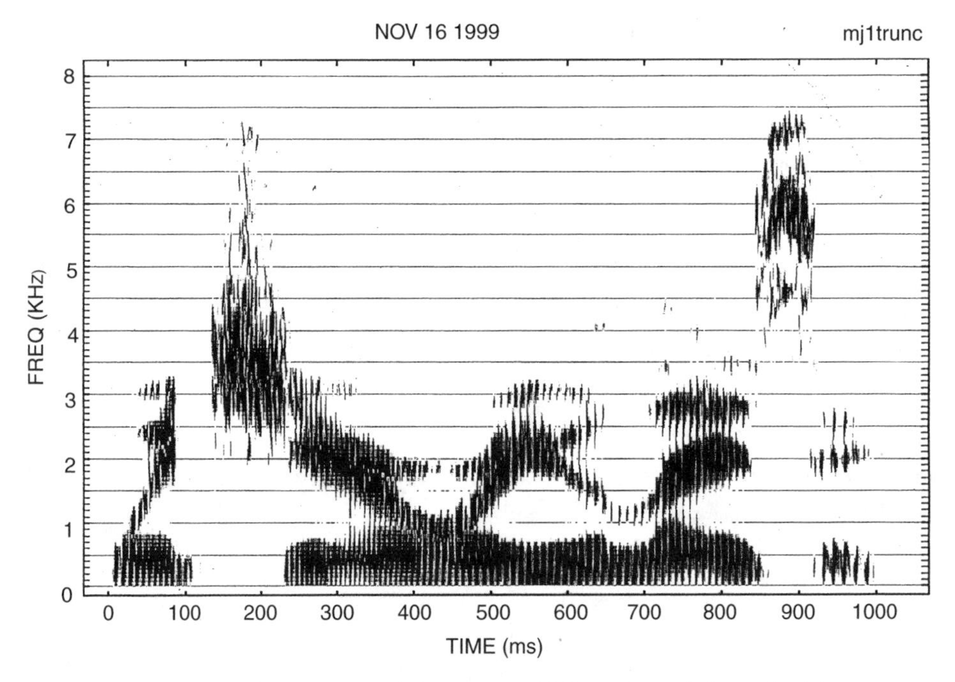

Figure 2.7 Spectrogram of the sentence "When the sunlight stri(kes)"

2.3.2 *Linear Prediction and the Webster Equation*

The method of linear prediction provides a particularly appropriate representation of the power spectrum of the speech signal. If we assume that the speech signal at time n is well predicted by a linear combination of p previous samples, we may write

$$x_n = \sum_{k=1}^{p} a_k x_{n-k} + e_n, \tag{2.43}$$

where the weights, a_k, of the linear combination are called the linear prediction coefficients (LPCs) and the error at time n, e_n, is small with respect to x_n when averaged over time. A comprehensive treatment of linear prediction of speech is given in Markel and Gray [211]; however, for the purposes of this book, the following summary will suffice.

Equation (2.43) is equivalent to the source–filter model in the sense that the source function, μ_n, is

$$\mu_n = Ge_n \tag{2.44}$$

for some constant gain, G. The filter is an all-pole filter with transfer function, $H(z)$, given by

$$H(z) = \frac{G}{1 - \sum_{k=1}^{p} a_k z^{-k}}, \tag{2.45}$$

where the a_k are just the LPCs from (2.43).

As in the case of Fourier analysis, (2.43)–(2.45) hold for short intervals of approximately 10 ms duration. For each such interval, we can obtain an optimal estimate of the a_k by a minimum mean square error (MMSE) technique.

The prediction error, E_n, is defined by

$$E_n = \sum_{m=0}^{N+p-1} e_{n+m}^2. \tag{2.46}$$

The MMSE is obtained by solving

$$\nabla_a E_n = 0, \tag{2.47}$$

which is equivalent to solving the linear system

$$\begin{pmatrix} R(0) & \cdots & R(p-1) \\ \vdots & \ddots & \vdots \\ R(p-1) & \cdots & R(0) \end{pmatrix} \begin{pmatrix} a_1 \\ a_2 \\ \vdots \\ a_p \end{pmatrix} = \begin{pmatrix} R(1) \\ \vdots \\ R(p) \end{pmatrix}, \tag{2.48}$$

where $R(k)$ is the autocorrelation function of $x(t)$ at time n given by

$$R_n(k) = \sum_{m=1}^{N+k-1} w_m w_{m+k} x_{n+m} x_{n+m+k} \tag{2.49}$$

with $N \cong 100$. Because of the Toeplitz property of the correlation matrix and the relationship of the right-hand side of (2.48) to that matrix, there is an efficient algorithm for solving for the a_k due to Durbin [265]. Let $E^0 = R(0)$. Then, for $1 \leq i \leq p$, compute the partial correlation coefficients (PARCORs) according to

$$k_i = \frac{1}{E^{(i-1)}} \left[R_n(i) - \sum_{j=1}^{i-1} a_j^{(i-1)} R_n(i-j) \right]. \tag{2.50}$$

Then, for $1 \leq i \leq p$, compute the LPCs from

$$a_i^{(i)} = k_i \tag{2.51}$$

and, for $1 \leq j \leq i-1$,

$$a_j^{(i)} = a_j^{(1-i)} - k_i a_{i-j}^{(i-1)}. \tag{2.52}$$

Then the residual error is updated from

$$E^{(i)} = (1 - k_i^2) E^{(i-1)}. \tag{2.53}$$

Finally, the desired LPCs for a pth-order predictor are

$$a_j = a_j^{(p)}, \quad \text{for} \quad 1 \leq j \leq p. \tag{2.54}$$

The PARCORs are the negatives of the reflection coefficients, that is,

$$k_i = -\frac{A_{i+1} - A_i}{A_{i+1} + A_i} \tag{2.55}$$

From (2.55) we see that the LPCs are related to the area function, $A(x)$, in the Webster equation. In fact, Wakita [325] has shown that the linear predictor of (2.43) is equivalent to the solution of the lossless Webster equation. Thus the very general method of linear prediction is actually a physical model of the speech signal. In addition, the poles, z_i of $H(z)$ in (2.45) are just the formants or resonances that appear in the solution to the Webster equation as indicated in Fig. 2.4. Write the poles in the form of

$$z_i = |z| e^{j\theta i}. \tag{2.56}$$

The formant frequencies, f_i, and bandwidths, σ_i, are determined by

$$\theta_i = \frac{2\pi f_i}{T} \tag{2.57}$$

and

$$|z| = e^{\sigma_i T} \tag{2.58}$$

respectively.

The LPCs characterize the power spectrum of the speech signal in the sense that

$$\lim_{p\to\infty} |H(\omega)|^2 = |S_n(\omega)|^2; \tag{2.59}$$

thus an alternative to the information-bearing features of the speech signal S_{kn}, defined in (2.41), is the set of LPCs defined in (2.54). In practice, the cepstral coefficients, c_n, defined by

$$c_n = \frac{1}{2\pi}\int_{-\pi}^{\pi} \log|H(\omega)|e^{j\omega n}\, d\omega, \tag{2.60}$$

are often used in preference to the a_i themselves. Equation (2.60) can be evaluated recursively from

$$c_0 = \log G \tag{2.61}$$

and, for $n = 1, 2, \ldots,$

$$c_n = a_n + \sum_{k=1}^{n-1} \frac{k}{n} a_{n-k} c_k. \tag{2.62}$$

Still other feature sets are obtained by taking time derivatives of the c_n and by applying a non-linear transformation to the frequency, ω, in (2.60). This modification is called the mel-scale cepstrum [59, 314]. For the purposes of this discussion, all such feature sets will be considered to be equivalent. We will refer to this assumption in Chapter 3.

2.4 Time–Frequency Representations

As discussed in Section 2.1.3, the speech signal is intrinsically non-stationary. Since all feature sets of the signal are derived from time–frequency distributions of its energy, it follows that there is some ambiguity in any such representation of the speech signal. The following analysis serves to explain and quantify the effects of non-stationarity.

In Section 2.3.1 we defined the spectrogram as a method for representing the time variation of energy in the speech signal. Rewriting (2.37) and (2.39) in continuous time, the spectrogram $S_x(\omega, t)$, of the signal, $x(t)$, observed through the window $g(\tau)$, becomes

$$S_x(\omega, t) = \left| \int_{-\infty}^{\infty} g(\tau)x(t+\tau)e^{-j\omega\tau}\, d\tau \right|^2 . \tag{2.63}$$

It is natural to ask whether or not we can find a better time–frequency representation in some well-defined sense. In particular, we seek another representation, $F_x(\omega, t)$, that will give better estimates in both time and frequency of the time variation of the spectrum due to non-stationarity. For example, the Wigner transform, $W_x(\omega, t)$, defined by

$$W_x(\omega, t) = \int_{-\infty}^{\infty} x\left(t + \frac{\tau}{2}\right) x^*\left(t - \frac{\tau}{2}\right) e^{-j\omega t}\, d\tau, \tag{2.64}$$

will give perfect resolution of the FM chirp, $x_c(t) = e^{j(\omega_0 t + \frac{\tau}{2} m t^2)}$, at the expense of a discontinuous component, $\delta(\omega)2\cos(2\omega_0 t)$. In comparison, $S_x(\omega, t)$ will give continuous but poor resolution of $x_i(t)$. Perhaps there is an optimal compromise.

We begin by imposing the weak constraint on $F_x(\omega, t)$ that it be shift invariant in both time and frequency. If this condition is satisfied then any $F_x(\omega, t)$ has the form

$$F_x(\omega, t) = \frac{1}{2\pi}\phi(\omega, t) ** W_x(\omega, t) \tag{2.65}$$

for some kernel $\psi(\omega, t)$. The symbol ** indicates convolution in time and frequency, so that (2.65) may be conveniently evaluated by means of the two-dimensional Fourier transform pair,

$$X(\nu, \tau) = \mathcal{F}\{x(\omega, t)\} = \iint_{-\infty}^{\infty} x(\omega, t)e^{-j(\nu t + \omega\tau)}\, d\omega\, dt \tag{2.66a}$$

and

$$x(\omega, t) = \mathcal{F}^{-1}\{X(\nu, \tau)\} = \frac{1}{2\pi} \iint_{-\infty}^{\infty} X(\nu, \tau)e^{j(\nu t + \omega\tau)}\, d\nu\, d\tau. \tag{2.66b}$$

Then (2.65) may be replaced by

$$F_x(\omega, t) = \mathcal{F}^{-1}\{\Phi(\nu, \tau)A_x(\nu, \tau)\}, \tag{2.67}$$

where

$$\Phi(\nu, \tau) = \mathcal{F}\{\phi(\omega, t)\} \tag{2.68a}$$

and

$$A_x(\nu, \tau) = \mathcal{F}\{W_x(\omega, t)\}. \tag{2.68b}$$

For reasons that will become clear, $\Phi(\nu, \tau)$ defined in (2.68a) is sometimes called the point-spread function, and $A_x(\nu, \tau)$ the ambiguity function.

From (2.67) and (2.68) we see that the spectrogram of (2.63) is a special case of (2.65) and can be written as

$$S_x(\omega, t) = W_g(\omega, t) **W_x(\omega, t), \tag{2.69}$$

where $W_g(\omega, t)$ is understood to be the Wigner transform of the window function, $g(\tau)$.

A consequence of (2.69) is that if $F_x(\omega, t)$ is shift invariant and positive, then any such distribution can be written as a superposition of spectrograms of the form

$$F_x(\omega, t) = \int_{-\infty}^{\infty} S_x(\omega, t, g_\alpha(\tau))\, d\alpha, \tag{2.70}$$

where the spectrogram is constructed from the window function $g_\alpha(\tau)$ depending only on the parameter, α.

Any $F_x(\omega, t)$ of the form of (2.70) will have some degree of localization and smoothness. That is, the effect of the kernel function, $\phi(\omega, t)$, will be to spread out the energy at (ω, t) over an ellipse centered at that point and with semi-axes θ_ω and θ_t. Perfect localization corresponds to $\theta_\omega = \theta_t = 0$. We also require that the distribution of energy be smooth, by which is meant that energy at point (ν, τ) in the transform domain is distributed over an ellipse centered at that point and with semi-axes Σ_ν and Σ_τ. Perfect smoothness corresponds to the condition that $\Sigma_\nu = \Sigma_\tau = 0$.

Riley [274] has shown that $F_x(\omega, t)$ is governed by an uncertainty principle according to which it is always the case that

$$\sigma_\omega \Sigma_\tau \geq \tfrac{1}{2} \tag{2.71a}$$

and

$$\sigma_t \Sigma_\nu \geq \tfrac{1}{2} \tag{2.71b}$$

with equality if and only if

$$\phi(\omega, t) = \beta e^{-t^2/2\theta_T^2} e^{-\omega^2/2\theta_\Omega^2}, \tag{2.72}$$

where β, σ_Ω, and σ_T are constants depending on $\phi(\omega, t)$.

There are two important implications of (2.71) and (2.72). First and foremost, any choice of $F_x(\omega, t)$ will cause some loss in resolution in both time and frequency. We cannot overcome this consequence of non-stationarity. However, we will, in Section 2.6, explore a different approach to the problem of non-stationarity based on (2.34).

Second, from (2.69), (2.70) and (2.72) it is clear that a useful $F_x(\omega, t)$ is a spectrogram based on a Gaussian window in time and frequency with $\theta_T \theta_\Omega = \frac{1}{2}$. The improvement resulting from this distribution may be judged by comparing the conventional spectrogram shown in Fig. 2.8a with the improved spectrogram of Fig. 2.8b. Due to computational complexity, the smoothed spectrogram is not used in practice. As a result, the features typically used may have somewhat higher variances than could otherwise be obtained. It is clear from Fig. 2.8 that the conventional spectrogram does provide reasonable information-bearing features.

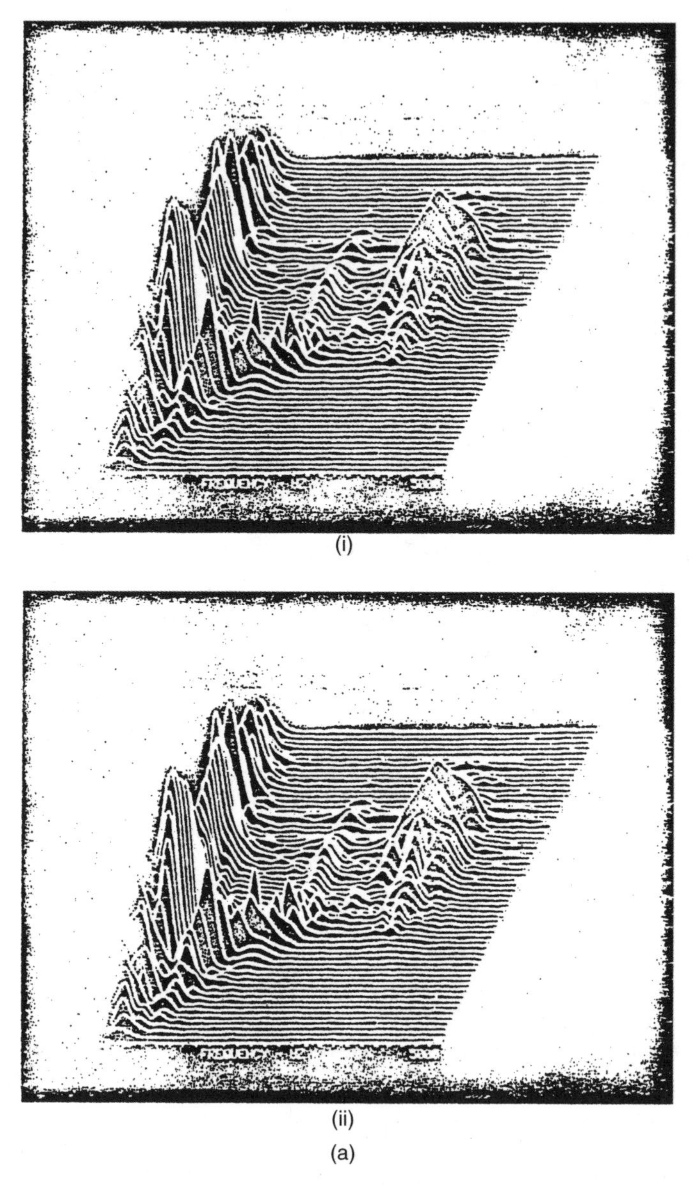

Figure 2.8 (a) (i) Spectrogram of the word "read" computed from contiguous 8.0 ms speech segments; (ii) pitch synchronous spectrogram of the word "read". (b) (i) Smoothed pitch synchronous spectrogram of the word "read"; (ii) smoothed pitch synchronous spectrogram of the word "read"

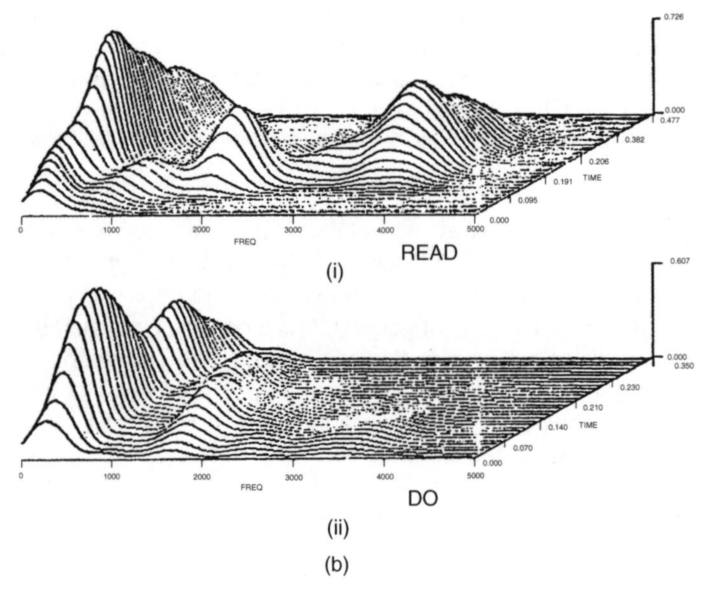

Figure 2.8 (*continued*)

2.5 Classification of Acoustic Patterns in Speech

Everyday experience confirms the highly variable nature of speech. We are aware of the wide ranges of voices, accents, and speaking styles present in ordinary discourse. Yet, virtually all people seem to correctly understand spoken messages effortlessly. One explanation for this remarkable ability is that speech is literate. That is, speech is characterized by a relatively small number of distinct and somehow invariant acoustic patterns. Were this not so, there would be little hope of learning to speak because, as Bruner *et al.* [40] observe:

> [W]ere we to utilize fully our capacity for registering the differences in things and to respond to each event encountered as unique, we would soon be overwhelmed by the complexity of our environment .... The learning and utilization of categories represents one of the most elementary and general forms of cognition by which man adjusts to his environment. [40]

The ability of humans to perform the kind of categorical perception described above has long been recognized as essential to our mental abilities. Plato [248] explained the phenomenon in his theory of forms as follows.

> You know that we always postulate in each case a single form for each set of particular things, to which we apply the same name. Then let us take any set you choose. For example there are many particular beds and tables. But there are only two forms, one of bed and one of table. If you look at a bed, or anything else, sideways or endways or from other angles, does it make any difference to the bed? Isn't it merely that it looks different without being different? And similarly with other things.... The apparent size of an object, as you know, varies with its distance from our eye. So also a stick will look bent if you put it in the

water, straight when you take it out and deceptive differences of shading can make the same surface seem to the eye concave or convex; and our minds are clearly liable to all sorts of confusions of this kind.... Measuring, counting and weighing have happily been discovered to help us out of these difficulties and to ensure that we should not be guided by apparent differences of size, quantity and heaviness, but by calculations of number, measurement and weight.

In common parlance, we speak of our human abilities to recognize patterns. If asked, we would almost certainly agree with the proposition that this ability we possess is a significant aspect of our intelligence. Upon further reflection, we would find the conventional meanings of the terms "pattern" and "intelligence" to be vague enough to cause us difficulty in stating precisely how they are related. We call many diverse objects, events and ideas "patterns". We refer, for example, to sequences of numbers as they occur in puzzles as having a "pattern". We notice the "pattern" of a word exactly contained in a longer one though the two may be unrelated in meaning. We speak of styles of musical compositions as displaying patterns which render the romantic easily distinguished from the classical. Certainly to solve a puzzle or discover a camouflaged sequence of letters or appreciate music requires intelligence. But what exactly does that entail?

The modern explanation of pattern recognition is that the variability of patterns in general and acoustic patterns in particular can be understood by a rigorous appeal to the theory of probability. In Chapter 10 we shall return to the question of pattern recognition and carefully compare the Platonic theory of forms with the modern mathematical theory which we review below.

2.5.1 Statistical Decision Theory

The problem treated in the statistical pattern recognition literature [241] is illustrated in Fig. 2.9. It is customary, though by no means necessary, to choose $\mathbb{R}^n$, the n-dimensional Euclidean space, as the vector space the points of which, $\mathbf{x}_j$, represent the "objects" under consideration. The vector $\mathbf{x}_j$ is the n-tuple $(x_{1j}, x_{2j}, x_{n_j})$ whose components x_{jk} are called features. Each coordinate axis is a scale on which its corresponding feature is measured, so the space is called the feature space.

In these pages, the features will be Fourier spectra as defined in (2.41), LPCs computed from (2.50)–(2.54), or cepstra derived from (2.60) or (2.62). Based on the argument made in (2.42), vectors of such features are measured approximately every 2.5 ms. Later, in Chapter 3, we will refer to these vectors as observations.

In general, we shall be interested in the N-class pattern recognition problem which is that of devising a decision rule, f, which classifies an unknown "object", $\mathbf{x}$, as a member of at most one of the N classes. We are required to say "at most one class" because it is often desirable to make no assignment of a particular vector $\mathbf{x}$. This choice is referred to as the "rejection option" and is but a technical matter which we shall not consider further here. The patterns are regions of the feature space and will be designated by ω_i for $1 \leq i \leq N$. The union of the ω_i is called Ω and

$$\{\omega_i\}_{i=1}^{N} = \Omega \subseteq \mathbb{R}^n. \tag{2.73}$$

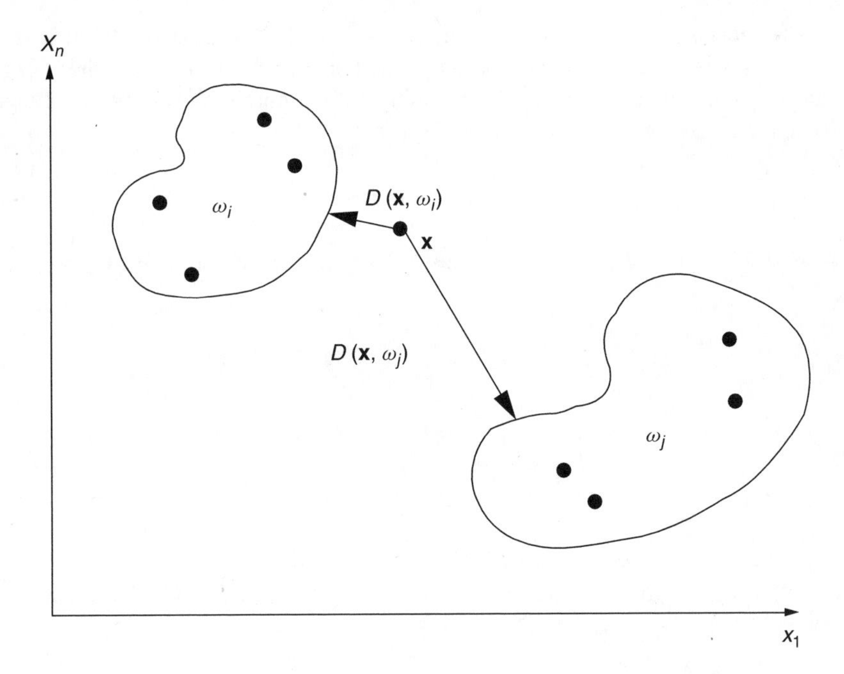

Figure 2.9 The point–set distance function as a measure of pattern dissimilarity

The decision rule, f, is written as

$$f(\mathbf{x}) = \omega_i, \tag{2.74}$$

meaning that the vector $\mathbf{x}$ is assigned to the ith class by the rule f. A rule of the form (2.74) must be constructed on the basis of a training set, $\mathcal{Y}$, comprising the set of vectors $\{\mathbf{y}_j\}_{j=1}^{M}$. Most often, though not necessarily, the training set wil be labeled, that is, each vector will be correctly marked with its class membership. The subset of a labeled training set corresponding to the ith class will be called $\mathcal{Y}_i$, so that $\mathbf{y}_j \in \mathcal{Y}_i \rightarrow \mathbf{y}_j \in \omega_i$. This will be denoted by writing $\mathbf{y}_j^{(i)}$. Obviously

$$\mathcal{Y} = \cup_{i=1}^{N} \mathcal{Y}_i, \tag{2.75}$$

where

$$\mathcal{Y}_i = \{\mathbf{y}_j^{(i)}\}_{j=1}^{m_i}, \tag{2.76}$$

and from (2.75) it is clear that

$$M = \sum_{i=1}^{N} m_i. \tag{2.77}$$

The performance of a decision rule will be evaluated on the basis of a test set $\mathcal{X} = \{\mathbf{x}_l\}_{l=1}^{L}$. This set is assumed to be correctly labeled, but we shall omit the superscript indicating class membership when the meaning is unambiguous without it. To get the truest measure of performance of a decision rule,

$$\mathcal{X} \cap \mathcal{Y} = \phi \tag{2.78}$$

should be strictly observed. The most direct measure of performance is the extent to which

$$f(\mathbf{y}_j^{(i)}) = \omega_i, \qquad \text{for } 1 \le i \le N, 1 \le j \le m_i, \tag{2.79}$$

and

$$f(\mathbf{x}_l^{(i)}) = \omega_i, \qquad \text{for } 1 \le l \le L. \tag{2.80}$$

In practice, (2.79) will more often be satisfied than (2.80).

It will often be useful to define a "prototype" of the ith class $\mathbf{y}_p^{(i)}$. The most common definition is

$$\mathbf{y}_p^{(i)} = \frac{1}{m_i} \sum_{j=1}^{m_i} \mathbf{y}_j^{(i)}, \tag{2.81}$$

which gives, in some sense, an "average" exemplar of the pattern ω_i.

2.5.2 *Estimation of Class-Conditional Probability Density Functions*

As indicated in Figure 2.9, non-parametric decision rules assign an unlabeled vector membership in the pattern to which it is closest in terms of a well-defined distance measure, $D(\mathbf{x}, \omega_i)$. This distance is the distance from a point to a set which must be constructed from an ordinary topological metric. Recall that in elementary topology one defines a metric $d(\mathbf{x}, \mathbf{y})$ on a vector space, say $\mathbb{R}^n$, as any function satisfying

$$d(\mathbf{x}, \mathbf{y}) \ge 0, \tag{2.82a}$$

$$d(\mathbf{x}, \mathbf{y}) = d(\mathbf{y}, \mathbf{x}), \tag{2.82b}$$

$$d(\mathbf{x}, \mathbf{z}) \le d(\mathbf{x}, \mathbf{y}) + d(\mathbf{y}, \mathbf{z}), \tag{2.82c}$$

the positivity, symmetry and triangle inequality conditions. In practice, conditions (2.82) are often not strictly observed.

Some well-known metrics are the following:

$$d(\mathbf{x}, \mathbf{y}) = \begin{cases} 0, & \text{if } \mathbf{x} = \mathbf{y}, \\ 1, & \text{otherwise.} \end{cases} \tag{2.83}$$

This seemingly trivial metric can be extremely useful in problems involving a finite number of attributes which are either present or not.

In continuous feature spaces the Minkowski p-metrics,

$$d_p(\mathbf{x}, \mathbf{y}) = \left[\sum_{i=1}^{N} | x_i - y_i |^p \right]^{\frac{1}{p}}, \tag{2.84}$$

are often used. There are three special cases of (2.84) in common usage: $p = 1$, $p = 2$, and $p = \infty$, giving rise to the Hamming, Euclidean and Chebyshev metrics, respectively. For $p = \infty$,

$$d(\mathbf{x}, \mathbf{y}) = \max_i \{| x_i - y_i |\}. \tag{2.85}$$

In the next section we shall see the importance of metrics of the form

$$d(\mathbf{x}, \mathbf{y}) = (\mathbf{x} - \mathbf{y})T(\mathbf{x} - \mathbf{y})', \tag{2.86}$$

where T is any positive definite $n \times n$ matrix and the prime denotes vector transpose.

Any of the metrics (2.83)–(2.86), or others still, may be used to define point–set distances. Perhaps the simplest of these is the distance to the prototype,

$$D(\mathbf{x}, \omega_i) = d\left(\mathbf{x}, \mathbf{y}_p^{(i)}\right), \tag{2.87}$$

where $\mathbf{y}_p^{(i)}$ is defined by (2.81). The family of "nearest-neighbor" distances is, for reasons which will be seen later, highly effective. Here we let

$$D(\mathbf{x}, \omega_i) = d\left(\mathbf{x}, \mathbf{x}_{[k]}^{(i)}\right), \tag{2.88}$$

where $\mathbf{x}_{[k]}^{(i)}$ is the kth nearest neighbor to $\mathbf{x}$ in the set ω_i. The kth nearest neighbor is usually found by sorting the distances $d(\mathbf{x}, \mathbf{y}_j^{(i)})$ for $1 \leq j \leq m_i$ so that

$$d\left(\mathbf{x}, \mathbf{y}_{j1}^{(i)}\right) \leq d\left(\mathbf{x}, \mathbf{y}_{j2}^{(i)}\right) \leq \ldots \leq d\left(\mathbf{x}, \mathbf{y}_{jm_i}^{(i)}\right). \tag{2.89}$$

The training vector in the kth term in the sequence (2.89) is, of course, $\mathbf{x}_{[k]}^{(i)}$. There is a "rule of thumb" which states that one should set $k \leq \sqrt{m_i}$.

There is an unusual distance which leads to what, for obvious reasons, is called the "majority vote rule":

$$D(\mathbf{x}, \omega_i) = (k_i)^{-1}, \tag{2.90}$$

where k_i is given by

$$k_i = \left| \mathcal{Y}_i \cap \{\mathbf{x}_{[\ell]}\}_{\ell=1}^{k} \right|. \tag{2.91}$$

We use $| S |$ to denote the cardinality of the set S. Hence (2.91) defines k_i as the number of members of $\mathcal{Y}_i$ which are among the K nearest neighbors of $\mathbf{x}$ without respect to class.

In a spirit similar to that of (2.88), one may define

$$D(\mathbf{x}, \omega_i) = \frac{1}{K} \sum_{k=1}^{K} d\left(\mathbf{x}, \mathbf{x}_{[k]}^{(i)}\right), \tag{2.92}$$

which is just the average distance from $\mathbf{x}$ to the K nearest members of $\mathcal{Y}_i$. Alternatively, we may weigh the distance by the number of samples of $\mathcal{Y}_i$ which are within the distance (2.92) of $\mathbf{x}$. Thus let

$$D(\mathbf{x}, \omega_i) = \frac{1}{\ell} \sum_{k=1}^{K} d\left(\mathbf{x}, \mathbf{x}_{[k]}^{(i)}\right), \tag{2.93}$$

where

$$\ell = \max_{j} \left\{ j \mid d\left(\mathbf{x}, \mathbf{x}_{[j]}^{(i)}\right) \leq \frac{1}{K} \sum_{k=1}^{K} d\left(\mathbf{x}, \mathbf{x}_{[k]}^{(i)}\right) \right\}. \tag{2.94}$$

From (2.87)–(2.94) we get the family of non-parametric decision rules

$$f(\mathbf{x}) = \omega_i \iff D(\mathbf{x}, \omega_i) \leq D(\mathbf{x}, \omega_j), \quad 1 \leq j \leq N. \tag{2.95}$$

Rules of the form (2.95) are closest to our intuitive sense of classification. They are easy to implement since the training process consists simply in collecting and storing the labeled feature vectors. These rules often outperform all others in terms of classification accuracy. They may, however, be quite costly to operate, especially in high-dimensional spaces.

Thus far, we have simply stated the rules, justifying them only by an appeal to intuition. We defer, until the end of the next section, a more rigorous analysis of their underlying principles.

Parametric Decision Rules

The primary method of classification to be considered is that in which an unknown vector is said to belong to the class of highest probability based on the values of the observed features. This is denoted by

$$f(\mathbf{x}) = \omega_i \iff P(\omega_i \mid \mathbf{x}) \leq P(\omega_j \mid \mathbf{x}), \quad 1 \leq j \leq N. \tag{2.96}$$

We can construct a classifier by treating feature vectors as random variables. A consequence of Bayes' law is that

$$P(\omega_i \mid \mathbf{x}) = \frac{p(\mathbf{x} \mid \omega_i) P(\omega_i)}{p(\mathbf{x})}, \tag{2.97}$$

where $P(\omega_i)$ is the "prior probability" of the ith class, that is, the probability of a vector coming from ω_i before it is observed; $p(\mathbf{x})$ is the probability density function of the feature vectors without respect to their class and $p(\mathbf{x} \mid \omega_i)$ is called the ith class-conditional probability density function, by which term is meant that $p(\mathbf{x} \mid \omega_i)$ is the probability density function of feature vectors which correspond to the ith pattern only.

Since the factor $p(\mathbf{x})$ is common to $p(\mathbf{x} \mid \omega_i)$ for all i and assuming for a moment that $p(\omega_i) = \frac{1}{N}$ for all i, making the patterns equally likely a priori, we can rewrite (2.96) in view of (2.97) as

$$f(\mathbf{x}) = \omega_i \iff p(\mathbf{x} \mid \omega_i) \geq p(\mathbf{x} \mid \omega_j), \qquad \text{for } 1 \leq j \leq N. \tag{2.98}$$

The decision rule (2.98) is called the maximum a posteriori probability (MAP) rule.

The MAP rule is sometimes augmented by prior probabilities and a loss or cost function to derive the "minimum risk rule". The total risk of deciding ω_i at $\mathbf{x}$, R, is the loss summed over all classes, which is computed from

$$R_i = \sum_{j=1}^{N} L_{ij} \frac{p(\mathbf{x} \mid \omega_i) P(\omega_i)}{p(\mathbf{x})}, \tag{2.99}$$

where the loss function L_{ij} is the penalty incurred for misclassifying ω_j as ω_i. It is customary to let

$$L_{ij} = c_{ij}(1 - \delta_{ij}), \tag{2.100}$$

where c_{ij} is a fixed cost for the ω_j to ω_i classification error and δ_{ij} is the Kronecker delta function. In this case we can substitute $f(\mathbf{x})$ for ω_i in (2.99) and solve it by finding that $f(\mathbf{x})$ which minimizes R. The solution

$$f(\mathbf{x}) = \omega_i \iff L_{ij} P(\omega_j) p(\mathbf{x} \mid \omega_j) \leq L_{kj} P(\omega_k) p(\mathbf{x} \mid \omega_k), \qquad \text{for } 1 \leq k \leq N, \tag{2.101}$$

is called the "minimum risk" decision rule.

If we set $c_{ij} = 1$, then the risk becomes P_e, the probability of classification error. For this particular loss function, called the zero–one loss function, (2.101) becomes

$$f(\mathbf{x}) = \omega_i \iff p(\mathbf{x} \mid \omega_i) P(\omega_i) \geq p(\mathbf{x} \mid \omega_j) P(\omega_j), \qquad \text{for } 1 \leq j \leq N. \tag{2.102}$$

Rule (2.102), therefore, minimizes P_e, meaning that if the $p(\mathbf{x} \mid \omega_i)$ and $P(\omega_i)$ are known exactly, then no other classification scheme relying only on the feature vectors $\mathcal{Y}$, can yield a lower rate of incorrect classifications as the number of trials tends to infinity.

A standard proof of the optimality of (2.102) is given by Patrick [241]. The practical meaning of the theoretical result is that we assume that $p(\mathbf{x}|\omega_i)$ can be estimated to any desired degree of accuracy and that good estimates will provide a asymptotically optimal performance in the limit as M gets large (see (2.77)).

This important result can be demonstrated as follows. Let us define the point risk of a decision rule, $r(f(\mathbf{x}))$, by

$$r(f(\mathbf{x})) = \sum_{j \neq i} p(\mathbf{x} \mid \omega_j)P(\omega_j)/p(\mathbf{x}) \tag{2.103}$$
$$= 1 - p(\mathbf{x} \mid \omega_i)P(\omega_i)/p(\mathbf{x})$$
$$= P_e.$$

From (2.103) it is clear that P_e is minimized by the decision rule of (2.101), which we shall designate as $f^*(\mathbf{x})$. The global risk of a decision rule, $R(f(\mathbf{x}))$ is defined as

$$R(f(\mathbf{x})) = E_{\mathbb{R}^n}\{r(f(x))\} \tag{2.104}$$
$$= \int_{\mathbb{R}^n} r(f(\mathbf{x}))p(\mathbf{x})\,d\mathbf{x}$$
$$= \int_{\mathbb{R}^n} \sum_{i=1}^{N} L(f(\mathbf{x}), i)p(\mathbf{x} \mid \omega_i).$$

Now note that all terms in the integrand of (2.104) are positive so it must be minimized pointwise. Since we know from eq. (2.103) that the point risk is minimized by $f^*(\mathbf{x})$, so must the global risk. Thus $f^*(\mathbf{x})$ is the optimal decision rule.

In many cases of practical importance, the physics of the process under study will dictate the form of $p(\mathbf{x} \mid \omega_i)$ to be a member of a parametric family of densities. The values of the parameters can then be estimated from the training data.

Perhaps the most useful parametric form is the multivariate normal or Gaussian density function,

$$p(\mathbf{x} \mid \omega_i) = \frac{1}{(2\pi)^{n/2} \mid U_i \mid^{1/2}} E^{-\frac{1}{2}(\mathbf{x}-\boldsymbol{\mu}_i)\mathbf{U}_i^{-1}(\mathbf{x}-\boldsymbol{\mu}_i)'}, \tag{2.105}$$

where $\boldsymbol{\mu}_i$ is the mean vector defined by

$$\boldsymbol{\mu}_i = E\{\mathbf{x}^{(i)}\} \tag{2.106}$$

and $\mathbf{U}_i$ is the covariance matrix whose jkth entry u_{ijk} is given by

$$u_{ijk} = E\left\{\left(x_j^{(i)} - \mu_{ij}\right)\left(x_k^{(i)} - \mu_{ik}\right)\right\}. \tag{2.107}$$

The expectation operator, E, appearing in (2.106) and (2.107) computes the expected value of $p(\mathbf{x})$ from

$$E\{g(\mathbf{x})\} = \int_{\mathbb{R}^n} g(\mathbf{x})p(\mathbf{x})\,d\mathbf{x}. \tag{2.108}$$

The utility of the Gaussian density function stems from two facts. First, quite often the actually observed features are really linear combinations of independent non-normal random variables which are either unknown to us or not directly measurable. The well-known central limit theorem tells us that such features are well described as Gaussian random variables in the following sense. If x_k are independent and identically distributed random variables for $1 \leq k \leq K$ and having finite mean μ and variance σ^2, then the sum, $S_k = \sum_{k=1}^{K} x_k$, is a random variable such that

$$\lim_{k \to \infty} p \left[\frac{S_k - k\mu}{\sqrt{k\sigma^2}} < \theta \right] = \int_{-\infty}^{\theta} e^{-\frac{y^2}{2}} \, dy. \tag{2.109}$$

In other words, as k increases, S_k becomes more nearly Gaussian.

Second, as is clear from (2.105), the normal density is completely specified by its first- and second-order moments, which are easily estimated from training data by

$$\hat{\boldsymbol{\mu}}_i = \frac{1}{m_i} \sum_{j=1}^{m_i} \mathbf{y}_j^{(i)} = \mathbf{y}_p^{(i)} \tag{2.110}$$

and

$$\hat{\mathbf{U}}_i = \frac{1}{m_i} \mathbf{Y}'\mathbf{Y} - \boldsymbol{\mu}_i' \boldsymbol{\mu}_i, \tag{2.111}$$

where the jth row of $\mathbf{Y}_i$ is $\mathbf{y}_j^{(i)}$ for $1 \leq j \leq m_i$.

There are other closed-form density functions which can be used in the same way. The more common ones are chi-square, beta, gamma, binomial, Poisson, Cauchy and Laplace. If neither these nor any other closed-form density is appropriate, then one can resort to several methods of approximating the class-conditional densities in terms of orthogonal polynomials, potential functions, Parzen estimators or Gaussian mixtures.

Equivalence of the Basic Decision Rules

We are now in a position to observe several respects in which parametric and non-parametric classifiers are equivalent. First, it is immediately clear that both methods have a geometrical interpretation in that they implicity partition the feature space into disjoint regions. In the non-parametric case, the decision boundary between ω_i and ω_j is just

$$\left\{ \mathbf{X} \in \mathbb{R}^n \mid D(\mathbf{x}, \omega_i) = D(\mathbf{x}, \omega_j) \right\} \tag{2.112}$$

for whatever distance measure is chosen. Similarly in the parametric case, the decision boundary is the locus

$$\left\{ \mathbf{x} \in \mathbb{R}^n \mid p(\mathbf{x} \mid \omega_i) = p(\mathbf{x} \mid \omega_j) \right\}. \tag{2.113}$$

Clearly (2.113) can be modified to accommodate (2.101) and (2.102) in an analogous way.

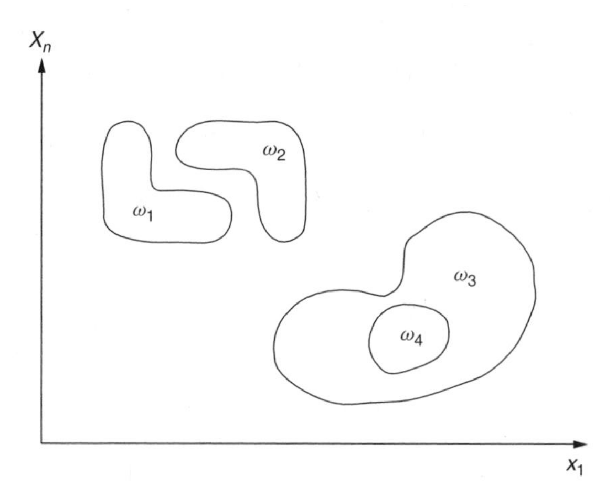

Figure 2.10 Non-linearly separable pattern classes

Note that, in general, the loci defined by (2.112) and (2.113) are not hyperplanes, and can easily separate the pathological cases depicted in Fig. 2.10.

The relationship between the two decision rules is somewhat less obvious so that it is helpful to motivate the analysis by considering the special case of normal densities. Taking the natural log of both sides of (2.105), we get

$$\log_e(p(\mathbf{x} \mid \omega_i)) = -\frac{1}{2}\left[n \log_e(2\pi) + \log \mid \mathbf{U}_i \mid + (\mathbf{x} - \boldsymbol{\mu}_i)\mathbf{U}_i^{-1}(\mathbf{x} - \boldsymbol{\mu}_i)'\right]. \quad (2.114)$$

Neither the first nor second terms on the right-hand side of (2.114) is dependent on $\mathbf{x}$, and $\mathbf{U}_i^{-1}$ is positive definite, so that

$$D(\mathbf{X}, \omega_i) = (\mathbf{x} - \boldsymbol{\mu}_i)\mathbf{U}_i^{-1}(\mathbf{x} - \boldsymbol{\mu}_i)' \quad (2.115)$$

is a well-formed metric. In fact it is a special case of (2.86) which is often referred to as the Mahalanobis distance. Clearly, then, given values for $\boldsymbol{\mu}_i$ and $\mathbf{U}_i$, either a probability density or a distance function is equally well determined. Furthermore, the natural logarithm is monotonic in its argument so that distance and density are inversely monotonically related in the sense that

$$p(\mathbf{x}_1 \mid \omega_i) < p(\mathbf{x}_2 \mid \omega_i) \Longleftrightarrow D(\mathbf{x}_1, \omega_i) > D(\mathbf{x}_2, \omega_i). \quad (2.116)$$

The inverse variation of the Mahalanobis distance and the normal density function extends to arbitrary probability densities and metrics. The general relationship is elucidated by the theory of non-parametric density estimation a central result of which is the following theorem due to Fix and Hodges [241], stated here without proof.

Theorem 1 (Fix and Hodges). Let $\mathbf{x} \in \mathbb{R}^n$ be a random variable of probability density function $p(\mathbf{x})$ continuous at $\mathbf{x}$ and let $\{\mathbf{y}_j\}_{j=1}^{M}$ be a set of M observations of $\mathbf{x}$. Define a sequence of convex tolerance regions of $\mathbf{x}$, $\{\mathcal{T}_i\}_{i=1}^{\infty}$, such that exactly k_i of the $\mathbf{y}_j$ lie inside $\mathcal{T}_i$, a region enclosing volume Φ_i. Further, let the tolerance regions be such that they shrink rapidly enough around $\mathbf{x}$ so that

$$\lim_{M\to\infty} \sup_{\mathbf{y}_j \in \mathcal{T}_M}\{\| \mathbf{x} - \mathbf{y}_j \|\} = 0 \tag{2.117}$$

but slowly enough so that

$$\lim_{M\to\infty} \{M\Phi_M\} = \infty \tag{2.118}$$

and

$$\lim_{M\to\infty} k_M = \infty. \tag{2.119}$$

Finally, define the sequence of functions

$$\hat{p}_M(\mathbf{x}) = \frac{k_M}{M\Phi_M(x)}. \tag{2.120}$$

Then $\hat{p}_M(\mathbf{x})$ is a consistent estimator of $p(\mathbf{x})$ in the sense that

$$\lim_{M\to\infty} E_{\mathbb{R}^n}\{p(\mathbf{x}) - \hat{p}_M(\mathbf{x})\} = 0. \tag{2.121}$$

Equation (2.120) exactly captures our physical intuition of the inverse relationship of density to volume. As illustrated in Figure 2.11, the $\{\mathcal{T}_k\}$ are of arbitrary convex shape. Fraser [241] asserts that it is useful to let $\mathcal{T}_k$ be the sphere of radius $\| \mathbf{x} - \mathbf{x}_{[k]} \|$ centered at $\mathbf{x}$.

The utility of the spherical tolerance regions is made manifest as follows. Let us define the coverage, C_k, of the kth tolerance region as the probability that some observation $\mathbf{x}$ lies interior to $\mathcal{T}_k$. Thus

$$C_k = \int_{\mathcal{T}_k} p(\mathbf{x})\, d\mathbf{x}. \tag{2.122}$$

Since $\mathcal{T}_k$ is dependent on $\{\mathbf{y}_j\}_{j=1}^{M}$, C_k is itself a random variable on the unit interval having probability density function $g(C_k)$. Following Wilks [241], we observe that the probability of exactly m out M new observations falling in $\mathcal{T}_k$ has the binomial distribution

$$p(|\{\mathbf{x}, \in \mathcal{T}_k\}| = m) = \frac{M!}{m!(M-m)!} C_k^m (1 - C_k)^{M-m}, \tag{2.123}$$

from which it follows that $g(C_k)$ is the beta density $\beta(m, M - m)$ independent of $p(\mathbf{x})$. In other words, the coverages of the spherical tolerance regions are not determined by the underlying statistics of the observations. Such tolerance regions are said to be "distribution-free", the significance of which property is that the estimator (2.120) is equally valid for all random variables. Of course, as is clear from (2.117)–(2.120) and (2.122), the theory is an asymptotic one while, in practice, we never have infinitely many samples. Yet, in

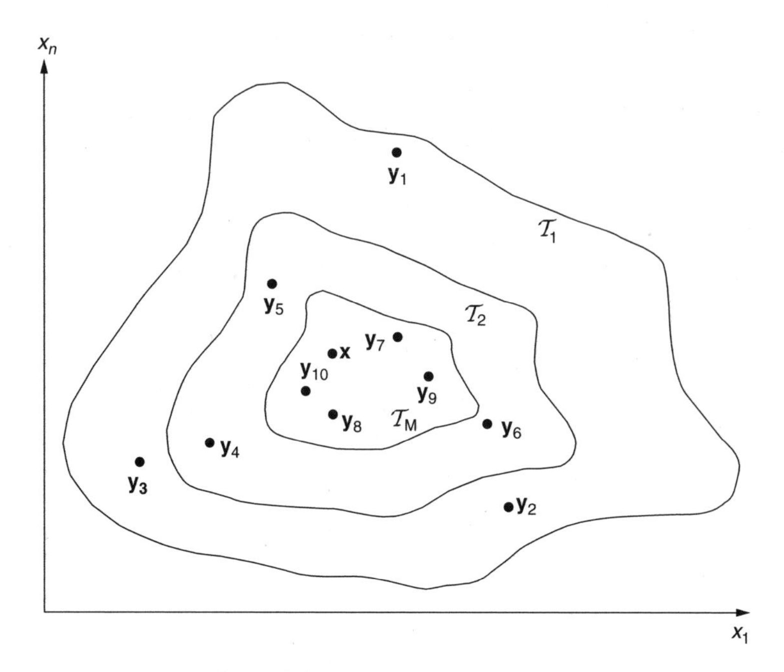

Figure 2.11 Tolerance regions in $\mathbb{R}^n$

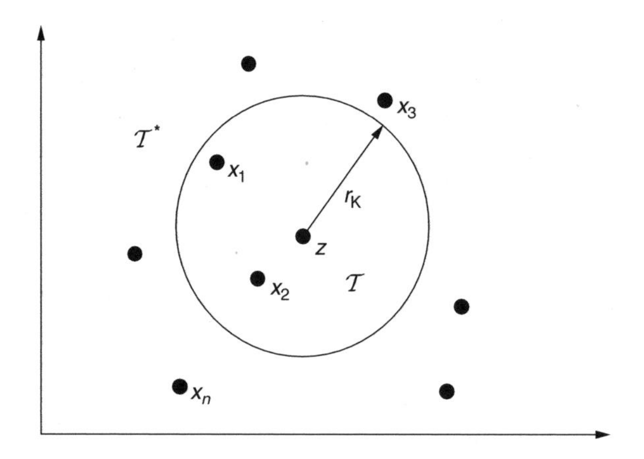

Figure 2.12 Points in the observation space $\{x_1, x_2, \ldots, x_n\}$ drawn from class ω_i having density function $f(z)$

practice, one collects as large a set of training data as possible and most often achieves good results.

The actual reduction to practice of the Fix and Hodges theorem rests on the observation that in $\mathbb{R}^n$, for spherical $\{\mathcal{T}_k\}$, Φ_k is proportional to $d(\mathbf{x}, \mathbf{x}_{[k]})^n$ for any proper metric $d(\mathbf{x}, \mathbf{y})$. Thus from (2.120) and Figure 2.12, it is clear the $p(\mathbf{x})$ and $d(\mathbf{x}, \mathbf{x}_{[k]})$ bear an inverse relationship to each other.

Loftsgaarten and Quesenberry [204] derived (2.121) for the special case $k = 1$ from which we get the nearest neighbor rule (2.88). For arbitrary k we get the generalized kth nearest neighbor rule (2.88) of Patrick and Fischer [242]. Whereas the rules based on (2.88) derive from applying the theory independently to each $\mathcal{Y}_i$, Cover and Hart [53] used but a single sequence of tolerance regions on $\mathcal{Y}$ to show that $\frac{k_i}{M}$, where k_i is as defined in (2.91), is a consistent estimator of $p(\mathbf{x} \mid \omega_i)$ in the sense of (2.122). From this result comes the "majority vote" classifier (2.90). The rules (2.92) and (2.93) do not admit of distribution-free tolerance regions but are in the same spirit as those cited above and have, in fact, been observed to outperform them [189].

2.5.3 Information-Preserving Transformations

As we noted earlier, the physical signal that we measure may be quite complex but encoded in it is one of N patterns whose identity is our only concern. The signal is transformed, in the sense that its representation is changed, by the measurement and feature extraction process until it is finally decoded by the classifier. In this context, the transformations involved will be useful to us to the extent that they leave the identity of the patterns unaltered. In this sense these transformations may be said to be information-preserving.

Mathematically, the process of changing the representation of a signal will be accomplished by the abstract transformation, $\mathcal{T}$, which maps the abstract space of $\mathcal{A}$ into another abstract space $\mathcal{B}$, which we signify by writing

$$\mathcal{T} : \mathcal{A} \to \mathcal{B}; \tag{2.124}$$

for any $a \in \mathcal{A}$ there will be some $b \in \mathcal{B}$, in which case we say that b is the image of a under $\mathcal{T}$ and write $\mathcal{T}(a) = b$.

Let us suppose that we are using a particular pattern recognition system which manifests classification error probability

$$p_e = \sum_{i=1}^{N} \int_{E_i} p(\mathbf{x} \mid \omega_i)\, d\mathbf{x}, \tag{2.125}$$

where

$$E_i = \left\{ \mathbf{x} \mid p(\mathbf{x} \mid \omega_i) < p(\mathbf{x} \mid \omega_j),\ i \neq j \right\}. \tag{2.126}$$

If we now introduce the transformation $\mathcal{T}$ into the process we would like to observe that

$$p_e = \sum_{i=1}^{N} \int_{\mathcal{T}(E_i)} p(\mathcal{T}(\mathbf{x}) \mid \omega_i) d(\mathcal{T}(\mathbf{x})), \tag{2.127}$$

where

$$d(\mathcal{T}(\mathbf{x})) = \sum_{i=1}^{n} \frac{\partial \mathcal{T}}{\partial x_i} dx_i. \tag{2.128}$$

If (2.127) is satisfied then $\mathcal{T}$ will be considered to be an information-preserving transformation.

It is easy enough to either measure p_e empirically or compute it from (2.125) in order to evaluate the quality of a transformation. It is more difficult to use p_e in order to design an information-preserving transformation. For this purpose it will often be desirable to design $\mathcal{T}$ so that it preserves the topology of the space in the sense that if

$$D(\mathbf{x}, \omega_i) > D(\mathbf{x}, \omega_j) \tag{2.129}$$

then

$$D(\mathcal{T}(\mathbf{x}), \omega_i) > D(\mathcal{T}(\mathbf{x}), \omega_j). \tag{2.130}$$

If (2.130) holds whenever (2.129) does, then $\mathcal{T}$ will not alter the order of the distances from $\mathbf{x}$ to the patterns and hence, by using decision rules (2.95), will not alter p_e.

The method may be adapted to the discrete problem of feature extraction. In so doing we seek a linear transformation, $\mathcal{T}$, such that

$$\mathbf{y}'_{m\times 1} = \mathcal{T}_{m\times n}\mathbf{x}'_{n\times 1}, \tag{2.131}$$

thus $\mathcal{T}$ maps an n-dimensional feature space onto an m-dimensional one with $m \leq n$.

We want $\mathcal{T}$ to be optimal in the sense that it minimizes E, the norm squared difference between the n-dimensional feature space and its m-dimensional image under $\mathcal{T}$. Let

$$E = \sum_{i=1}^{M} \frac{\|\mathbf{y} - \sum_{k=1}^{m} a_{ik}\mathbf{t}_k\|^2}{\| \mathbf{y}_i \|^2}, \tag{2.132}$$

where a_{ik} (for $1 \leq i \leq M$, $1 \leq k \leq m$) are constants to be determined and $\mathbf{t}_k$ is the kth row of $\mathcal{T}$ which we seek.

Setting $\frac{\partial E}{\partial a_{ik}} = 0$ for $1 \leq i \leq N$, $1 \leq k \leq m$, and solving for a_{ik} results in

$$\sum_{j=1}^{m} a_{ij}(\mathbf{t}_j \cdot \mathbf{t}_k) = \mathbf{y}_i \cdot \mathbf{t}_k. \tag{2.133}$$

If we require that $\mathcal{T}$ be orthonormal then (2.131) becomes simply

$$a_{ik} = \mathbf{y}_i \cdot \mathbf{t}_k. \tag{2.134}$$

Substituting (2.134) in (2.132) and expanding the numerator, we observe that we can minimize E by maximizing

$$\beta = \sum_{k=1}^{m} \mathbf{t}_k \mathbf{U} \mathbf{t}'_k \tag{2.135}$$

subject to the m orthonormality constraints

$$\gamma_k = \|\mathbf{t}_k\|^2 = 1, \qquad k = 1, 2, \dots, m. \tag{2.136}$$

The desired maximum can be calculated by the method of Lagrange multipliers according to which we set

$$d\beta + \sum_{k=1}^{m} \lambda_k \gamma_k = 0, \tag{2.137}$$

which reduces to

$$\mathbf{t}_k'(\mathbf{U} - \lambda_k \mathbf{I}) = 0, \qquad \text{for } 1 \le k \le m, \tag{2.138}$$

where $\mathbf{U}$ is the covariance matrix of the training set $\mathcal{Y}$ and, by definition, λ_k is an eigenvalue with corresponding eigenvector $\mathbf{t}_k$. $\mathbf{U}$ is obtained from

$$\mathbf{U} = \frac{1}{M}\mathbf{Y}'\mathbf{Y} - \mathbf{u}'\mathbf{u}, \tag{2.139}$$

$\mathbf{u}$ being the mean vector of $\mathcal{Y}$, namely,

$$\mathbf{u} = \frac{1}{M}\sum_{i=1}^{N}\sum_{j=1}^{m_i} \mathbf{y}_j^{(i)}. \tag{2.140}$$

Note that (2.139) and (2.140) are formally identical to (2.111) and (2.110), respectively.
Substituting (2.138) into (2.135), we find that at the maximum

$$\beta = \sum_{k=1}^{m} \lambda_k. \tag{2.141}$$

Thus the maximum possible value of β is the sum of the m largest eigenvalues of $\mathbf{U}$. This occurs when the rows of $\mathcal{T}$ are the eigenvectors corresponding to those eigenvalues. The significance of the eigenvalues is that they are the variances of $\mathcal{Y}$ in the directions of the corresponding eigenvectors which are actually the coordinate axes of the new feature space obtained from the transformation $\mathcal{T}$.

The fidelity of the transformation $\mathcal{T}$ is just the error E, which can be shown to be

$$E = \sum_{k=m+1}^{n} \lambda_k, \tag{2.142}$$

where the summation assumes that the eigenvalues are sorted in descending order of magnitude. The smaller the value of E, the more information is preserved by the transformation $\mathcal{T}$.

Typically, if we plot the magnitude of the eigenvalues against their descending rank in magnitude, we get the behavior shown in Fig. 2.13. When the magnitude of successive eigenvalues drops sharply, we obtain an indication of the intrinsic dimensionality, n^*, of the feature space.

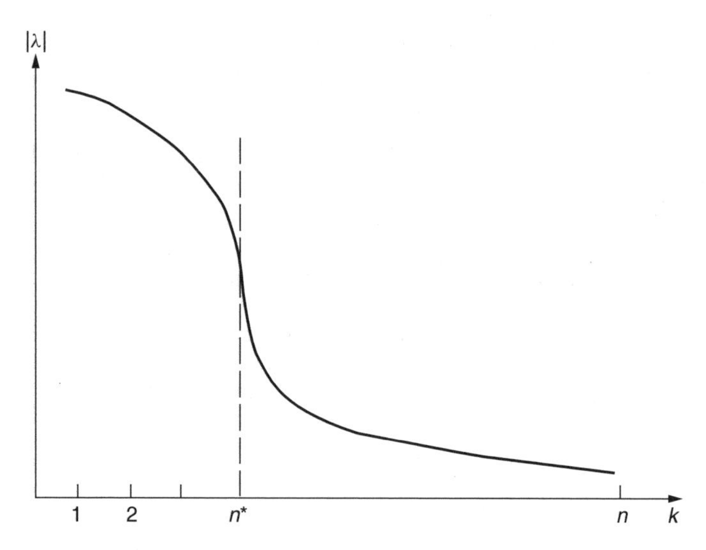

Figure 2.13 Plot of eigenvalues showing the intrinsic dimensionality n^*

2.5.4 *Unsupervised Density Estimation – Quantization*

Unsupervised density estimation locates the more populated regions of the space and labels them as patterns. Algorithms which perform this function are often called clustering procedures or quantizers. We shall consider a method for locating the clusters in unlabeled data based on a generalization of the Lloyd–Max [203] quantizer. The method will appear in Section 3.1.1.

In what follows, we shall assume we are given a finite set, Ω, of N observations,

$$\Omega = \{\mathbf{x}_1, \mathbf{x}_2, \dots, \mathbf{x}_N\}. \tag{2.143}$$

Let us assume that Ω contains samples of $M > 1$ different pattern classes. The observations are not assigned a priori to a class, nor are the classes themselves named. We simply wish to be able to distinguish among the observations in a reasonable way, trusting that the significance of the distinction will emerge from another source of information.

Finally, then, we are left to estimate the density $F(\mathbf{z})$, by some robust but necessarily suboptimal technique. Many such procedures are discussed in the literature [121]. Because of its generality and utility, we describe here one particular method called the k-means algorithm [340].

The k-means procedure is an automatic iteration scheme which will quite reliably find any specified number of clusters. Its properties are well understood [198]. The iteration consists of three basic steps: classification, computation of cluster centers, and convergence testing.

Assuming that we wish to find M clusters, we choose M arbitrary tokens to serve as initial cluster centers. For simplicity, we set

$$\mathbf{x}_p^{(i)} = \mathbf{x}_i, \qquad \text{for } 1 \le i \le M. \tag{2.144}$$

Classification then proceeds on the basis of the nearest neighbor rule, namely,

$$\mathbf{x}_j \in \omega_i \iff \delta(\mathbf{x}_j, \mathbf{x}_p^{(i)}) \leq \delta(\mathbf{x}_j, \mathbf{x}_p^{(k)}), \qquad 1 \leq k \leq M. \tag{2.145}$$

After (2.145) has been applied for $1 \leq j \leq N$, we recompute the cluster centers using a minimax criterion. That is, we let

$$\mathbf{x}_p^{(i)} = \mathbf{x}_j^{(i)} \qquad \text{such that } \max_k\{\delta(\mathbf{x}_j^{(i)}, \mathbf{x}_k^{(i)})\} \text{ is minimized for } 1 \leq i \leq M. \tag{2.146}$$

The convergence test consists of checking whether or not the same tokens are designated as cluster centers as in the previous iteration. If not, another iteration is performed. If the $\mathbf{x}_i$ themselves were given in terms of coordinates instead of in terms of pairwise dissimilarity values, then (2.146) could be replaced by

$$\mathbf{x}_p^{(i)} = \frac{1}{m_i}\sum_{j=1}^{m_i} \mathbf{x}_j^{(i)}. \tag{2.147}$$

2.5.5 A Note on Connectionism

During the past decade, there has been a resurgence of interest in connectionist theories.

Recent research on this topic can trace its origins to the work of McCulloch and Pitts [215] who suggested that what was important about the central nervous system was not the details of its electrochemical operation, but rather its implementation of Boolean logic. Specifically, they showed that combinations of binary "neurons" could be used to form a model of universal computation, the significance of which we will examine closely in Chapter 9.

The "neural networks" of McColloch and Pitts influenced Rosenblatt [280] to propose linear threshold devices or perceptrons, and Hebb [123] to devise still other neural models that could adapt to external stimuli and recognize patterns. Their research garnered considerable attention because of the similarity of their models to real biological systems, the likeness strengthened by the advent of the Hodgkin and Huxley [129] model of the non-linear behavior of nerve cells.

The perceptrons of Rosenblatt came under attack on two fronts. Neurophysiologists argued that they are too simple and monolithic to be considered serious models of the structure of real brains. Then, mathematicians, especially Minsky and Papert [221], demonstrated that they are computationaly weak models unable to implement logical functions such as the XOR. However, a straightforward generalization of the perceptron removed the mathematical objection. By cascading perceptrons to form the multi-layer perceptron (MLP) of Fig. 2.14, general pattern recognition could be achieved. The training algorithm for the MLP was known to Rosenblatt but did not gain wide recognition until Rumelhart [282] devised an algorithm based on propagating the classification errors backward through the MLP. The method is, in fact, just a classical gradient optimization procedure.

The power of the MLP to discriminate between patterns rests on the Kolmogorov representation theorem [205] which states that any continuous function of N variables can be computed using only linear combinations of non-linear continuous increasing

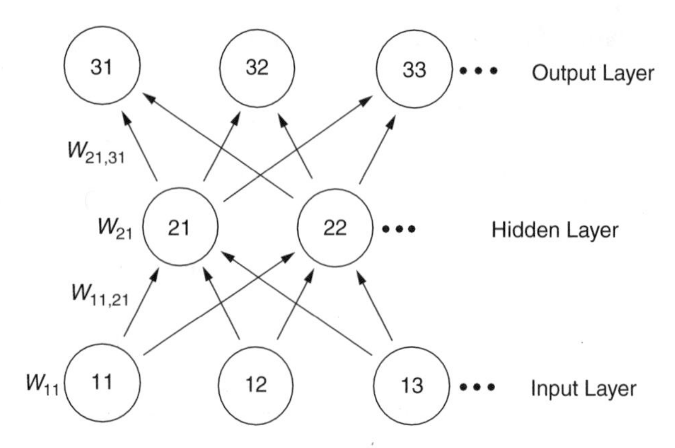

Figure 2.14 A multi-layer perceptron neural network

functions of only one variable. This theorem is a response to the 13th problem of Hilbert (cf. Reid [273]) regarding the solutions of general equations of seventh degree using functions of only two variables. Its relevance to the MLP is that the activation equations are exactly of the form considered by Kolmogorov and hence any arbitrary decision boundary can be computed from them.

If the MLP or other related connectionist circuits are both powerful enough computationally and faithful enough biologically [134], then some technical questions must be addressed. Given that the classifiers of (2.98)–(2.102) are optimal, what additional information is captured in the biologically inspired theories but not available to statistical decision-theoretic methods that could provide for superior performance of the former? Second, how are dynamics represented in neural models? Time is intrinsic in the classical methods but not in the connectionist approach. Time is an essential element of all pattern recognition in language. Finally, what benefit, if any, accrues to a relatively uniform architecture composed of simple elements? Does it allow for integration of diverse signals? Does it provide for robust performance in the presence of component failures?

There are presently no definitive answers to these questions. It will be interesting to see if answers emerge in the decades ahead.

2.6 Temporal Invariance and Stationarity

The motivation for the non-parametric methods is found in the observation that whereas "instantaneous" spectra or other primitive measurements of the speech signal can be compared without regard to their temporal location, sequences of these measurements, such as are required to represent speech signals of greater temporal extent, must, due to the non-stationarity of speech signals, take account of time to be meaningfully compared. In particular, in order to compare word-length utterances, we shall require a distance measure that is invariant to those local changes of time scale which are not associated with the identity of the utterance. Conversely, a distance measure useful for classification must be sensitive to other temporal variations that are information-bearing.

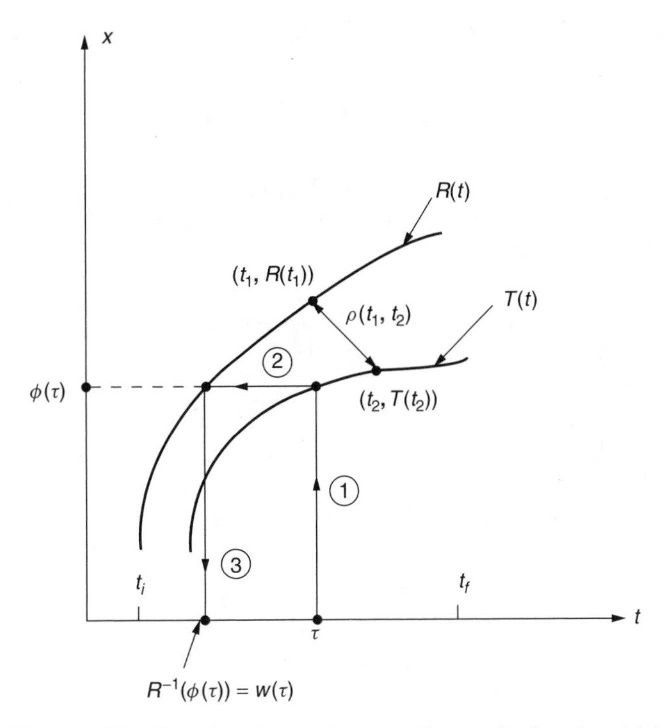

Figure 2.15 Changing time scales according to the function $\phi(\tau)$

A method for constructing just such a distance function was suggested by Vintsyuk [322]. The reasoning is best understood by referring to Fig. 2.15 in which two words, v and $w \in V$, are represented by the continuous functions $x(t)$ and $y(t)$, respectively. Suppose that for any $t \in [0, T]$, the local or pointwise distance between $x(t)$ and $y(t)$ is computed by evaluating the known function $\rho(x(t), y(t))$. To measure $D(u, v)$, the distance between $x(t)$ and $y(t)$, and hence between v and w, we must integrate the local metric over some appropriate interval.

2.6.1 A Variational Problem

To achieve some measure of invariance with respect to local time scale, it is natural to perform the integration in such a way that its minimum value is always obtained for the particular x and y. Vintsyuk proposed that

$$D(v, w) = \min_{\{\phi(t)\}} \int_0^T \rho(x(t), y(\phi(t)))g(t, \phi(t), \dot{\phi}(t))\, dt, \tag{2.148}$$

where $\phi(t)$ is any function that maps the natural time scale of $y(t)$ monotonically onto that of $x(t)$, and g is a weighting function which is small when $\phi(t) \cong t$ and large otherwise. Intuitively speaking, the effect of g is to prevent $\phi(t)$ from causing too abrupt or too non-linear a change of time scale.

Now note that (2.146) is an instance of the classical variational problem $\delta J = 0$, where

$$J = \int_{t_1}^{t_2} F(\tau, \phi(\tau), \dot{\phi}(\tau))\, d\tau. \tag{2.149}$$

In (2.149), F is a known function and $\dot{\phi}$ signifies, as usual, differentiation with respect to time. The solution to (2.149) is found by solving the well-known Euler–Lagrange equation (cf. [329, pp. 20ff.]).

$$\frac{\partial F}{\partial \phi} - \frac{d}{d\tau}\frac{\partial F}{\partial \dot{\phi}} = 0. \tag{2.150}$$

If

$$F(\tau, \phi(\tau), \dot{\phi}(\tau)) = \rho(x(\tau), y(\tau)) = [x(\tau) - y(\phi(\tau))]^2 \tag{2.151}$$

and $g = 1$ for all $t \in [t_1, t_2]$, so that there are no constraints other than the monotonicity of ϕ, then (2.148) reduces to

$$\frac{\partial}{\partial \phi}\|x(\tau) - y(\phi(\tau))^2\| = 0. \tag{2.152}$$

If y is strictly monotonic increasing, (2.152) will be satisfied if $\|x(\tau) - y(\phi(\tau))\|$ is identically zero, from which it follows that

$$\phi(\tau) = y^{-1}(x(\tau)), \qquad \forall \tau \in [t_1, t_2], \tag{2.153}$$

and hence $D(v, w) = 0$. A graphical construction of this solution is illustrated by the numbered lines in Fig. 2.15.

Analytical solutions such as (2.153) are available only in special cases. Unfortunately, several properties of the speech signal conspire to preclude such a simple solution. First, (2.151) is not an appropriate local metric for speech. Based on evidence [110], we require, at the very least, something of the form

$$\rho(x(\tau), y(\tau)) = \log \frac{||\mathbf{X}(s, \tau)||^2}{||\mathbf{Y}(s, \tau)||^2}, \tag{2.154}$$

where $\mathbf{X}(s, \tau)$ and $\mathbf{Y}(s, \tau)$ are, respectively, the "instantaneous", that is, short-duration, spectra of $x(\tau)$ and $y(\tau)$ at time τ and the norm is understood to be taken with respect to the complex frequency s. In fact, the problem is somewhat more complicated than indicated by (2.154) which accounts only for spectral information and ignores such prosodic features as pitch and intensity.

Second, we must account for local variations in time scale that actually are information-bearing. Thus $g(\tau) \equiv 1$ and $\phi(\tau)$ essentially unconstrained represent intolerable oversimplifications. More realistically, we should set

$$D(v, w) = \min_{\{\phi(t)\}} \int_0^T \log \frac{||\mathbf{X}(s, \tau)||^2}{||\mathbf{Y}(s, \tau)||^2} g(\tau)\, d\tau, \tag{2.155}$$

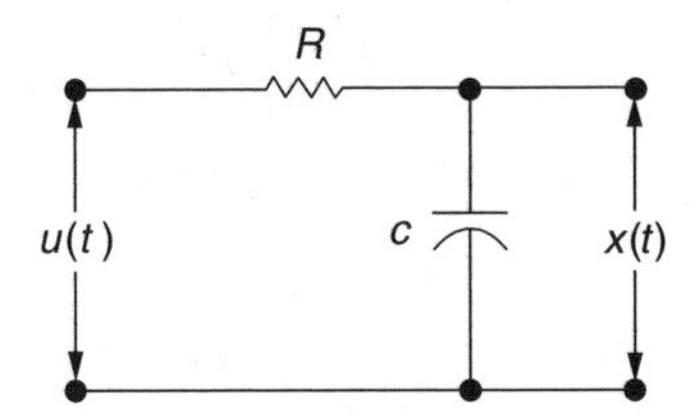

Figure 2.16 A simple RC circuit as a problem in optimal control

in which $\phi(t)$ is subject to linear inequality constraints and $g(t)$ is such that only mildly non-linear solutions to (2.150) are allowed. Under such conditions, (2.155) will not be analytically tractable.

In fact, however, equations similar to (2.155) also appear in optimal control problems and were found by Bellman [31] to be amenable to numerical solution by a method he called dynamic programming.

Consider the example of the simple circuit shown in Fig. 2.16. Suppose it is desired to charge the capacitor C by means of the control voltage $u(t)$ from initial voltage x_i at time $t = 0$ to some final voltage x_t at time t_t in such a way that the minimum amount of energy J is dissipated by the resistor R. This is recognized as the variational problem $\delta J = 0$; that is, minimize J with respect to $u(t)$, where

$$J = RC^2 \int_0^{t_f} \dot{x}^2 \, dt = \frac{1}{R} \int_0^{t_f} (u(t) - x(t))^2 \, dt. \tag{2.156}$$

In the notation of (2.149), $F(t, u, \dot{u}) = (1/R)(u(t) - x(t))^2$, and (2.150) reduces to $(d/dt)(u - x) = 0$, the closed-form solution for which forces $u(t) - x(t)$ to be constant. For given values of x_0, x_f, and t_f we can compute the exact $u(t)$ needed.

2.6.2 A Solution by Dynamic Programming

This very simple problem can also be solved numerically by the method of dynamic programming. First, the problem must be discretized so that the integral in (2.156) is approximated by the summation

$$J = \sum_{l=0}^{n} \Delta J_l = \frac{1}{R} \sum_{l=0}^{n} [u(l\Delta t) - x(l\Delta t)]^2 = \frac{1}{R} \sum_{l=0}^{n} [u_l - x_l]^2, \tag{2.157}$$

where $\Delta t = t_f / n$ for some suitably large n and the boundary conditions are that $u_0 = u(0), x_0 = x_i$, and $x_n = x_f$.

Next, let $x(t)$ be quantized so that it takes one of only m values $\{l\Delta x\}_{l=1}^{m}$ with Δx and m chosen to provide adequate resolution over a sufficiently large range of values of x. When the voltage drop across C is $l\Delta x$, the circuit is said to be in the lth state. Now let the state be changed every Δt seconds by changing the voltage u_l for $l = 1, 2, \ldots, n$. Then for any particular program $\{u_l\}_{l=1}^{n}$, the corresponding state sequence can be computed from principles of elementary circuit theory. Since the circuit can be in but m different states at each of n different times, there are exactly m^n distinct programs for $u(t)$. In principle, each one can be independently evaluated according to (2.157) and the best one, in the sense of minimizing J, chosen. The chosen program is thus a discrete time approximation to the solution of the continuous variational problem of finding $u(t)$ that minimizes (2.156).

Of course, for realistic values of m and n, the required number of evaluations of (2.157) precludes the use of the naive exhaustive technique. Bellman's well-known insight which obviates the need for brute force is embodied by his optimality principle, which notes that for problems such as minimizing (2.156), any program optimal for $0 \leq l \leq n$ is optimal over any subinterval $r \leq l \leq n$ for $0 \leq r \leq n$.

The application of this principle to the solution of the RC circuit problem is illustrated in Fig. 2.17. Let J_{kr} be the minimum value of J after any sequence of k state changes terminating in state r. Let ΔJ_{lr} be the incremental energy dissipated in changing the state of the circuit from l to r by changing u_{k-1} to u_k at time $k\Delta t$. Then according to the principle of optimality it follows that

$$J_{kr} = \min_{1 \leq l \leq m} \left\{ J_{k-1,l} + \Delta J_{lr} \right\}, \tag{2.158}$$

for $1 \leq k \leq n$ and $1 \leq r \leq m$. To start the procedure the boundary values are used to set J_{0l}, and thereafter only nm^2 operations are needed rather than m^n for the naive method. After n state changes we will find $J_{\min} = J_{nr^*}$, where r^* is the state corresponding to the desired x_f. The sequence $\{u_k\}_{k=0}^{n}$ is the sought-after optimal program. The technique is called dynamic programming because it computes an optimal control policy or program by adjusting the policy dynamically, that is, at each state change.

The purpose of the foregoing somewhat lengthy digression is to set the non-parametric methods in their proper historical context. Vintsyuk noticed that the then new technique of dynamic programming was precisely what was needed to evaluate the metric (2.155) with appropriate constraints on $\phi(t)$ which, for the speech problem, plays the role of the control $u(t)$ in the above example. Since the original contribution by Vintsyuk, the technique of dynamic programming has been independently applied to the automatic speech recognition problem by several researchers [35, 37, 140, 288, 289, 290, 321] who

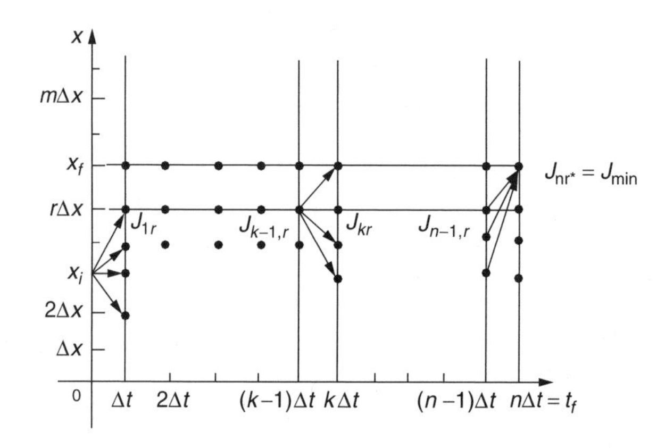

Figure 2.17 Dynamic programming solution of the optimal control of an RC circuit

chose not to motivate their reports with the variational problem but instead focused on the solution $\phi(t)$ and its effect of non-linear transformation of the time scale of one signal to that of the other. Thus rather than accentuating the essential feature of the method, the time-invariant distance measure $D(v, w)$, they have obscured its significance by giving primacy to the "alignment function", $\phi(t)$, which is merely a by-product of the solution and of considerably less importance in the actual classification process. As a result of this emphasis, the method has come to be known as dynamic time warping (DTW): "dynamic" no doubt from dynamic programming, and "time warping" not from science fiction's rendition of relativistic physics but rather to indicate non-linear transformation of time scale.

A complete, and in several respects innovative, application of the method of Vintsyuk was given by Itakura [140]. He advocated the use of dynamic programming with a particular set of features, linear prediction coefficients, a particular metric, the logarithm of the ratio of prediction residuals, and a particular set of constraints on the derivative of the time scale transformation. Since this method has gained many adherents, we shall briefly describe it and use it as a generic non-parametric method.

As was the case with the optimal control problem, we begin by discretizing the problem in time by sampling $x(t)$ and $y(t)$ every Δt seconds, forming the sequences $\{x(n\Delta t)\}_{n=1}^{N}$ and $\{y(m\Delta t)\}_{m=1}^{M}$. At the first level, the sequences are blocked into frames of S samples over which a pth-order autocorrelation analysis is performed from which the LPCs are computed as in (2.50)–(2.54).

The crucial characteristic of the Itakura method is the choice of metric in the feature space, namely

$$\rho(\mathbf{x}(n\Delta t), \mathbf{y}(m\Delta t)) = \log \frac{\mathbf{a_x R_y a_x'}}{\mathbf{a_y R_y a_y'}}, \tag{2.159}$$

where $\mathbf{x}(n\Delta t)$ and $\mathbf{y}(m\Delta t)$ are the frames centered about times $n\Delta t$ and $m\Delta t$, respectively; $\mathbf{a_x}$ and $\mathbf{a_y}$ are the p-dimensional LPC vectors derived from $\mathbf{x}$ and $\mathbf{y}$, respectively, and $\mathbf{R_y}$ denotes the pth-prder autocorrelation matrix of the frame $\mathbf{y}$.

Itakura showed that the ratio of quadratic forms in (2.159) can be efficiently computed by a p-dimensional inner product. Other properties of (2.159) are derived in [110], two of which are relevant here: the metric captures the spirit of (2.82) but is not, strictly speaking, a metric. While it is true that $\rho(\mathbf{x}, \mathbf{y}) \geq 0$ for all $\mathbf{x}, \mathbf{y}$, it deviates slightly from being symmetric and obeying the triangle inequality and is thus more correctly called a distortion function. The non-negativity property, however, is sufficient to justify invoking the Loftsgaarden and Quesenberry result.

Also important to Itakura's method was his choice of constraints on $\phi(t)$ which are introduced via the function $g(t)$. First, there are global constraints

$$g(n\Delta t) = \begin{cases} \infty, & \text{if } \dot{\phi}(n\Delta t) < \frac{1}{2}, \\ 1, & \text{if } \frac{1}{2} \leq \dot{\phi}(n\Delta t) \leq 2, \\ \infty, & \text{if } \dot{\phi}(n\Delta t) > 2, \end{cases} \tag{2.160}$$

which have the effect of limiting the overall compression or dilation of the time scale of $y(t)$ to a factor of $\frac{1}{2}$ or 2, respectively. Second, certain local constraints are incorporated into $g(t)$ so that the change of time scale is restricted locally to the same range as it is globally. This is achieved by requiring

$$g(n\Delta t) = \begin{cases} \infty, & \text{if } \phi((n-1)\Delta t) = \phi((n-2)\Delta t) = \phi(n\Delta t), \\ 1, & \text{if } 0 \le \phi(n\Delta t) - \phi((n-1)\Delta T) < 2, \\ \infty, & \text{otherwise.} \end{cases} \tag{2.161}$$

The choice of an LPC parameterization of the speech signal in conjunction with the distance function of (2.159) and the constraints (2.160) and (2.161) lead to the following discrete analog of the variational problem (2.155):

$$D(v, w) = \min_{\phi} \sum_{n=1}^{N} \rho(\mathbf{x}(n\Delta t), \mathbf{y}(n\Delta t)) \cdot g(n\Delta t). \tag{2.162}$$

The combinatorial optimization problem (2.162) can be solved by a dynamic programming algorithm analogous to (2.158). Let us omit the symbol Δt so that, for any $n \in [1, N]$, $\phi(n) = m$ for some $m \in [0, M]$ consistent with (2.160) and (2.161). Let us further simplify the notation by letting $d(n, m)$ stand for $\rho(\mathbf{x}(n\Delta t), \mathbf{y}(m\Delta t))$, so that $d(n, m)$ is the incremental distance associated with identifying frame n of $\mathbf{x}$ with frame m of $\mathbf{y}$. The minimum cumulative distance of the first n frames of $\mathbf{x}$ being mapped onto $\mathbf{m}$ frames of $\mathbf{y}$ is designated by $D(n, m)$. Then the minimization in (2.162) can be accomplished by computing

$$D(n, m) = d(n, m) + \min\{D(n-1, m)g(n-1, m), D(n-1, m-1), D(n-1, m-2)\}, \tag{2.163}$$

for $1 \le n \le N$, $1 \le m \le M$, where $\phi(1) = 1$ and $\phi(N) = M$. Note that the boundary conditions on ϕ are those that would be required to solve the Euler–Lagrange equation corresponding to (2.159). At the end of recursive application of (2.163), $D(v, w) = D(N, M)$. In the process, we simultaneously constructed a function ϕ which relates the time scales of v and w in such a way that $D(v, w)$ is insensitive to their differences. If so desired, we could explicitly recover ϕ from the execution of the procedure defined by (2.163).

The procedure is illustrated in Fig. 2.18 in which the interior of the parallelogram is the feasible region for the combinatorial optimization problem (2.162) defined by the global constraints (2.160) and the arrows indicate the local constraints (2.161) which restrict ϕ to only those mild non-linearities which would be expected to result from changes in speaking rate. A complete solution is shown in Fig. 2.19.

To use (2.163) to recognize an unlabeled utterance w as some word $v_i \in V$, we perform the analog of (2.95). We assume that we have available at least one labeled sample of each $v_i \in V$. We compute $D(v_i, w)$ for all $v_i \in V$ and decide that $w \in v^*$, where

$$v^* = \arg\min_{v \in V} \{D(v, w)\}, \tag{2.164}$$

which, by analogy with (2.95), is the Bayesian decision based upon a nonparametric estimate of the class-conditional density functions for the set of vocabulary words.

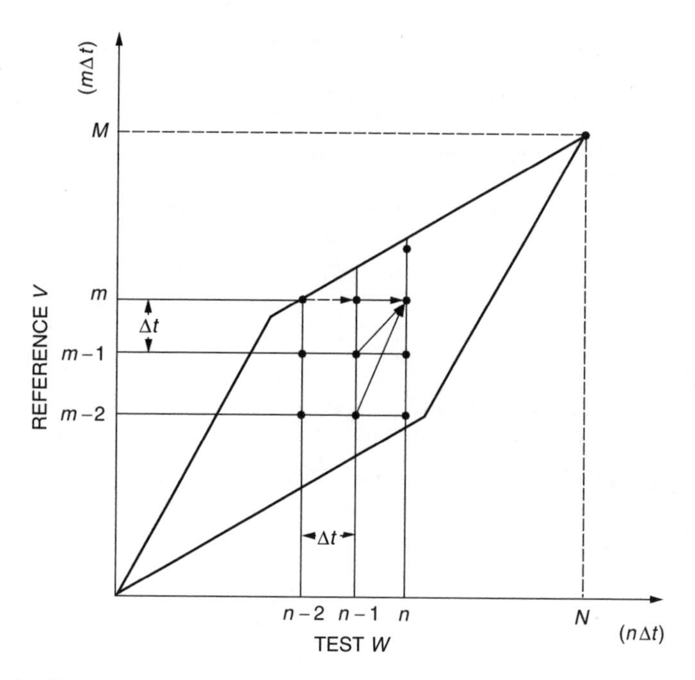

Figure 2.18 Dynamic programming solution of the acoustic pattern recognition problem

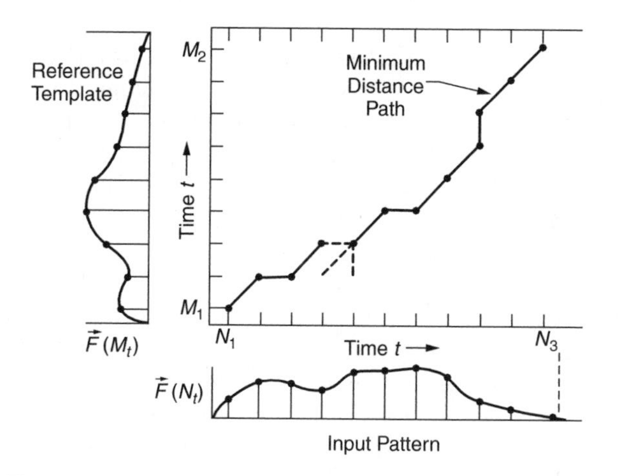

Figure 2.19 Dynamic time warping with speech templates

2.7 Taxonomy of Linguistic Structure

Sections 2.1 through 2.6 explain how information is encoded in the speech signal. The remainder of this book is devoted to the analysis of the organization of the information-bearing features according to various aspects of linguistic structure. We define the three primary components of language: symbols, grammar, and meaning. The symbols of the

Table 2.1 Taxonomy of linguistic structure

Symbols	Grammar	Meaning
acoustic-phonetic	phonological	morphological
prosodic	phonotactic	lexical
	morphological	semantic
	syntactic	pragmatic
	prosodic	prosodic

speech code are arbitrary, subject only to physical and articulatory constraints. Grammar refers to the rules for making allowable sequences of symbols. Grammatical rules are the arbitrary relationships of the symbols to each other without regard for meaning. However, some aspects of grammar provide a framework for the expression of meaning. Finally, meaning is the relationship between symbols and reality. Meaningful messages describe the physical world and our mental image of it.

Each of these elements of language is composed of particular aspects of linguistic structure. This is summarized in Table 2.1.

2.7.1 Acoustic Phonetics, Phonology, and Phonotactics

Linguistic analysis ultimately rests on the assumption that speech is literate. That is, all of the acoustic patterns in the speech signal are well described by a small set of phonetic symbols. There are four broad phonetic categories, vowels, fricatives, plosives, and nasals. Vowels are produced by vocal cord vibration modulating the air stream forced out of the lungs and exciting the acoustic resonances of the vocal tract. Fricatives result from an air jet, created by an articulatory constriction of the vocal tract, impinging on one of its fixed surfaces. Plosives are generated by a sudden release of pressure built up behind an articulatory closure of the vocal tract. Finally, nasals are those sounds produced by connecting the nasal cavity to the vocal tract allowing sound to be radiated from the nostrils. Each of the broad phonetic categories has a characteristic acoustic spectrum. Vowel spectra have prominent resonances called formants. Fricatives have high-pass spectra. Plosives are identified by short bursts of broadband energy. Nasals are distinguished by the presence of zeros in their spectra.

The broad phonetic categories may be divided into smaller classes according to the place and manner of articulation. For vowels, place refers to the position of the tongue body from front to back and from high to low. The front of the vocal tract is that part near the mouth. The lower part of the vocal tract is the part nearest the movable part of the jaw. Different positions of the tongue body correspond to different area functions which, as shown in Section 2.1, result in different solutions of (2.2) each having a characteristic spectrum. We will be more specific about this effect in a moment.

For phonetic categories other than vowels, place refers to the location of the constriction of the vocal tract created by positions of the articulators. In these cases, place is defined as labial (lips), dental (teeth), alveolar (ridge behind the teeth), palatal (hard palate), velar (soft palate), pharyngeal (throat), and glottal (larynx).

For all phonetic categories, manner refers method of generating the sound. The principal methods are voicing (vibrating vocal cords), plosion (complete closure), frication (partial closure), nasalization (opening the nasal cavity with the velum), and liquid (smooth slow motion of the articulators). A typical phonetic inventory for English is shown in Table 2.2. Each entry in the table has a distinct place and manner of articulation, resulting in a distinct acoustic signal with a distinct spectrum. The symbols in the table stand for the sound corresponding to the capitalized portion of the adjacent English word.

Sounds in the phonetic inventory in Table 2.2 can be defined in terms of their place and manner of articulation. The vowels are, of course, all voiced. Their articulatory places are shown in Table 2.3.

The vowels have distinct spectra and, in particular, characteristic formant frequencies. These are shown in Table 2.4.

For broad phonetic categories other than vowels, the phonetic units can be arranged as shown in Table 2.5.

The acoustic-phonetic symbols described above are joined in sequences according to the grammatical rules of phonology and phonotactics. Phonology describes the ways in which the sounds typically associated with a symbol are changed by the phonetic context

Table 2.2 The ARPAbet phonetic inventory

Symbol	Sound	Symbol	Sound
i	hEEd	v	Voice
I	hId	T	THick
e	hAy	D	THose
E	bEd	s	See
@	hAd	z	Zoo
a	cOd	S	meSH
c	pAW	Z	meaSUre
o	hOE	h	Heat
U	hOOd	m	Mother
u	hOOt	n	North
R	hEARd	G	riNG
x	Ahead	l	Loop
A	bUd	L	battLE
Y	hIde	M	bottOM
W	hOW	N	buttON
O	bOY	F	baTTer
X	rosEs	Q	baTman
p	Pond	w	Won
b	Bond	y	You
t	Tug	r	Rope
d	Dug	C	CHild
k	Kit	J	Jug
g	Got	H	WHere
f	File	#	(silence)

Table 2.3 Place of the vowels

	Back	Middle	Front
High			i
	U	a	I
Middle	o	R	E
		A	
Low	c		@

Table 2.4 Typical formant frequencies (Hz) for the vowels

Vowel	f1	f2	f3
i	255	2330	3000
I	350	1975	2560
E	560	1875	2550
@	735	1625	2465
a	760	1065	2550
A	640	1250	2610
c	610	865	2540
U	475	1070	2410
u	290	940	2180

Table 2.5 Place and manner for the broad phonetic categories

Voiced	Labial	Dental	Alveolar	Palatal	Velar	Glottal
Plosive	b		d	J	g	
Fricative	v	D	z	Z		
Nasal	m		n	G		
Liquid	w		l	y		h
Unvoiced						
Plosive	p		t	C	k	
Fricative	f	T	s	S		

in which they appear. There are two primary reasons for the phonological variations of sounds. First is the phenomenon of coarticulation of a sound due to the motions of the articulators as they move from one typical place of articulation to another. Another way to describe the coarticulatory effect is as a set of constraints on the ways $A(x, t)$ in (2.2) can change in time due to physiological and mechanical factors that determine the way it changes in space. The second consideration is both aesthetic and practical. That is, it is often desirable to change sounds in context so they will be more pleasing to the ear and/or easier to produce by simplifying their articulatory motions.

Phonotactics determines the allowable sequences of phonetic units. To some extent, the order of sounds in an utterance is constrained by the physiology and mechanics of the vocal apparatus. These factors are primarily reflected in the phonology of a language. However, there are sound sequences in any language that are euphonious and easy to articulate, yet they do not appear. We shall treat these restrictions as arbitrary rules that must, however, be observed if the language is to be faithfully described.

2.7.2 Morphology and Lexical Structure

Morphology is another aspect of grammar that places further restrictions on sequences of sounds used to form words, that is, on lexical structure. For example, specific sounds may be appended to nouns to indicate case or number or to verbs in order to indicate mood or tense. Also, a specific sound may be attached to a root word as a prefix or suffix in order to alter the logical or spatial meaning of a word. For example, the prefix "un" in English is the logical equivalent of negation and the suffix "er" changes the name of an activity into the name of one who engages in that activity.

Two things are clear about morphology. First, it is widely used to increase the lexicon of a language. Second, it is a grammatical function but it has significant consequences for the meanings of words, hence its appearance in two columns of Table 2.1.

2.7.3 Prosody, Syntax, and Semantics

Prosody is that aspect of speech that a listener perceives as melody and emotion. The physical correlates of prosody are f0, the vibration frequency of the vocal cords that a listener perceives as pitch, energy as measured by the term G in (2.45), and duration of the phonetic units which is reflected in speaking rate. Listeners are sensitive to words and/or phonetic units that are pronounced faster or slower than average, the contrast marking other aspects of linguistic structure.

Prosody is a suprasegmental feature of speech. That is, it is not a property of individual acoustic-phonetic units, even though it may significantly affect their spectra, but rather is properly associated with phrases, sentences, and longer discourses. Throughout this book we shall ignore the suprasegmental effects of prosody, treating it instead as a special set of symbols (i.e. segments) that have syntactic and semantic implications. Specifically, prosodic contrasts are used as markers of phrase structure that have effects on meaning.

Syntax is the linguistic structure that describes the composition of sentences from words without regard for meaning. There are two principal aspects of syntax. First is the designation of a fixed set of syntactic constituents. At the lexical level, each word is assigned to at least one such constituent, depending on its role in the sentence. Typically, these constituents are the so-called parts of speech such as noun and verb. Second, the parts of speech are organized into larger constituents called phrases according to well-defined rules. Although these rules are independent of meaning, the determination of the phrase structure of a sentence by parsing it into its syntactic constituents is a necessary first step in the extraction and/or representation of meaning.

The linguistic structure that addresses the meanings of utterances is called semantics. It also has two significant aspects. First is lexical semantics, which associates individual words to the objects, qualities, and actions they represent. As mentioned in Section 2.7.2, lexical semantics is partially dependent upon morphology. The second aspect of semantics with which we shall be concerned is compositional in nature, describing the way that lexical semantics and phrase structure cooperate to express the variety of much richer meanings of sentences.

2.7.4 Pragmatics and Dialog

Pragmatics refers to the relationship between utterances and speakers. Different speakers may use the same words to express different semantics. They may also use the same words differently in different social contexts, thereby conveying different meanings.

Dialog refers to the expression of meaning in ordinary conversation as opposed to isolated utterances. Dialog conventions may substantially alter both syntactic rules and semantic content. Dialog also poses the difficult problems of ellipsis, the omission of words or phrases when they are understood by context, and the resolution of reference, the determination of the proper attachment of modifiers to the words or phrases they modify.

These aspects of linguistic structure are listed simply to acknowledge their existence. They will not be explicitly discussed in these pages. The omission is not intended to diminish their importance. As we shall later see, there is simply a lack of mathematical machinery that can be brought to bear on these topics.

We will now proceed to the main concern of this book, the mathematical treatment of phonology, phonotactics, and syntax in Chapters 3 through 7 and of semantics in Chapters 8 through 10. Our analysis is neither standard nor exhaustive. Traditional examinations of linguistic structure may be found in Chomsky and Halle [47], Jakobson [143], Fant [83] and Oshika *et al.* [239]. These pages are concerned with the expression of the essential aspects of linguistic structure in mathematical models.

3

Mathematical Models of Linguistic Structure

3.1 Probabilistic Functions of a Discrete Markov Process

An important implication of the analysis of Section 2.1 is the non-stationarity of the speech signal. In Section 2.4 we observed that a consequence of the non-stationarity is an intrinsic uncertainty in computing the spectrum of the speech signal. In Section 2.6 we made an implicit model of the non-stationarity by forming a distance measure that is insensitive to local variations of time scale. We will now develop an explicit method of accounting for non-stationarity based on a powerful stochastic model. For the most part, except for Section 3.1.5, we will assume that the speech signal is piecewise stationary. As illustrated in Fig. 3.1, we will say that the segment, w_k, on the interval from t_{k-1} to t_k is stationary. This is reasonable in view of equation (2.34).

A probabilistic function of a (hidden) Markov chain is a stochastic process generated by two interrelated mechanisms, an underlying Markov chain having a finite number of states, and a set of random functions, one of which is associated with each state. At discrete instants of time, the process is assumed to be in some state and an observation is generated by the random function corresponding to the current state. The underlying Markov chain then changes states according to its transition probability matrix. The observer sees only the output of the random functions associated with each state and cannot directly observe the states of the underlying Markov chain; hence the term hidden Markov model.

In principle, the underlying Markov chain may be of any order and the outputs from its states may be multivariate random processes having some continuous joint probability density function. In this discussion, however, we shall restrict ourselves to consideration of Markov chains of order one, that is, those for which the probability of transition to any state depends only upon that state and its predecessor.

3.1.1 The Discrete Observation Hidden Markov Model

We shall limit the discussion to processes whose observations are drawn from a discrete finite alphabet according to discrete probability distribution functions associated with the states. It is quite natural to think of the speech signal as being generated by such a

Mathematical Models for Speech Technology. Stephen Levinson
 ISBN: 0-470-84407-8

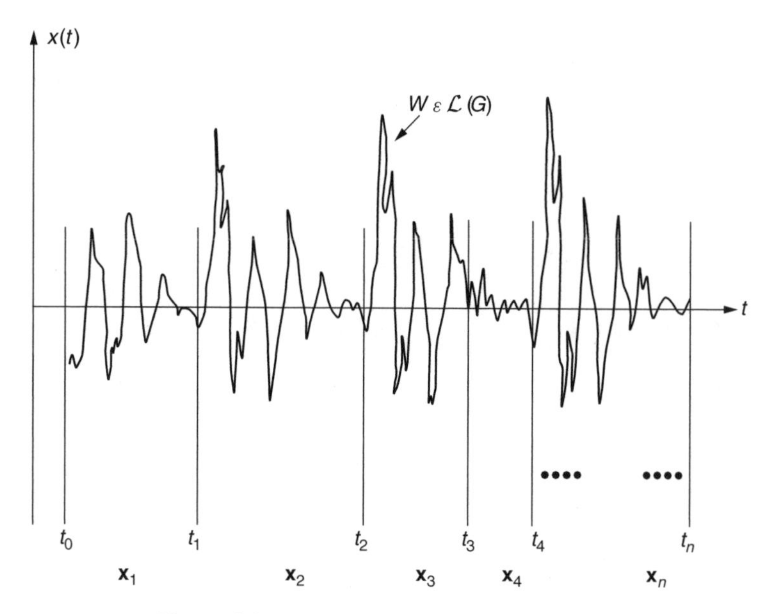

Figure 3.1 Segmenting the speech waveform

process. We can imagine the vocal tract as being in one of a finite number of articulatory configurations or states. In each state a short (in time) signal is produced that has one of a finite number of prototypical spectra depending, of course, on the state. Thus, the power spectra of short intervals of the speech signal are determined solely by the current state of the model, while the variation of the spectral composition of the signal with time is governed predominantly by the probabilistic state transition law of the underlying Markov chain. For speech signals derived from a small vocabulary of isolated words, the model is reasonably faithful. The foregoing is, of course, an oversimplification intended only for the purpose of motivating the following theoretical discussion.

Let us say that the underlying Markov chain has N states $q_1, q_2, \ldots, q_N$ representing the stationary intervals, and the observations are drawn from an alphabet, $\mathbf{V}$, of M prototypical spectra $v_1, v_2, \ldots, v_M$. The underlying Markov chain can then be specified in terms of an initial state distribution vector $\boldsymbol{\pi}' = (\pi_1, \pi_2, \ldots, \pi_N)$ and a state transition matrix, $\mathbf{A} = [a_{ij}]$, $1 \leq i, j \leq N$. Here, π_i is the probability of q_i at some arbitrary time, $t = 0$, and a_{ij} is the probability of transiting to state q_j given current state, q_i, that is, $a_{ij} = \text{Prob}(q_j \text{ at } t + 1 | q_i \text{ at } t)$.

The random processes associated with the states can be collectively represented by another stochastic matrix $\mathbf{B} = [b_{jk}]$ in which, for $1 \leq j \leq N$ and $1 \leq k \leq M$, b_{jk} is the probability of observing symbol v_k given current state q_j. The v_k are obtained by quantizing the feature space using the algorithm of (2.144)–(2.147) in Section 2.5.4. We denote this as $b_{jk} = \text{Prob}(v_k \text{ at } t | q_j \text{ at } t)$. Thus a hidden Markov model, $\mathbf{M}$, shown in Fig. 3.2, is identified with the parameter set $(\boldsymbol{\pi}, \mathbf{A}, \mathbf{B})$.

To use hidden Markov models to perform speech recognition we must solve two specific problems: observation sequence probability estimation, which will be used for classification of an utterance; and model parameter estimation, which will serve as a procedure for

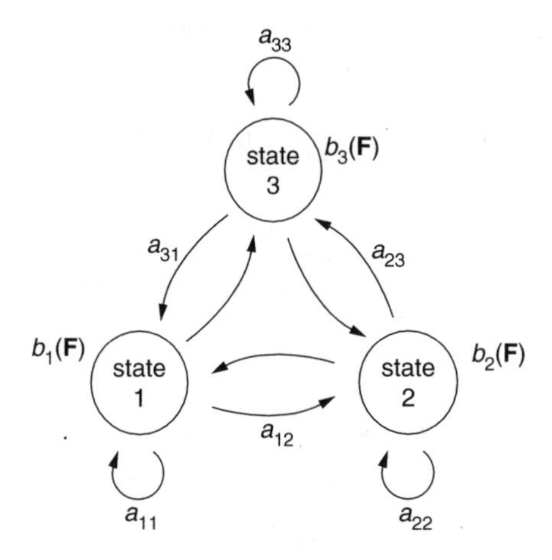

Figure 3.2 A three-state hidden Markov model

training models for each vocabulary word. Both problems proceed from a sequence, $\mathbf{O}$, of observations $O_1 O_2 \cdots O_T$ where each O_t, for $1 \leq t \leq T$, is some $v_k \in \mathbf{V}$.

Our particular classification problem is as follows. We wish to recognize utterances known to be selected from some vocabulary, W, of words $w_1, w_2, \ldots, w_v$. We are given an observation sequence, $\mathbf{O}$, derived from the utterance of some unknown $w_i \in W$ and a set of V models $\mathbf{M}_1, \mathbf{M}_2, \ldots, \mathbf{M}_V$. We must compute $P_i = \text{Prob}(\mathbf{O}|\mathbf{M}_i)$ for $1 \leq i \leq V$. We will then classify the unknown utterance as w_i if and only if $P_i \geq P_j$ for $1 \leq j \leq V$.

The training problem is simply that of determining the models $\mathbf{M}_i = (\pi_i, A_i, B_i)$ for $1 \leq i \leq V$ given training sequences $\mathbf{O}^{(1)}, \mathbf{O}^{(2)}, \ldots, \mathbf{O}^{(V)}$, where $\mathbf{O}^{(i)}$ is known to have been derived from an utterance of word w_i for $1 \leq i \leq V$.

One could, in principle, compute $\text{Prob}(\mathbf{O}|\mathbf{M})$ by computing the joint probability $\text{Prob}(\mathbf{O}, s|\mathbf{M})$ for each state sequence, s, of length T, and summing over all state sequences. Obviously this is computationally intractable. Fortunately, however, there is an efficient method for computing P. Let us define the function $\alpha_t(i)$, for $1 \leq t \leq T$, as Prob $(O_1 O_2 \ldots O_t$ and q_i at $t|\mathbf{M})$. According to the definition $\alpha_1(i) = \pi_i b_i(O_1)$, where $b_i(O_1)$ is understood to mean b_{ik} if and only if $O_1 \equiv v_k$; then we have the following recursive relationship for the "forward probabilities":

$$\alpha_{t+1}(j) = \left[\sum_{i=1}^{N} \alpha_t(i) a_{ij}\right] b_j(O_{t+1}), \qquad 1 \leq t \leq T-1. \tag{3.1}$$

Similarly, we define another function, $\beta_t(j) = \text{Prob}(O_{t+1} O_{t+2} \ldots O_T | q_j$ at t and $\mathbf{M})$. We set $\beta_T(j) = 1$ for all j and then use the backward recursion

$$\beta_t(i) = \sum_{j=1}^{N} a_{ij} b_j(O_{t+1}) \beta_{t+1}(j), \qquad T-1 \geq t \geq 1, \tag{3.2}$$

to compute the "backward probabilities".

The two functions can be used to compute P according to

$$P = \text{Prob}(\mathbf{O}|\mathbf{M}) = \sum_{i=1}^{N}\sum_{j=1}^{N} \alpha_t(i)a_{ij}b_j(O_{t+1})\beta_{t+1}(j) \tag{3.3}$$

for any t such that $1 \leq t \leq T-1$. Equations (3.1) to (3.3) are from Baum [25] and are sometimes referred to as the "forward–backward" algorithm.

Setting $t = T-1$ in (3.3) gives

$$P = \sum_{i=1}^{N} \alpha_T(i) \tag{3.4}$$

so that P can be computed from the forward probabilities alone. A similar formula for P can be obtained from the backward probabilities by setting $t = 1$. These and several other formulas in this section may be compactly written in matrix notation. For instance,

$$P = \boldsymbol{\pi}'\mathbf{B}_1\mathbf{A}\mathbf{B}_2\mathbf{A}\cdots\mathbf{A}\mathbf{B}_T\mathbf{1}, \tag{3.5}$$

where $\mathbf{1}$ is the N-vector $(1, 1, 1, \ldots, 1)'$ and

$$\mathbf{B}_t = \begin{pmatrix} b_1(O_t) & & & \mathbf{O} \\ & b_2(O_t) & & \\ & & \ddots & \\ \mathbf{0} & & & b_N(O_t) \end{pmatrix} \tag{3.6}$$

for $1 \leq t \leq T$. From (3.5) it is clear that P is a homogeneous polynomial in the π_i, a_{ij}, and b_{jk}. Any of (3.3) through (3.5) may be used to solve the classification problem. The forward and backward probabilities will prove to be convenient in other contexts.

When we compute P with the forward–backward algorithm, we are including the probabilities of all possible state sequences that may have generated $\mathbf{O}$. Alternatively, we may define P as the maximum over all state sequences $\mathbf{i} = (i_0, i_1, \ldots, i_T)$ of the joint probability $P(\mathbf{O}, \mathbf{i})$. This distinguished state sequence and the corresponding probability of the observation sequence can be simultaneously computed by means of the Viterbi [323] algorithm. This dynamic programming technique proceeds as follows: Let $\phi_1(i) = \pi_i b_i(O_1)$ for $1 \leq i \leq N$. Then we can perform the following recursion for $2 \leq t \leq T$ and $1 \leq j \leq N$:

$$\phi_t(j) = \max_{1\leq i\leq N} [\phi_{t-1}(i)a_{ij}]b_j(O_t) \tag{3.7}$$

and

$$\Psi_t(j) = i^*, \tag{3.8}$$

where i^* is a choice of an index i that maximizes $\phi_t(i)$.

The result is that $P = \max_{1 \le i \le N}[\phi_T(i)]$. Also the maximum likelihood state sequence can be recovered from Ψ as follows. Let $q_T = i^*$, where i^* maximizes P. Then for $T \ge t \ge 2, q_{t-1} = \Psi_t(q_t)$. If one only wishes to compute P, the linked list, Ψ, need not be maintained as in (3.8). Only the recursion (3.7) is required.

The problem of training a model, unfortunately, does not have such a simple solution. In fact, given any finite observation sequence as training data, we cannot optimally train the model. We can, however, choose $\boldsymbol{\pi}$, $\mathbf{A}$, and $\mathbf{B}$ such that Prob($\mathbf{O}|\mathbf{M}$) is locally maximized. For an asymptotic analysis of the training problem the reader should consult Baum and Petrie [27].

We can use the forward and backward probabilities to formulate a solution to the problem of training by parameter estimation. Given some estimates of the parameter values, we can compute, for example, that the expected number of transitions, γ_{ij}, from q_i to q_j, conditioned on the observation sequence is just

$$\gamma_{ij} = \frac{1}{P}\sum_{t=1}^{T-1} \alpha_t(i)a_{ij}b_j(O_{t+1})\beta_{t+1}(j). \tag{3.9}$$

Then, the expected number of transitions, γ_i out of q_i, given $\mathbf{O}$, is

$$\gamma_i = \sum_{j=1}^{N} \gamma_{ij} = \frac{1}{P}\sum_{t=1}^{T-1} \alpha_t(i)\beta_t(i), \tag{3.10}$$

the last step of which is based on (3.2). The ratio γ_{ij}/γ_i is then an estimate of the probability of state q_j, given that the previous state was q_i. This ratio may be taken as a new estimate, $\overline{a}_{ij}$, of a_{ij}. That is,

$$\overline{a}_{ij} = \frac{\gamma_{ij}}{\gamma_i} = \frac{\sum_{t=1}^{T-1} \alpha_t(i)a_{ij}b_j(O_{t+1})\beta_{t+1}(j)}{\sum_{t=1}^{T-1} \alpha_t(i)\beta_t(i)}. \tag{3.11}$$

Similarly, we can make a new estimate of b_{jk} as the frequency of occurrence of v_k in q_j relative to the frequency of occurrence of any symbol in state q_j. Stated in terms of the forward and backward probabilities, we have

$$\overline{b}_{jk} = \frac{\sum_{t \ni O_t = v_k} \alpha_t(j)\beta_t(j)}{\sum_{t=1}^{T} \alpha_t(j)\beta_t(j)}. \tag{3.12}$$

Finally, new values of the initial state probabilities may be obtained from

$$\overline{\pi}_i = \frac{1}{P}\alpha_1(i)\beta_1(i). \tag{3.13}$$

As we shall see in the next section, the reestimates are guaranteed to increase P, except at a critical point.

Proof of the Reestimation Formula

The reestimation formulas (3.11), (3.12), and (3.13) are instances of the Baum–Welch algorithm. Although it is not at all obvious, each application of the formulas is guaranteed to increase P except if we are at a critical point of P, in which case the new estimates will be identical to their current values. Several proofs of this rather surprising fact are given in the literature; see [25] [29]. Because we shall need to modify it later to cope with the finite sample size problem, we shall briefly sketch Baum's proof [25] here. The proof is based on the following two lemmas:

Lemma 1. Let $u_i, i = 1, \ldots, S$ *be positive real numbers, and let* $v_i, i = 1, \ldots, S$ *be non-negative real numbers such that* $\sum_i v_i > 0$. *Then from the concavity of the log function it follows that*

$$\ln\left(\frac{\sum v_i}{\sum u_i}\right) = \ln\left[\sum_i \left(\frac{u_i}{\sum_k u_k}\right) \cdot \frac{v_i}{u_i}\right] \tag{3.14}$$

$$\geq \sum_i \frac{u_i}{\sum_k u_k} \ln\left(\frac{v_i}{u_i}\right) \tag{3.15}$$

$$= \frac{1}{\sum_k u_k}\left[\sum_i (u_i \ln v_i - u_i \ln u_i\right]. \tag{3.16}$$

Here every summation is from 1 to S.

Lemma 2. If $c_i > 0$ $i = 1, \ldots, N$, *then, subject to the constraint* $\sum_i x_i = 1$, *the function*

$$F(\mathbf{x}) = \sum_i c_i \ln x_i \tag{3.17}$$

attains its unique global maximum when

$$x_i = \frac{c_i}{\sum_i c_i}. \tag{3.18}$$

The proof follows from the observation that by the Lagrange method

$$\frac{\partial}{\partial x_i}\left[F(\mathbf{x}) = \lambda \sum_i x_i\right] = \frac{c_i}{x_i} - \lambda = 0. \tag{3.19}$$

Multiplying by x_i and summing over i gives $\lambda = \sum_i c_i$, hence the result.

Now in Lemma 1, let S be the number of state sequences of length T. For the ith sequence, let u_i be the joint probability

$$u_i = \text{Prob[state sequence } i, \text{ observation } \mathbf{O}|\text{model } \mathbf{M}]$$

$$= P(i, \mathbf{O}|\mathbf{M}).$$

Let v_i be the same joint probability conditioned on model $\overline{\mathbf{M}}$. Then

$$\textstyle\sum_i u_i = p(\mathbf{O}|\mathbf{M}) \stackrel{\triangle}{=} P(\mathbf{M}), \quad (3.20)$$

$$\textstyle\sum_i v_i = p(\mathbf{O}|\overline{\mathbf{M}}) \stackrel{\triangle}{=} P(\overline{\mathbf{M}}), \quad (3.21)$$

and the lemma gives

$$\ln \frac{P(\overline{\mathbf{M}})}{P(\mathbf{M})} \geq \frac{1}{P(\mathbf{M})} \cdot [Q(\mathbf{M}, \overline{\mathbf{M}}) - Q(\mathbf{M}, \mathbf{M})], \quad (3.22)$$

where

$$Q(\mathbf{M}, \overline{\mathbf{M}}) \stackrel{\triangle}{=} \sum_i u_i \ln v_i. \quad (3.23)$$

Thus, if we can find a model $\overline{\mathbf{M}}$ that makes the right-hand side of (3.22) positive, we have a way of improving the model $\mathbf{M}$. Clearly, the largest guaranteed improvement by this method results for $\overline{\mathbf{M}}$, which maximizes $Q(\mathbf{M}, \overline{\mathbf{M}})$, and hence maximizes the right-hand side of (3.22). The remarkable fact proven in Baum & Eagon [26] is that $Q(\mathbf{M}, \overline{\mathbf{M}})$ attains its maximum when $\overline{\mathbf{M}}$ is related to $\mathbf{M}$ by the reestimation formulas (3.11) through (3.13). To show this let the sth-state sequence be $s_0, s_1, \ldots, s_T$, and the given observation sequence be $O_{k_1}, \ldots, O_{k_T}$. Then

$$\ln v_s = \ln P(s, \mathbf{O}|\overline{\mathbf{M}}) = \ln \overline{\pi}_{s_0} + \sum_{t=0}^{T-1} \ln \overline{a}_{s_t s_{t+1}} + \sum_{t=0}^{T-1} \ln \overline{b}_{s_{t+1}}(O_{t+1}). \quad (3.24)$$

Substituting this in (3.22) for $Q(\mathbf{M}, \overline{\mathbf{M}})$, and regrouping terms in the summations according to state transitions and observed symbols, it can be seen that

$$Q(\mathbf{M}, \overline{\mathbf{M}}) = \sum_{i=1}^{N}\sum_{j=1}^{N} c_{ij} \ln \overline{a}_{ij} + \sum_{j=1}^{N}\sum_{k=1}^{M} d_{jk} \ln \overline{b}_j(k) + \sum_{i=1}^{N} e_i \ln \overline{\pi}_i. \quad (3.25)$$

Here

$$c_{ij} = \sum_{s=1}^{S} p(s, \mathbf{O}|\mathbf{M}) n_{ij}(s), \quad (3.26)$$

$$d_{jk} = \sum_{s=1}^{S} p(s, \mathbf{O}|\mathbf{M}) m_{jk}(s), \quad (3.27)$$

$$e_i = \sum_{s=1}^{S} p(s, \mathbf{O}|\mathbf{M}) r_i(s), \quad (3.28)$$

and for the sth-state sequence

$$n_{ij}(s) = \text{number of transitions from state } q_i \text{ to } q_j,$$

$$m_{jk}(s) = \text{number of times symbol } k \text{ is generated in state } q_j,$$

$$r_i(s) = \begin{cases} 1, & \text{if initial state is } q_i \\ 0, & \text{otherwise.} \end{cases}$$

Thus, c_{ij}, d_{jk}, and e_i are the expected values of n_{ij}, m_{jk}, r_i, respectively, based on model $\mathbf{M}$.

Expression (3.25) is now a sum of $2N + 1$ independent expressions of the type maximized in Lemma 2. Hence, $Q(\mathbf{M}, \overline{\mathbf{M}})$ is maximized if

$$\overline{a}_{ij} = \frac{c_{ij}}{\sum_j c_{ij}}, \tag{3.29}$$

$$\overline{b}_j(k) = \frac{d_{jk}}{\sum_k d_{jk}}, \tag{3.30}$$

$$\overline{\pi}_i = \frac{e_i}{\sum_i e_i}. \tag{3.31}$$

These are recognized as the reestimation formulas (3.11) through (3.13).

Solution by Optimization Techniques

Lest the reader be led to believe that the reestimation formulas are peculiar to stochastic processes, we shall examine them briefly from several different points of view. Note that the reestimation formulas update the model in such a way that the constraints

$$\sum_{i=1}^{N} \pi_i = 1, \tag{3.32}$$

$$\sum_{j=1}^{N} a_{ij} = 1, \qquad \text{for } 1 \leq i \leq N, \tag{3.33a}$$

and

$$\sum_{k=1}^{M} b_{jk} = 1, \qquad \text{for } 1 \leq j \leq N, \tag{3.33b}$$

are automatically satisfied at each iteration. The constraints are, of course, required to make the hidden Markov model well defined. It is thus natural to look at the training problem as a problem of constrained optimization of P and, at least formally, solve it by the classical method of Lagrange multipliers. For simplicity, we shall restrict the discussion to optimization with respect to **A**. Let Q be the Lagrangian of P with respect to the constraints (3.33a). We see that

$$Q = P + \sum_{i=1}^{N} \lambda_i \left(\sum_{j=1}^{N} a_{ij} - 1 \right), \tag{3.34}$$

where the λ_i are the as yet undetermined Lagrange multipliers.

At a critical point of P on the interior of the manifold defined by (3.32) and (3.33), it will be the case that, for $1 \leq i, j \leq N$,

$$\frac{\partial Q}{\partial a_{ij}} = \frac{\partial P}{\partial a_{ij}} + \lambda_i = 0. \tag{3.35}$$

Multiplying (3.35) by a_{ij} and summing over j, we get

$$\sum_{j=1}^{N} a_{ij} \frac{\partial P}{\partial a_{ij}} = -\left[\sum_{j=1}^{N} a_{ij} \right] \lambda_i = -\lambda_i = \frac{\partial P}{\partial a_{ij}}, \tag{3.36}$$

where the right-hand side of (3.36) follows from substituting (3.33a) for the sum of a_{ij} and then replacing λ_i according to (3.35). From (3.36) it may be seen that P is maximized when

$$a_{ij} = \frac{a_{ij} \frac{\partial P}{\partial a_{ij}}}{\sum_{k=1}^{N} a_{ik} \frac{\partial P}{\partial a_{ik}}}. \tag{3.37}$$

A similar argument can be made for the π and **B** parameters.

While it is true that solving (3.37) for a_{ij} is analytically intractable, it can be used to provide some useful insights into the Baum–Welch reestimation formulas and alternatives to them for solving the training problem. Let us begin by computing $\partial P / \partial a_{ij}$ by differentiating (3.3), according to the formula for differentiating a product,

$$\frac{\partial P}{\partial a_{ij}} = \sum_{t=1}^{T-1} \alpha_t(i) b_j(O_{t+1}) \beta_{t+1}(j). \tag{3.38}$$

Substituting the right-hand side of (3.38) for $\partial P / \partial a_{ij}$ in (3.37), we get

$$a_{ij} = \frac{\sum_{t=1}^{T-1} \alpha_t(i) a_{ij} b_j(O_{t+1}) \beta_{t+1}(j)}{\sum_{j=1}^{N} \sum_{t=1}^{T-1} \alpha_t(i) a_{ij} b_j(O_{t+1}) \beta_{t+1}(j)}. \tag{3.39}$$

Then changing the order of summation in the denominator of (3.39) and substituting in the right-hand side of (3.2), we get

$$a_{ij} = \frac{\sum_{t=1}^{T-1} \alpha_t(i) a_{ij} b_j(O_{t+1}) \beta_{t+1}(j)}{\sum_{t=1}^{T-1} \alpha_t(i) \beta_t(i)}. \tag{3.40}$$

The right-hand side of (3.40) is thus seen to be identical to that of the reestimation formula (3.11). Thus, at a critical point, the reestimation formula (3.11) solves the equations (3.40). Similarly, if we differentiate (3.3) with respect to π_i and b_{jk}, we get

$$\frac{\partial P}{\partial \pi_i} = \sum_{j=1}^{N} b_i(O_1)a_{ij}b_j(O_2)\beta_2(j) = b_i(O_1)\beta_1(i) \tag{3.41}$$

and

$$\frac{\partial P}{\partial b_{jk}} = \sum_{t \ni O_t = v_k} \sum_{i=1}^{N} \alpha_t(i)a_{ij}\beta_{t+1}(j) + \delta(O_1, v_k)\pi_j\beta_1(j), \tag{3.42}$$

respectively. In (3.42), δ is understood to be the Kronecker δ function.

By substituting (3.41) and (3.42) into their respective analogs of (3.37), we obtain the reestimation formulas (3.13) and (3.12), respectively, at a critical point. Thus it appears that the reestimation formulas may have more general applications than might appear from their statistical motivation.

Equation (3.37) suggests that we define a transformation, $\mathcal{T}$, of the parameter space onto itself as

$$\mathcal{T}(\mathbf{x})_{ij} = \frac{x_{ij}\frac{\partial P}{\partial x_{ij}}}{\sum_{k=1}^{N} x_{ik}\frac{\partial P}{\partial x_{ik}}}, \tag{3.43}$$

where $\mathcal{T}(\mathbf{x})_{ij}$ is understood to mean the ijth coordinate of the image of $\mathbf{x}$ under $\mathcal{T}$. The parameter space is restricted to be the manifold such that $x_{ij} \geq 0$ for $1 \leq i, j \leq N$ and $\sum_{j=1}^{N} x_{ij} = 1$ for $1 \leq i \leq N$. Thus the reestimation formulas (3.11), (3.12), and (3.13) are a special case of the transformation (3.43), with P a particular homogeneous polynomial in the x_{ij} having positive coefficients. Here the x_{ij} include the π_i, the a_{ij}, and the b_{jk}. Baum and Eagon [26] have shown that for any such polynomial $P[\mathcal{T}(\mathbf{x})] > P(\mathbf{x})$ unless $\mathbf{x}$ is a critical point of P. Thus the transformation, $\mathcal{T}$, is appropriately called a growth transformation. The conditions under which $\mathcal{T}$ is a growth transformation were relaxed by Baum and Sell [29] to include all polynomials with positive coefficients. They further proved that P increases monotonically on the segment from $\mathbf{x}$ to $\mathcal{T}(\mathbf{x})$. Specifically, they showed that $P[\eta\mathcal{T}(\mathbf{x}) + (1 - \eta)\mathbf{x}] \geq P(\mathbf{x})$ for $0 \leq \eta \leq 1$. Other properties of the transformation (3.43) have been explored by Passman [240] and Stebe [307]. There may be still less restrictive general criteria on P for $\mathcal{T}$ to be a growth transformation.

We can give $\mathcal{T}(\mathbf{x})$ a simple geometric interpretation. For the purposes of this discussion we shall restrict ourselves to $\mathbf{x} \in \mathbb{R}^N$, $x_i \geq 0$ for $1 \leq i \leq N$, and the single constraint $G(\mathbf{x}) = \sum_{i=1}^{N} x_i - 1 = 0$. We do so without loss of generality, since constraints such as those of (3.33) are disjoint, that is, no pair of constraints has any common variables. As shown in Fig. 3.3, given any $\mathbf{x}$ satisfying $G(\mathbf{x}) = 0$, $\mathcal{T}(\mathbf{x})$ is the intersection of the vector $\mathbf{X}$, or its extension, with the hyperplane $\sum_{i=1}^{N} x_i - 1 = 0$, where $\mathbf{X}$ has components $x_i\frac{\partial P}{\partial x_i}$ for $1 \leq i \leq N$.

This may be shown by observing that a line in the direction of $\mathbf{X}$ passing through the origin has the equation $\mathbf{y} = r\mathbf{X}$, where r is a non-negative scalar.

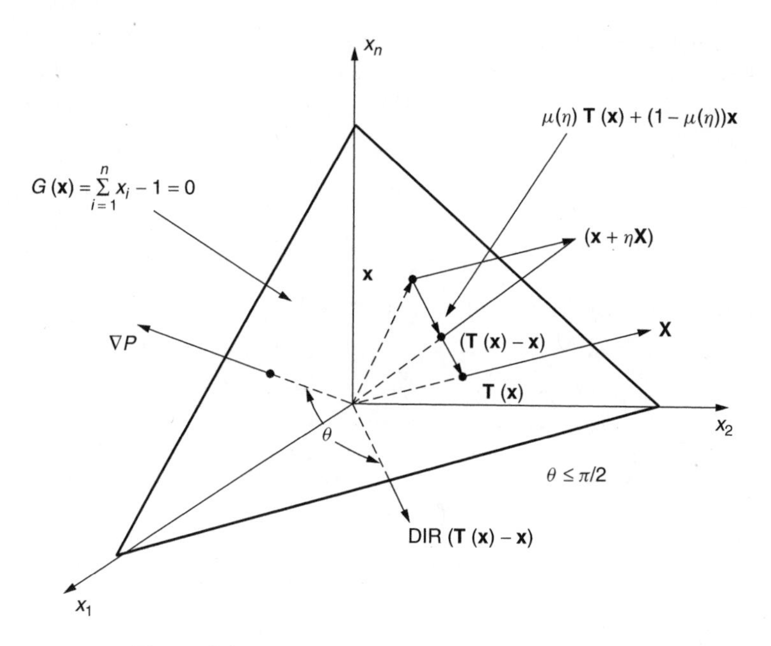

Figure 3.3 The geometry of the Baum algorithm

Componentwise this is equivalent to

$$y_i = r x_i \frac{\partial P}{\partial x_i}, \qquad \text{for } 1 \leq i \leq N. \tag{3.44}$$

We can find that r for which $\mathbf{y}$ intersects the hyperplane $G(\mathbf{x}) = 0$ by summing over i. Thus

$$\sum_{i=1}^{N} y_i = r \sum_{i=1}^{N} x_i \frac{\partial P}{\partial x_i} = 1, \tag{3.45}$$

since $\mathbf{y}$ lies on the hyperplane $G(\mathbf{x}) = 0$. Rearranging (3.45), we have

$$r = \frac{1}{\sum_{i=1}^{N} x_i \frac{\partial P}{\partial x_i}} \tag{3.46}$$

and

$$y_i = \frac{x_i \frac{\partial P}{\partial x_i}}{\sum_{j=1}^{N} x_j \frac{\partial P}{\partial x_j}}. \tag{3.47}$$

Furthermore, as also shown in Fig. 3.3, the vector $(\mathcal{T}(\mathbf{x}) - \mathbf{x})$ is the set of intersections of the vector $(\mathbf{x} + \eta\mathbf{X})$ with the hyperplane $G(\mathbf{x}) = 0$ for $0 \leq \eta \leq +\infty$ with $\mathcal{T}(\mathbf{x})$ corresponding to $\eta = +\infty$ and $\mathbf{x}$ to $\eta = 0$.

Finally, in view of the result of Baum and Sell, quoted above, the vector $(\mathcal{T}(\mathbf{x}) - \mathbf{x})$ must also have a positive projection on ∇P. This, too, is easily seen. If P is a polynomial with positive coefficients, then $\partial P/\partial x_i \geq 0$ for $1 \leq i \leq N$. From the definition of $\mathcal{T}$ it is clear that

$$\mathcal{T}(\mathbf{x})_i \geq x_i \Leftrightarrow \frac{\partial P}{\partial x_i} \geq \sum_{j=1}^{N} x_j \frac{\partial P}{\partial x_j} = r, \tag{3.48}$$

where r is some constant. Then it must be true that

$$\sum_{i=1}^{N} [\mathcal{T}(\mathbf{x})_i - x_i] \left(\frac{\partial P}{\partial x_i} - r \right) \geq 0, \tag{3.49}$$

since both factors in each summand are of the same sign. Rearranging (3.49), we have

$$\sum_{i=1}^{N} [\mathcal{T}(\mathbf{x})_i - x_i] \frac{\partial P}{\partial x_i} \geq r \sum_{i=1}^{N} [\mathcal{T}(\mathbf{x})_i - x_i] = 0. \tag{3.50}$$

The right-hand side is zero since $\sum_{i=1}^{N} \mathcal{T}(\mathbf{x})_i = \sum_{i=1}^{N} x_i = 1$. Thus $(\mathcal{T}(\mathbf{x}) - \mathbf{x}) \cdot \nabla P \geq 0$, proving that a step of the transformation has a positive projection along the gradient of P.

This merely guarantees that we can move an infinitesimal amount in the direction of $(\mathcal{T}(\mathbf{x}) - \mathbf{x})$ while increasing P. The theorem of Baum and Eagon, however, guarantees much more, namely that we can take a finite step and be assured of increasing P. A geometrical interpretation and proof of this result is as follows.

We begin by recalling two simple properties of homogeneous polynomials. A well-known theorem due to Euler states that if $P(\mathbf{x})$ is a homogeneous polynomial with positive coefficients,

$$\sum_{L=1}^{N} x_i \frac{\partial P}{\partial x_i} = w_p P(\mathbf{x}), \tag{3.51}$$

where the constant w_p is the weight of $P(\mathbf{x})$ and is equal the degree of each term in $P(\mathbf{x})$. This result is easily derived by differentiating $P(\mathbf{x})$ and collecting terms.

The second result is that at a critical point of P subject to the constraint that $\sum_{i=1}^{N} x_i = 1$,

$$\frac{\partial P}{\partial x_i} = w_p P(\mathbf{x}). \tag{3.52}$$

This follows from the method of Lagrange multipliers according to which

$$\nabla \left(P + \lambda \left[\sum_{i=1}^{N} x_i - 1 \right] \right) = 0 \tag{3.53}$$

at a critical point. Performing the differentiation indicated in (3.53), we find that the Lagrange multiplier is just

$$\lambda = -\frac{\partial P}{\partial x_i}, \qquad 1 \leq i \leq N. \tag{3.54}$$

Multiplying by x_i and summing over i, we obtain

$$\sum_i \left[x_i - \frac{\partial P}{\partial x_i} + x_i \lambda \right] = 0. \tag{3.55}$$

Then, recalling the constraint on x_i, (3.55) reduces to

$$\sum_i x_i \frac{\partial P}{\partial x_i} = -\lambda \tag{3.56}$$

and, invoking Euler's result (3.51), (3.56) becomes

$$-\lambda = w_p P(\mathbf{x}). \tag{3.57}$$

Substituting the value of λ into (3.54) gives the result of (3.52). Note that (3.57) implies that the gradient of P with respect to the constraint is normal to the plane $G(\mathbf{x})$ specified by the constraint. Note also that at a critical point

$$\mathcal{T}(\mathbf{x})_i = \frac{x_i \dfrac{\partial P}{\partial x_i}}{\sum_j x_j \dfrac{\partial P}{\partial x_j}} = \frac{x_i w_p P(\mathbf{x})}{w_p P(\mathbf{x})} = x_i \tag{3.58}$$

so that a critical point of P is a fixed point of the mapping $\mathcal{T}(\mathbf{x})$.

We can now use these results to extend (3.50) to the full theorem of Baum and Eagon. Either (3.50) holds for some finite step Δx in the direction of $\mathcal{T}(\mathbf{x}) - \mathbf{x}$ or $\nabla P \cdot (\mathcal{T}(\mathbf{x}) - \mathbf{x})$ becomes negative at that distance. If the latter is true then there must be a point, $\mathbf{z}$, between $\mathbf{x}$ and $\mathbf{x} + \Delta\mathbf{x}$ at which $\nabla P \cdot (\mathcal{T}(\mathbf{x}) - \mathbf{x}) = 0$. This point will be a critical point of $P(\mathbf{x})$ with respect to the constraint $G(\mathbf{x})$.

This critical point is the intersection of the vector $\mathbf{x} + \eta\mathbf{X}$ for some $\eta > 0$, with the vector $(\mathcal{T}(\mathbf{x}) - \mathbf{x})$. Because it is a critical point it is also the intersection of $\mathbf{z} + \mu\mathbf{Z}$ with $(\mathcal{T}(\mathbf{x}) - \mathbf{x})$. Thus we have two equations for the unique line passing through the origin and the point $\mathbf{z}$, namely,

$$\mathbf{y} = r\mathbf{X} \tag{3.59}$$

and

$$\mathbf{y} = r\mathbf{Z}. \tag{3.60}$$

Thus the directions of lines (3.59) and (3.60) must be identical or

$$\frac{\partial P}{\partial x_i} = \text{constant}, \qquad \text{for all } i. \tag{3.61}$$

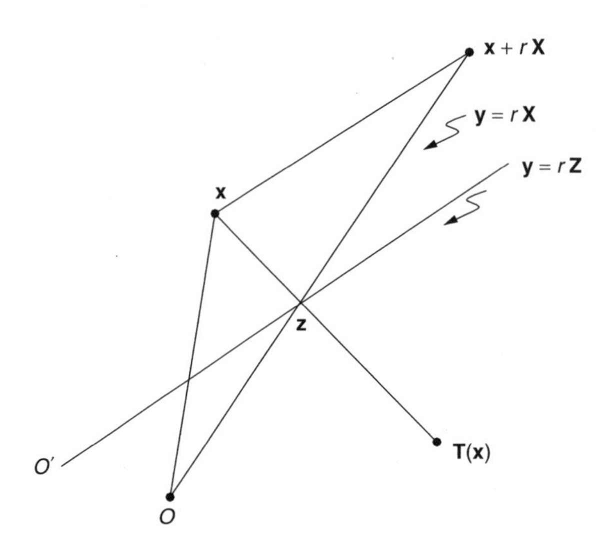

Figure 3.4 The local maximum of the likelihood function cannot lie on the line segment **T(x)−x**.

But as shown in Fig. 3.4, this is impossible since if (3.61) were true, $\mathbf{x}$ would be a critical point of $P(\mathbf{x})$ and hence a fixed point of $\mathcal{T}(\mathbf{x})$ in which case $\mathbf{x} = \mathcal{T}(\mathbf{x})$, a contradiction.

From this contradiction we conclude that $\mathbf{z}$ is not a critical point and $\nabla P \cdot (\mathcal{T}(\mathbf{x}) - \mathbf{x}) > 0$ everywhere on the vector $(\mathcal{T}(\mathbf{x}) - \mathbf{x})$. And finally $P(\overline{\mathbf{x}}) \geq P(\mathbf{x})$ everywhere on $G(\mathbf{x})$ with equality only at a critical point of $P(\mathbf{x})$ corresponding to a fixed point of $\mathcal{T}(\mathbf{x})$. This is the Baum *et al.* result.

While the reestimation formulas provide an elegant method for maximizing P, their success depends critically on the constraint set (3.32)–(3.34). As we will suggest later, in some cases there may be advantages in using classical optimization methods.

The principle of the classical methods is to search along the projection of ∇P on the constraint space, G, for a local maximum. The method of Rosen [277], for example, uses only ∇P and a crude search strategy. The method of Davidon is one of many quasi-Newton techniques that uses the Fletcher–Powell [90] approximation to the inverse of the Hessian of P and an exact line search with adaptive step size. There are many collections of general purpose subroutines for constrained optimization that can be used to solve the training problem. We have successfully used a version of the Davidon procedure from the Harwell Subroutine Library [135]. However, for the constraints that $\boldsymbol{\pi}$, $\mathbf{A}$, and $\mathbf{B}$ be stochastic, the computation can be greatly simplified.

We illustrate this by outlining the gradient search algorithm for the case where P is a function of the variables $x_1, \ldots, x_N$ subject to the constraints $x_i \geq 0$ for $1 \leq i \leq N$ and $\sum_{i=1}^{N} x_i - 1 = 0$. For convenience we will call the last constraint G_1, and the inequality constraints on $x_1, \ldots, x_N$ as $G_2, \ldots, G_{N+1}$, respectively.

An initial starting point $\mathbf{x}$ is chosen and the "active" constraints identified. For our case G_1 is always active. For $i > 1$, G_i is active if $x_{i-1} = 0$. Let G_{n_j}, $j = 1, \ldots, \ell$ be the active constraints (with $n_1 = 1$) at the initial point. Let $Q = P + \sum_{j=1}^{\ell} \lambda_j G_{n_j}$. Then according to the Kuhn–Tucker theorem [125], the Lagrange multipliers, λ_j, are determined by demanding that ∇Q be orthogonal to ∇G_{n_j} for $1 \leq j \leq \ell$. Now

$$\nabla Q = \nabla P + \sum_{j=1}^{\ell} \lambda_j \nabla G_{n_j} = \nabla P + \mathbf{\Gamma}\boldsymbol{\lambda}, \tag{3.62}$$

where $\mathbf{\Gamma}$ is the $N \times \ell$ matrix with $\Gamma_{ij} = (\nabla G_{n_j})_i = \partial G_{n_j}/\partial x_i$, and $\boldsymbol{\lambda}$ is the vector with components λ_j for $j = 1, \ldots, \ell$. Thus the Kuhn–Tucker requirement is equivalent to

$$\mathbf{\Gamma}' \nabla Q = 0 \tag{3.63}$$

or, from (3.62),

$$\boldsymbol{\lambda} = -(\mathbf{\Gamma}'\mathbf{\Gamma})^{-1}\mathbf{\Gamma}'\nabla P. \tag{3.64}$$

For our special constraints we have

$$\Gamma_{i1} = 1, \qquad \text{for } 1 \leq i \leq N, \tag{3.65}$$

and, for $j \neq 1$,

$$\Gamma_{ij} = \begin{cases} 1, & \text{if } i = n_j - 1, \\ 0, & \text{otherwise.} \end{cases} \tag{3.66}$$

With $\mathbf{\Gamma}$ defined this way,

$$\mathbf{\Gamma}'\mathbf{\Gamma} = \begin{pmatrix} N & 1 & \ldots & 1 \\ 1 & 1 & & \mathbf{O} \\ \vdots & \mathbf{O} & \ddots & \\ 1 & & & 1 \end{pmatrix} \tag{3.67}$$

and $(\mathbf{\Gamma}'\mathbf{\Gamma})^{-1}$ may be shown to be

$$(\mathbf{\Gamma}'\mathbf{\Gamma})^{-1} = \frac{1}{N - \ell + 1} \begin{pmatrix} 1 & -1 & -1 & \cdots & -1 \\ -1 & N-\ell & 1 & \cdots & 1 \\ -1 & 1 & N-\ell & 1 & 1 \\ \vdots & & & \ddots & 1 \\ -1 & 1 & 1 & & N-\ell \end{pmatrix}. \tag{3.68}$$

Substituting (3.68) into (3.64) gives $\boldsymbol{\lambda}$. When this $\boldsymbol{\lambda}$ is substituted back into (3.62), it turns out that the resulting vector ∇Q can be computed by the following simple steps:

(i) Compute ∇P and let S be the sum of all components of ∇P *except* $(\nabla P)_{n_{j-1}}$, $j = 2, \ldots, \ell$.

(ii) Then

$$(\nabla Q)_i = 0, \qquad i = n_j, j = 2, \ldots, l \tag{3.69}$$

$$= (\nabla P)_i - \frac{S}{N - l + 1} \qquad \text{otherwise.} \tag{3.70}$$

Finally, the values of P are searched along the line

$$\mathbf{x}(\eta) = \mathbf{x} + \eta \frac{\nabla Q}{\| \nabla Q \|} \tag{3.71}$$

for a maximum with respect to η. The procedure is repeated at this new point.

In applying this technique to the actual training problem, there will be $2N + 1$ stochasticity constraints analogous to G_1 and a corresponding number of positivity constraints analogous to $G_2, G_3, \ldots, G_{N+1}$. In this case we have the option of treating all the parameters and their associated constraints together, or we may divide them into disjoint subsets and determine search directions for each subset independently.

Notice that this derivation does not require P to be of any special form. This may prove to be an advantage since the Baum–Welch algorithm is not applicable to all P. Furthermore, the constraints may be changed. Although, as we shall see later, the Baum–Welch algorithm can be somewhat generalized in this respect, it does not generalize to work with arbitrary linear constraints.

Considerations for Implementation

From the foregoing discussion it might appear that solutions to the problems of hidden Markov modeling can be obtained by straightforward translation of the relevant formulas into computer programs. Unfortunately, for all but the most trivial problems, the naive implementation will not succeed for two principal reasons. First, any of the methods of solution presented here for either the classification or the training problem require evaluation of $\alpha_t(i)$ and $\beta_i(i)$ for $1 \leq t \leq T$ and $1 \leq i \leq N$. From the recursive formulas for these quantities, (3.1) and (3.2), it is clear that as $T \to \infty, \alpha_T(i) \to 0$ and $\beta_1(i) \to 0$ in exponential fashion. In practice, the number of observations necessary to adequately train a model and/or compute its probability will result in underflow on any real computer if (3.1) and (3.2) are evaluated directly. Fortunately, there is a method for scaling these computations that not only solves the underflow problem but also greatly simplifies several other calculations.

The second problem is more serious, more subtle, and admits of a less gratifying, though still effective, solution. Baum and Petrie [27] have shown that the maximum likelihood estimates of the parameters of a hidden Markov process are consistent estimates of the parameters (converge to the true values as $T \to \infty$). The practical implication of the theorem is that, in training, one should use as many observations as possible which, as we have noted, makes scaling necessary. In reality, of course, the observation sequence will always be finite. Then the following situation can arise. Suppose a given training sequence of length T results in $b_{jk} = 0$. (It is, in fact, possible for a local maximum of P to lie on a boundary of the parameter manifold.) Suppose, further, that we are subsequently asked to compute the probability that a new observation sequence was generated by our model. Even if the new sequence was actually generated by the model, it can be such that $\alpha_{t-1}(i)a_{ij}$ is non-zero for only one value of j and that $O_t = v_k$, whence $\alpha_t(j) = 0$ and the probability of the observation then becomes zero. This phenomenon is fatal to a classification task; yet, the smaller T is, the more likely is its occurrence. Here, we offer a general solution of constraining the parameter values so that $x_{ij} \geq \varepsilon_{ij} > 0$. Jelinek and Mercer [150] have shown how to set the value of ε.

Finally, in this section we discuss the related problem of model stability. Baum and Eagon [26] note that successive applications of the reestimation formulas converge to a connected component of the local maximum set of P. If there is only a finite number of such extrema, the point of convergence is unique to within a renaming of the states. The component of the local maximum set to which the iteration converges as well as which of the $N!$ labelings of the states is determined by the initial estimates of the parameters. If we wish to average several models resulting from several different starting points to achieve model stability, we must be able to match the states of models whose states are permuted. We have devised a solution to this problem based on a minimum-weight bipartite matching algorithm [174].

Scaling

The principle on which we base our scaling is to multiply $\alpha_i(i)$ by some scaling coefficient independent of i so that it remains within the dynamic range of the computer for $1 \leq t \leq T$. We propose to perform a similar operation on $\beta_i(i)$ and then, at the end of the computation, remove the total effect of the scaling.

We illustrate the procedure for (3.11), the reestimation formula for the state transition probabilities. Let $\alpha_t(i)$ be computed according to (3.1) and then multiplied by a scaling coefficient, c_t, where, say,

$$c_t = \left[\sum_{i=1}^{N} \alpha_t(i) \right]^{-1} \tag{3.72}$$

so that $\sum_{i=1}^{N} c_t \alpha_t(i) = 1$ for $1 \leq t \leq T$. Then, as we compute $\beta_t(i)$ from (3.2), we form the product $c_t \beta_t(i)$ for $T \geq t \geq 1$ and $1 \leq i \leq N$. In terms of the scaled forward and backward probabilities, the right-hand side of (3.11) becomes

$$\frac{\sum_{t=1}^{T-1} C_t \alpha_t(i) a_{ij} b_j(O_{t+1}) \beta_{t+1}(j) D_{t+1}}{\sum_{t=1}^{T-1} \sum_{\ell=1}^{N} C_t \alpha_t(i) a_{i\ell} b_\ell(O_{t+1}) \beta_{t+1}(\ell) D_{t+1}}, \tag{3.73}$$

where

$$C_t = \prod_{\tau=1}^{t} c_\tau \tag{3.74}$$

and

$$D_t = \prod_{\tau=t}^{T} c_\tau. \tag{3.75}$$

This results from the individual scale factors being multiplied together as we perform the recursions of (3.1) and (3.2).

Now note that each summand in both the numerator and the denominator has the coefficient $C_t D_{t+1} = \prod_{\tau=1}^{T} c_\tau$. These coefficients can be factored out and canceled so that (3.73) has the correct value $\overline{a}_{ij}$ as specified by (3.11). The reader can verify that this technique may be equally well applied to the reestimation formulas (3.12) and (3.13). It should also be obvious that, in practice, the scaling operation need not be performed at every observation time. One can use any scaling interval for which underflow does not occur. In this case, the scale factors corresponding to values of t within any interval are set to unity.

While the above described scaling technique leaves the reestimation formulas invariant, (3.3) and (3.4) are still useless for computing P. However, $\log P$ can be recovered from the scale factors as follows. Assume that we compute c_t according to (3.72) for $t = 1, 2, \ldots, T$. Then

$$C_T \sum_{i=1}^{N} \alpha_T(i) = 1, \tag{3.76}$$

and from (3.76) it is obvious that $C_T = 1/P$. Thus, from (3.74) we have

$$\prod_{t=1}^{T} c_t = \frac{1}{P}. \tag{3.77}$$

The product of the individual scale factors cannot be evaluated but we can compute

$$\log P = -\sum_{t=1}^{T} \log c_t. \tag{3.78}$$

If one chooses to use the Viterbi algorithm for classification, then $\log P$ can be computed directly from $\boldsymbol{\pi}$, $\mathbf{A}$, and $\mathbf{B}$ without regard for the scale factors. Initially, we let $\phi_1(i) = \log[\pi_i b_i(O_1)]$ and then modify (3.7) so that

$$\phi_t(j) = \max_{1 \le i \le N} [\phi_{t-1}(i) + \log a_{ij}] + \log[b_j(O_t)]. \tag{3.79}$$

In this case $\log P = \max_{1 \le i \le N}[\phi_T(i)]$.

If the parameters of the model are to be computed by means of classical optimization techniques, we can make the computation better conditioned numerically by maximizing $\log P$ rather than P. The scaling method of (3.72) makes this straightforward.

First note that if we are to maximize $\log P$, then we will need the partial derivatives of $\log P$ with respect to the parameters of the model. So, for example, we will need

$$\frac{\partial}{\partial a_{ij}}(\log P) = \frac{1}{P}\frac{\partial P}{\partial a_{ij}} = C_T \frac{\partial P}{\partial a_{ij}}. \tag{3.80}$$

Substituting the right-hand side of (3.38) for $\partial P/\partial a_{ij}$ in the right-hand side of (3.80) yields

$$\frac{\partial}{\partial a_{ij}}(\log P) = C_T \sum_{t=1}^{T-1} \alpha_t(i) b_j(O_{t+1}) \beta_{t+1}(j) \tag{3.81}$$

$$= \sum_{t=1}^{T-1} C_t \alpha_t(i) b_j(O_{t+1}) \beta_{t+1}(j) D_{t+1}$$

$$= \sum_{t=1}^{T-1} \left(\prod_{\tau=1}^{t} c_\tau \right) \alpha_t(i) b_j(O_{t+1}) \beta_{t+1}(j) \left(\prod_{\tau=t+1}^{T} c_\tau \right).$$

Thus if we evaluate (3.38) formally, using not the true values of the forward and backward probabilities but the scaled values, then we will have the correct value of the partial derivatives of $\log P$ with respect to the transition probabilities. A similar argument can be made for the other parameters of the model and, thus, the scaling method of (3.72) provides a means for the direct evaluation of $\nabla(\log P)$, which is required for the classical optimization algorithms. Later we shall see that the combination of maximizing $\log P$ and this scaling technique simplifies the solution of the left-to-right Markov modeling problem as well.

Finite Training Sets

The second point is complementary to the first in the sense that it arises exactly because we can never have T large enough. The consequence of this unpleasant reality is that there may be certain events of low probability that will not be manifest in a finite observation sequence. Should such a sequence be used to estimate $\boldsymbol{\lambda}$, these events will be assigned probability zero, which values may later have catastrophic results when the parameter estimates are used in classification and an event of low probability is actually encountered. Three methods for mitigation of this difficulty have been used in automatic speech recognition. The method of Jelinek and Mercer [150] uses information other than that in the observation sequence to estimate small probabilities. The method employed by Nadas [227] is a different smoothing technique.

We now turn our attention to solving the problems created by finite training-set size. As we noted earlier, the effect of this problem is that observation sequences generated by a putative model will have zero probability conditioned on the model parameters. Since the cause of the difficulty is the assignment of zero to some parameters, usually one or more symbol probabilities b_{jk}, it is reasonable to try to solve the problem by constraining the parameters to be positive.

We can maximize P subject to the new constraints $a_{ij} \geq \varepsilon > 0$, $b_{jk} \geq \varepsilon > 0$, most easily using the classical methods. In fact, the algorithm described earlier based on the Kuhn–Tucker theorem is unchanged except that the procedure for determining the active constraints is based on ε rather than zero.

While the Lagrangian methods are perfectly adequate, it is also possible to build the new constraints into the Baum–Welch algorithm. We can show how this is done by making a slight modification to the proof of the algorithm given earlier. Recall that the proof of the Baum–Welch algorithm was based on maximization of $2N + 1$ expressions of the type maximized in (3.17) of Lemma 2. Since these expressions involve disjoint sets of variables chosen from $\mathbf{A}, \mathbf{B}, \boldsymbol{\pi}$, it suffices to consider any one of the maximizations. In fact, it suffices to show how Lemma 2 gets modified. Thus we wish now to maximize

$$F(\mathbf{x}) = \sum_i c_i \ln x_i \tag{3.82}$$

subject to the constraints

$$\sum_i x_i = 1 \tag{3.83}$$

and

$$x_i \geq \varepsilon, \qquad i = 1, \cdots N. \tag{3.84}$$

(From the following discussion it will be obvious that a trivial generalization allows ε to depend on i.)

Now without the inequality constraints (3.84), Lemma 2 showed that $F(\mathbf{x})$ attains its unique global maximum when $x_i = c_i / \sum_i c_i$. Suppose now that this global maximum occurs outside the region specified by the inequality constraints (3.84). Specifically, let

$$\overline{x}_i = \frac{c_i}{\sum_{j=1}^{N} c_j} \begin{cases} \geq \varepsilon, & \text{for } i = 1, \ldots, N - l, \\ < \varepsilon, & \text{for } i = N - l + 1, \ldots, N. \end{cases} \tag{3.85}$$

From the concavity of $F(\mathbf{x})$ it follows that the maximum, subject to the inequality constraints, must occur somewhere on the boundary specified by the violated constraints (3.85). Now it is easily shown that if $\overline{x}_i$ for some $i > N - \ell$ is replaced by ε, then the global maximum over the rest of the variables occurs at values *lower* than those given above. From this we conclude that we must set

$$\overline{x}_i = \varepsilon, \qquad \text{for } i > N - l, \tag{3.86}$$

and maximize

$$\tilde{F}(\mathbf{x}) = \sum_{i=1}^{N-l} c_i \ln x_i \tag{3.87}$$

subject to the constraint $\sum_{i=1}^{N-l} x_i = 1 - \ell\varepsilon$. But this, analogously to Lemma 2, occurs when

$$\overline{x}_i = (1 - l\varepsilon)\frac{c_i}{\sum_{j=1}^{N-l} c_j}, \qquad i \leq N - l. \tag{3.88}$$

If these new values of $\overline{x}_i$ satisfy the constraints, we are done. If one or more become lower than ε, they too must be set equal to ε, and l augmented appropriately.

Thus the modified Baum–Welch algorithm is as follows. Suppose we wish to constrain $b_{jk} \geq \varepsilon$ for $1 \leq j \leq N$ and $1 \leq k \leq M$. We first evaluate $\mathbf{B}$ using the reestimation formulas. Assume that some set of the parameters in the jth row of $\mathbf{B}$ violates the constraint so that $b_{jk_i} < \varepsilon$ for $1 \leq i \leq l$. Then set $\tilde{b}_{jk_i} = \varepsilon$ for $1 \leq i \leq l$ and readjust the remaining parameters according to (3.89) so that

$$\tilde{b}_{jk} = (1 - l\varepsilon)\frac{b_{jk}}{\sum_{i=1}^{N-l} b_{ji}} \qquad \forall k \notin \{k_i | 1 \leq i \leq l\}. \tag{3.89}$$

After performing the operation of (3.89) for each row of $\mathbf{B}$, the resulting $\tilde{\mathbf{B}}$ is the optimal update with respect to the desired constraints. The method can be extended to include the state transition matrix if so desired. There is no advantage to treating π in the same manner since, for any single observation sequence, $\overline{\pi}$ will always be a unit vector with exactly one non-zero component. In any case, (3.89) may be applied at each iteration of the reestimation formulas, or once as a post-processing stage after the Baum–Welch algorithm has converged.

Non-Ergodic Hidden Markov Models

For the purposes of isolated word recognition, it is useful to consider a special class of absorbing Markov chains that leads to what we call left-to-right models. These models have the following properties:

(i) The first observation is produced while the Markov chain is in a distinguished state called the starting state, designated q_1.
(ii) The last observation is generated while the Markov chain is in a distinguished state called the final or absorbing state, designated q_N.
(iii) Once the Markov chain leaves a state, that state cannot be revisited at a later time.

The simplest form of a left-to-right model is shown in Fig. 3.5, from which the origin of the term left-to-right becomes clear.

In this section we shall consider two problems associated with these special hidden Markov models. Note that a single, long-observation sequence is useless for training such models, because once the state q_N is reached, the rest of the sequence provides no further

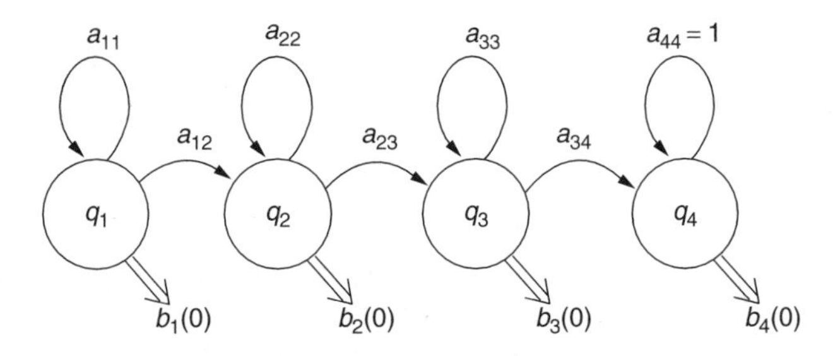

Figure 3.5 The left-to-right (non-ergodic) HMM.

information about earlier states. The appropriate training data for such a model is a *set* of observation sequences obtained by several starts in state q_1. In the case of isolated word recognition, for instance, several independent utterances of the same word provide such a set. We wish, therefore to modify the training algorithm to handle such training data. We also wish to compute the probability that a single given observation sequence, $O_1, O_2 \ldots, O_T$, was produced by the model, with the assumption that O_1 was produced in state q_1 and O_T in state q_N. The three conditions mentioned above can be satisfied as follows:

Condition (i) will be satisfied if we set $\boldsymbol{\pi} = (1, 0, \cdots, 0)$ and do not reestimate it. Condition (ii) can be imposed by setting

$$\beta_T(j) = \begin{cases} 1, & \text{for } j = N, \\ 0, & \text{otherwise.} \end{cases} \tag{3.90}$$

Condition (iii) can be guaranteed in the Baum–Welch algorithm by initially setting $a_{ij} = 0$ for $j < i$ (and in fact for any other combination of indices that specify transitions to be disallowed). It is clear from (3.43) that any parameter once set to zero will remain zero. For the gradient methods the appropriate a_{ij} are just set to zero and only the remaining parameters are adjusted.

The modification of the training procedure is as follows. Let us denote by $\mathbf{O} = [\mathbf{O}^{(1)}, \mathbf{O}^{(2)}, \ldots, \mathbf{O}^{(K)}]$ the set of observation sequences, where $\mathbf{O}^{(k)} = O_1^{(k)} O_2^{(k)} \cdots O_{T_k}^{(k)}$ is the kth sequence. We treat the observation sequences as independent of each other and then we adjust the parameters of the model $\mathbf{M}$ to maximize

$$P = \prod_{k=1}^{K} \text{Prob}(\mathbf{O}^{(k)}|\mathbf{M}) = \prod_{k=1}^{K} P_k. \tag{3.91}$$

Since the Baum–Welch algorithm computes the frequency of occurrence of various events, all we need to do is to compute these frequencies of occurrence in each sequence separately and add them together. Thus the new reestimation formulas may be written as

$$\bar{a}_{ij} = \frac{\sum_{k=1}^{K} \sum_{t=1}^{T_k - 1} \alpha_t^k(i) a_{ij} b_j(O_{t+1}^{(k)}) \beta_{t+1}^k(j)}{\sum_{k=1}^{K} \sum_{t=1}^{T_k - 1} \alpha_t^k(i) \beta_t^k(i)} \tag{3.92}$$

and

$$\bar{b}_{ij} = \frac{\sum_{k=1}^{K} \sum_{t \ni O_t(k) = v_j} \alpha_t^k(i) \beta_t^k(i)}{\sum_{k=1}^{K} \sum_{t=1}^{T_k} \alpha_t^k(i) \beta_t^k(i)}. \tag{3.93}$$

As noted above, $\boldsymbol{\pi}$ is not reestimated.

Scaling these computations requires some care since the scale factors for each individual set of forward and backward probabilities will be different. One way of circumventing the problem is to remove the scale factors from each summand before adding. We can accomplish this by returning the $1/P$ factor, which appears in (3.9) and (3.10) and was cancelled to obtain (3.11), to the reestimation formula. Using the reestimation formula for the transition probabilities as an example, (3.92) becomes

$$\bar{a}_{ij} = \frac{\sum_{k=1}^{K} \frac{1}{P_k} \sum_{t=1}^{T_k-1} \alpha_t^k(i) a_{ij} b_j(O_{t+1}^{(k)}) \beta_{t+1}^k(j)}{\sum_{k=1}^{K} \frac{1}{P_k} \sum_{t=1}^{T_k-1} \alpha_t^k(i) \beta_t^k(i)}. \tag{3.94}$$

If the right-hand side of (3.94) is evaluated using the scaled values of the forward and backward probabilities, then each term in the inner summation will be scaled by $C_t^k D_{t+1}^k$, which will then be canceled by the same factor which multiplies P_k. Thus, using the scaled values in computing (3.92) results in an unscaled $\bar{a}_{ij}$. The procedure is easily extended to computation of the symbol probabilities. Also note that for the purposes of classification only one subsequence is to be considered so that either (3.78) or (3.79) may be used unaltered to compute P.

To apply Lagrangian techniques to left-to-right models we note that, upon taking logarithms of (3.91), we have

$$\log P = \sum_{k=1}^{K} \log P_k. \tag{3.95}$$

The derivatives needed to maximize $\log P$ in (3.95) can be obtained by evaluating expressions for the derivatives of each individual subsequence and summing. For example, for a_{ij} we have (cf. (3.80) and (3.81))

$$\frac{\partial}{\partial a_{ij}}(\log P) = \sum_{k=1}^{K} \frac{\partial}{\partial a_{ij}}(\log P) = C_T^k \sum_{t=1}^{T_k-1} \alpha_t^k(i) b_j(O_{t+1}^{(k)}) \beta_{t+1}^k(j). \tag{3.96}$$

As in all previous cases, an analogous formula may be derived for the other parameters.

In practice, **A** and **B** for left-to-right models are especially sparse. Some of the zero values are so by design, but others are dependent on **O**. Parameters of this type will be found one at a time by standard line search strategies. We have found that the convergence of the Lagrangian techniques can be substantially accelerated by taking large enough steps so that several positivity constraints become binding. The corresponding variables are then clamped and (3.89) is applied before beginning the next iteration.

Matrix Notation

Several of the formulas derived above are much more compact in matrix notation. Let $'$ denote matrix transposition, as usual, and let the column vectors $\boldsymbol{\pi}$ and $\mathbf{1}$, and the matrices, $\mathbf{A}, \mathbf{B}_t, t = 1, \ldots, T$, be defined as above. Also let $\boldsymbol{\alpha}_t$ and $\boldsymbol{\beta}_t$ be column vectors with components $\alpha_t(i), i = 1, \ldots, N$ and $\beta_t(i), i = 1, \ldots, N$, respectively. Then the recursion for α_t is

$$\boldsymbol{\alpha}_{t+1} = \mathbf{B}_{t+1}\mathbf{A}'\boldsymbol{\alpha}_t, \qquad t = 1, \ldots, T-1. \tag{3.97}$$

The recursion for $\boldsymbol{\beta}_t$ is

$$\boldsymbol{\beta}_t = \mathbf{A}\mathbf{B}_{t+1}\boldsymbol{\beta}_{t+1}, \qquad t = T-1, \ldots, 1. \tag{3.98}$$

The starting values are

$$\boldsymbol{\alpha}_1 = \mathbf{B}_1\boldsymbol{\pi}, \tag{3.99}$$

$$\boldsymbol{\beta}_T = \mathbf{1}. \tag{3.100}$$

The probability P is given by

$$P = \boldsymbol{\beta}_t'\boldsymbol{\alpha}_t, \qquad \text{for any } t \text{ in } (1, T). \tag{3.101}$$

The special cases $t = 1$ and $t = T$ give

$$P = \boldsymbol{\pi}'\mathbf{B}_1\boldsymbol{\beta}_1 \tag{3.102}$$

and

$$P = \mathbf{1}'\boldsymbol{\alpha}_T = \mathbf{1}'\mathbf{B}_T\mathbf{A}'\mathbf{B}_{T-1}\cdots\mathbf{A}'\mathbf{B}_1\boldsymbol{\pi}. \tag{3.103}$$

In each of these formulas P can be regarded as the trace of a 1×1 matrix, which (as expanded in (3.103)) is a product of several matrices. The fact that the trace of a product of matrices is invariant to a cyclic permutation of the matrices can be used to advantage in finding the gradient of P. Define $\nabla_{\mathbf{A}}P$ as the matrix whose ijth component is $\partial P/\partial a_{ij}$. Similarly, define $\nabla_{\mathbf{B}}P$ and $\nabla_{\pi}P$. Then it is straightforward to show that

$$\nabla_{\pi}P = \mathbf{B}_1\boldsymbol{\beta}_1, \tag{3.104}$$

$$\nabla_{\mathbf{A}}P = \sum_{t=1}^{T-1}\boldsymbol{\alpha}_t\boldsymbol{\beta}_{t+1}'\mathbf{B}_{t+1}, \tag{3.105}$$

$$(\nabla_{\mathbf{B}}P)_{jk} = \sum_{t\ni O_t=k}(\mathbf{A}'\boldsymbol{\alpha}_{t-1})_j(\boldsymbol{\beta}_t)_j. \tag{3.106}$$

In the last equation, if $O_1 = v_k$ then the corresponding term in the sum is just $\boldsymbol{\pi}'\boldsymbol{\beta}_1$.

3.1.2 The Continuous Observation Case

The analyses described in Section 3.1.1 have been applied to several types of HMM other than the discrete symbol model with which we began this chapter. The most obvious generalization is that of replacing the discrete probability distributions with continuous multivariate density functions. In these cases, we omit the vector quantization stage and use the measurements, $\mathbf{x}_t \in \mathbb{R}^d$, directly. For compatibility with the notation used for the discrete case, we shall call the observations $\mathbf{O}_t$, bearing in mind that they are just the d-dimensional primary measurements that were called $\mathbf{x}_t$ in Section 2.5.4.

In addition to the discrete symbol case, Baum also considered models in which the observations drawn from q_j are distributed according to

$$b_j(O_t) = \mathcal{N}(O_t, \mu_j, \sigma_j), \tag{3.107}$$

where $\mathcal{N}(x, \mu, \sigma)$ denotes the univariate Gaussian density function of mean μ and variance σ^2. Later, Liporace [200] analyzed the d-dimensional multivariate problem. For this case, the parameter space $\mathbf{\Lambda} = \{\mathcal{A}^N \times \{\mathbb{R}^d\}^N \times \{\mathcal{U}^d\}^N\} \cdot \mathcal{A}^N$ is the set of all $N \times N$ rowwise stochastic matrices and $\mathcal{U}^d$ is the set of all real symmetric positive definite, $N \times N$ matrices. Following the precise strategy outlined above, the reestimation formulas are derived:

$$\overline{\mu}_{jr} = \frac{\sum_{t=1}^{T} \alpha_i(j)\beta_t(j)O_{tr}}{\sum_{t=1}^{T} \alpha_t(j)\beta_t(j)} \tag{3.108a}$$

for $1 \leq j \leq N$ and $1 \leq r \leq d$, where $\overline{\mu}_{jr}$ and O_{tr} are the rth components of the reestimate of the jth mean vector and the tth observation, respectively; similarly

$$\overline{u}_{jrs} = \frac{\sum_{t=1}^{T} \alpha_t(j)\beta_t(j)(O_{tr} - \mu_{jr})(O_{ts} - \mu_{js})}{\sum_{t=1}^{T} \alpha_t(j)\beta_t(j)} \tag{3.108b}$$

for $1 \leq j \leq N$ and $1 \leq r, s \leq d$, where u_{jrs} is the entry of $\mathbf{U}_j$ in the rth row and sth column. The formula for the state transition matrix for this case is identical to (3.11) with the replacement of O_t by $\mathbf{O}_t$.

Gaussian Mixtures

Consider an unobservable n-state Markov chain with state transition matrix $\mathbf{A} = [a_{ij}]_{n\times n}$. Associated with each state j of the hidden Markov chain is a probability density function, $b_j(\mathbf{x})$, of the observed d-dimensional random vector $\mathbf{x}$. Here we shall consider densities of the form

$$b_j(\mathbf{x}) = \sum_{k=1}^{m} c_{jk}\mathcal{N}(\mathbf{x}, \boldsymbol{\mu}_{jk}, \mathbf{U}_{jk}), \tag{3.109}$$

where m is known; $c_{jk} \geq 0$ for $1 \leq j \leq n$, $1 \leq k \leq m$; $\sum_{k=1}^{m} c_{jk} = 1$ for $1 \leq j \leq n$; and $\mathcal{N}(\mathbf{x}, \boldsymbol{\mu}, \mathbf{U})$ denotes the d-dimensional normal density function of mean vector $\boldsymbol{\mu}$ and covariance matrix $\mathbf{U}$.

It is convenient then to think of our hidden Markov chains as being defined over a parameter manifold $\Lambda = \{\mathcal{A}^n \times \mathcal{C}^m \times \mathbb{R}^d \times \mathcal{U}^d\}$, where $\mathcal{A}^n$ is the set of all $n \times n$ rowwise stochastic matrices; $\mathcal{C}^m$ is the set of all $m \times n$ rowwise stochastic matrices; $\mathbb{R}^d$ is the usual d-dimensional Euclidean space; and $\mathcal{U}^d$ is the set of all $d \times d$ real symmetric positive definite matrices. Then, for a given sequence of observations, $\mathbf{O} = \mathbf{O}_1, \mathbf{O}_2, \ldots, \mathbf{O}_T$, of the vector $\mathbf{x}$ and a particular choice of parameter values $\boldsymbol{\lambda} \in \Lambda$, we can efficiently evaluate the likelihood function, $L_{\boldsymbol{\lambda}}(\mathbf{O})$, of the hidden Markov chain by the forward–backward method of Baum [25].

The forward and backward partial likelihoods, $\alpha_t(j)$ and $\beta_t(i)$, are computed recursively from

$$\alpha_t(j) = \left[\sum_{i=1}^{n} \alpha_{t-1}(i)a_{ij}\right] b_j(\mathbf{O}_t) \tag{3.110a}$$

and

$$\beta_t(i) = \sum_{j=1}^{n} a_{ij} b_j(\mathbf{O}_{t+1})\beta_{t+1}(j), \tag{3.110b}$$

respectively. The recursion is initialized by setting $\alpha_0(1) = 1$, $\alpha_0(j) = 0$ for $2 \le j \le n$, and $\beta_T(i) = 1$ for $1 \le i \le n$, whereupon we may write

$$\mathcal{L}_\lambda(\mathbf{O}) = \sum_{i=1}^{n}\sum_{j=1}^{n} \alpha_t(i) a_{ij} b_j(\mathbf{O}_{t+1})\beta_{t+1}(j) \tag{3.111}$$

for any t between 1 and $T-1$.

The Estimation Algorithm

The parameter estimation problem is then one of maximizing $\mathcal{L}_\lambda(\mathbf{O})$ with respect to $\boldsymbol{\lambda}$ for a given $\mathbf{O}$. One way to maximize $\mathcal{L}_\lambda$ is to use conventional methods of constrained optimization. Liporace, on the other hand, advocates a reestimation technique analogous to that of Baum *et al.* [25, 28]. It is essentially a mapping $\mathcal{T}\colon \Lambda \to \Lambda$ with the property that $\mathcal{L}_{\mathcal{T}(\lambda)}(\mathbf{O}) \ge \mathcal{L}_\lambda(\mathbf{O})$, with equality if and only if $\boldsymbol{\lambda}$ is a critical point of $\mathcal{L}_\lambda(\mathbf{O})$, that is, $\nabla\mathcal{L}_\lambda(\mathbf{O}) = 0$. Thus a recursive application of $\mathcal{T}$ to some initial value of $\boldsymbol{\lambda}$ converges to a local maximum (or possibly an inflection point) of the likelihood functions. Liporace's result [200] relaxed the original requirement of Baum *et al.* [28] that $b_j(\mathbf{x})$ be strictly log concave to the requirement that it be strictly log concave and/or elliptically symmetric. We will further extend the class of admissible pdfs to mixtures and products of mixtures of strictly log concave and/or elliptically symmetric densities.

For the present problem, we will show that a suitable mapping $\mathcal{T}$ is given by the following equations:

$$\bar{a}_{ij} = \mathcal{T}(a_{ij}) = \frac{\sum_{t=1}^{T-1} \alpha_t(i) a_{ij} b_j(\mathbf{O}_{t+1})\beta_{t+1}(j)}{\sum_{t=1}^{T-1} \alpha_t(i)\beta_t(i)}, \tag{3.112}$$

$$\bar{c}_{jk} = \mathcal{T}(c_{jk}) = \frac{\sum_{t=1}^{T} \rho_t(j,k)\beta_t(j)}{\sum_{t=1}^{T} \alpha_t(j)\beta_t(j)}, \tag{3.113}$$

$$\bar{\boldsymbol{\mu}}_{jk} = \mathcal{T}(\boldsymbol{\mu}_{jk}) = \frac{\sum_{t=1}^{T} \rho_t(j,k)\beta_t(j)\mathbf{O}_t}{\sum_{t=1}^{T} \rho_t(j,k)\beta_t(j)}, \tag{3.114}$$

and

$$\bar{\mathbf{U}}_{jk} = \mathcal{T}(\mathbf{U}_{jk}) = \frac{\sum_{t=1}^{T} \rho_t(j,k)\beta_t(j)(\mathbf{O}_t - \boldsymbol{\mu}_{jk})(\mathbf{O}_t - \boldsymbol{\mu}_{jk})'}{\sum_{t=1}^{T} \rho_t(j,k)\beta_t(j)} \tag{3.115}$$

for $1 \le i, j \le n$, $1 \le k \le m$ and $1 \le r, s \le d$. In (3.113)–(3.115),

$$\rho_t(j,k) = \begin{cases} c_{jk}\dfrac{\partial b_j}{\partial c_{jk}}|_{O_1}, & \text{for } t = 1, \\ \sum_{i=1}^{n} \alpha_{t-1}(i) a_{ij} c_{jk}\dfrac{\partial b_j}{\partial c_{jk}}|_{O_t}, & \text{for } 1 < t \le T. \end{cases} \tag{3.116}$$

Proof of the Formulas

A general strategy for demonstrating that reestimation formulas similar to (3.108) and (3.109) for several different families of HMMs both exist and have the above mentioned desirable behavior rests on certain properties of the function

$$\mathcal{Q}(\boldsymbol{\lambda}, \overline{\boldsymbol{\lambda}}) = \mathcal{L}(\mathbf{O}|\boldsymbol{\lambda}) \log[\mathcal{L}(\mathbf{O}|\overline{\boldsymbol{\lambda}})], \tag{3.117}$$

often referred to as the auxiliary function. It is closely related to and motivated by the Kullback–Leibler statistic [164] which, for the case of the HMM, is given by

$$I(\boldsymbol{\lambda}, \overline{\boldsymbol{\lambda}}) = \int_O \mathcal{L}(\mathbf{O}|\boldsymbol{\lambda}) \log\left[\frac{\mathcal{L}(\mathbf{O}|\boldsymbol{\lambda})}{\mathcal{L}(\mathbf{O}|\overline{\boldsymbol{\lambda}})}\right] d\mu(\mathbf{O}). \tag{3.118}$$

The significance of (3.118) is that it expresses the mean over the observation space $\mathbf{O}$ of the amount of information per observation for discrimination between $\boldsymbol{\lambda}$ and $\overline{\boldsymbol{\lambda}}$. According to the principle of minimum cross entropy [299], the best estimate, $\overline{\boldsymbol{\lambda}}$, for the true model parameter given prior estimate $\boldsymbol{\lambda}$ is the one that minimizes (3.118). Since the likelihood function is everywhere non-negative, this can be accomplished by maximizing $\mathcal{Q}(\boldsymbol{\lambda}, \overline{\boldsymbol{\lambda}})$ in (3.117) with respect to $\overline{\boldsymbol{\lambda}}$. A reestimation formula is a solution for $\overline{\boldsymbol{\lambda}}$ of

$$\nabla_{\overline{\lambda}} \mathcal{Q}(\boldsymbol{\lambda}, \overline{\boldsymbol{\lambda}}) = 0, \qquad \text{for any} \qquad \boldsymbol{\lambda} \in \boldsymbol{\Lambda}, \tag{3.119}$$

having the form $\mathcal{F}(\boldsymbol{\lambda}) = \overline{\boldsymbol{\lambda}}$, where $\mathcal{F}$ is a mapping of $\boldsymbol{\Lambda}$ onto itself, and possessing the property that

$$\mathcal{Q}(\boldsymbol{\lambda}, \overline{\boldsymbol{\lambda}}) \geq \mathcal{Q}(\boldsymbol{\lambda}, \boldsymbol{\lambda}). \tag{3.120}$$

That iterations of $\mathcal{F}$ will ultimately converge to a local maximum of the likelihood function follows from three properties of the auxiliary function. It is easily shown that (i) in general,

$$\mathcal{Q}(\boldsymbol{\lambda}, \overline{\boldsymbol{\lambda}}) \geq \mathcal{Q}(\boldsymbol{\lambda}, \boldsymbol{\lambda}) \Rightarrow \mathcal{L}(\mathbf{O}|\overline{\boldsymbol{\lambda}}) \geq \mathcal{L}(\mathbf{O}|\boldsymbol{\lambda}), \tag{3.121}$$

and (ii)

$$\overline{\boldsymbol{\lambda}} = \mathcal{F}(\overline{\boldsymbol{\lambda}}) \Rightarrow \nabla_{\overline{\lambda}} \mathcal{L}(\mathbf{O}|\overline{\boldsymbol{\lambda}}) = \mathbf{0}. \tag{3.122}$$

Also it may be possible to show on a case basis that (iii) for a given $\mathcal{L}(\mathbf{O}|\boldsymbol{\lambda}), \overline{\boldsymbol{\lambda}} = \mathcal{F}(\boldsymbol{\lambda})$ is the unique global maximum of $\mathcal{Q}(\boldsymbol{\lambda}, \overline{\boldsymbol{\lambda}})$.

In several cases of interest that will be listed later, property (iii) can be shown to hold by proving that, for $\overline{\boldsymbol{\lambda}} = \mathcal{F}(\boldsymbol{\lambda})$,

$$\nabla_{\overline{\lambda}} \mathcal{Q}(\boldsymbol{\lambda}, \overline{\boldsymbol{\lambda}}) = \mathbf{0} \tag{3.123}$$

and, for $\overline{\boldsymbol{\lambda}}$ satisfying (3.123), that the eigenvalues of

$$\frac{\partial^2 \mathcal{Q}}{\partial \boldsymbol{\lambda}_i \partial \boldsymbol{\lambda}_j} \tag{3.124}$$

for any pair, λ_i, λ_j, of components of $\bar{\lambda}$ are all negative, and finally, that $Q(\lambda, \bar{\lambda}) \to -\infty$ as $\bar{\lambda} \to \infty$ or $\bar{\lambda} \to \partial\Lambda$. If (3.123) obtains then $\bar{\lambda}$ is a critical point of Q. If the Hessian (3.124) is negative definite then $\bar{\lambda}$ is a local maximum. Finally, if the auxiliary function tends to a large negative number as $\bar{\lambda}$ approaches either the point at infinity or the finite boundary of the parameter space, if one exists, then $\bar{\lambda}$ is, in fact, a global maximum.

Generalization of the Proof in Section 2.1.1

To prove property (iii), one cannot deal directly with the auxiliary function but must first represent it as a linear combination of terms of the form $\mathcal{L}(\mathbf{O}, \mathbf{S}|\lambda) \log[\mathcal{L}(\mathbf{O}, \mathbf{S}|\bar{\lambda})]$ with positive coefficients. Each such term is the contribution to $Q(\lambda, \bar{\lambda})$ due to the state sequence $\mathbf{S} \in Q^T$.

If property (iii) is true then the maximization of $\mathcal{L}(\mathbf{O}|\lambda)$ is achieved as illustrated in Fig. 3.6. Starting at any $\lambda \in \Lambda$, property (iii) guarantees that $\bar{\lambda} = \mathcal{F}(\lambda)$ increases Q. Property (i), (3.121), ensures that this results in an increase in $\mathcal{L}$ unless $\bar{\lambda}$ is a fixed point of $\mathcal{F}$, in which case Q is unchanged and by property (ii), (3.122), $\bar{\lambda}$ is a critical point, that is, a local maximum or possibly an inflection point, of $\mathcal{L}$. Different initial points, under successive transformations, will be mapped onto different local maxima depending solely on their location in Λ with respect to the separatrices.

Equations (3.112) and (3.113) for the reestimation of a_{ij} and c_{jk} are identical to (3.11) and follow directly from a theorem of Baum and Sell [29] because the likelihood function $\mathcal{L}_\lambda(\mathbf{O})$ given in (3.111) is a polynomial with non-negative coefficients in the variables a_{ij}, c_{jk}, $1 \le i, j \le n$, $1 \le k \le m$.

To prove (3.114) and (3.115) our strategy, following Liporace, is to define an appropriate auxiliary function $Q(\lambda, \bar{\lambda})$. This function will have the property that $Q(\lambda, \bar{\lambda}) > Q(\lambda, \lambda)$ implies $\mathcal{L}_{\bar{\lambda}}(\mathbf{O}) > \mathcal{L}_\lambda(\mathbf{O})$. Further, as a function of $\bar{\lambda}$ for any fixed λ, $Q(\lambda, \bar{\lambda})$ will have a unique global maximum given by (3.114)–(3.116).

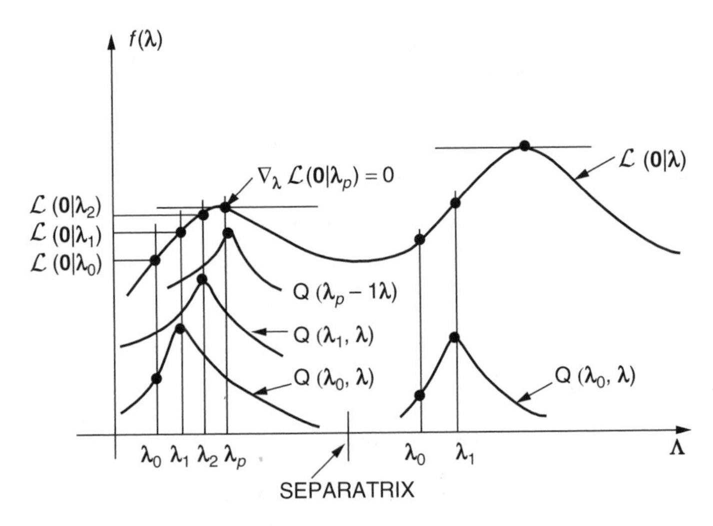

Figure 3.6 The role of the auxiliary function in the Baum algorithm

As a first step toward deriving such a function, we express the likelihood function as a sum over the set, $\mathcal{S}$, of all state sequences $\mathbf{S}$:

$$\mathcal{L}_\lambda(\mathbf{O}) = \sum_{\mathcal{S}} \mathcal{L}_\lambda(\mathbf{O}, \mathbf{S}) \tag{3.125}$$

$$= \sum_{\mathcal{S}} \prod_{t=1}^{T} a_{s_{t-1}s_t} \sum_{k=1}^{m} c_{s_t k} \mathcal{N}(\mathbf{O}_t, \mu_{s_t k}, \mathbf{U}_{s_t k}).$$

Let us partition the likelihood function further by choosing a particular sequence, $\mathbf{K} = (k_1, k_2, \ldots, k_T)$ of mixture densities. As in the case of state sequences we denote the set of all mixture sequences as $\mathcal{K} = \{1, 2, \ldots, m\}^T$. Thus for some particular $\mathbf{K} \in \mathcal{K}$ we can write the joint likelihood of $\mathbf{O}$, $\mathbf{S}$, and $\mathbf{K}$ as

$$\mathcal{L}_\lambda(\mathbf{O}, \mathbf{S}, \mathbf{K}) = \prod_{t=1}^{T} a_{s_{t-t}s_t} \mathcal{N}(\mathbf{O}_t, \mu_{s_t k_t}, \mathbf{U}_{s_t k_t}) c_{s_t k_t}. \tag{3.126}$$

We have now succeeded in partitioning the likelihood function as

$$\mathcal{L}_\lambda(\mathbf{O}) = \sum_{\mathbf{S} \in \mathcal{S}} \sum_{\mathbf{K} \in \mathcal{K}} \mathcal{L}_\lambda(\mathbf{O}, \mathbf{S}, \mathbf{K}). \tag{3.127}$$

In view of the similarity of the representation (3.127) to that of $\mathcal{L}_\lambda$ in [200], we now define the auxiliary function

$$Q(\lambda, \bar{\lambda}) = \sum_{\mathbf{S}} \sum_{\mathbf{K}} \mathcal{L}_\lambda(\mathbf{O}, \mathbf{S}, \mathbf{K}) \log \mathcal{L}_{\bar{\lambda}}(\mathbf{O}, \mathbf{S}, \mathbf{K}). \tag{3.128}$$

When the expressions for $\mathcal{L}_\lambda$ and $\mathcal{L}_{\bar{\lambda}}$ derived from (3.127) are substituted in (3.128), we get

$$Q(\lambda, \bar{\lambda}) = \sum_{\mathbf{S} \in \mathcal{S}} \sum_{\mathbf{K} \in \mathcal{K}} \sum_{t=1}^{T} \gamma_{s_t k_t t} \log \mathcal{N}(\mathbf{O}_t, \bar{\boldsymbol{\mu}}_{s_t k_t}, \bar{\mathbf{U}}_{s_t k_t}), \tag{3.129}$$

where $\gamma_{s_t k_t t} \geq 0$. The innermost summation in (3.129) is formally identical to that used by Liporace in his proof; therefore, the properties which he demonstrated for his auxiliary function with respect to $\boldsymbol{\mu}$ and $\mathbf{U}$ hold in our case as well, thus giving us (3.114) and (3.115). We may thus conclude that (3.126) is correct for $\mathcal{T}$ defined by (3.114)–(3.116). Furthermore, the parameter separation made explicit in (3.126)–(3.129) allows us to apply the same algorithm to mixtures of strictly log concave densities and/or elliptically symmetric densities as treated by Liporace in [200].

Discussion

Liporace [200] notes that by setting $a_{ij} = p_j$, $1 \leq j \leq n$, for all i, the special case of a single mixture can be treated. It is natural then to think of using a model with n clusters of m states, each with a single associated Gaussian density function, as a way of treating the Gaussian mixture problem considered here.

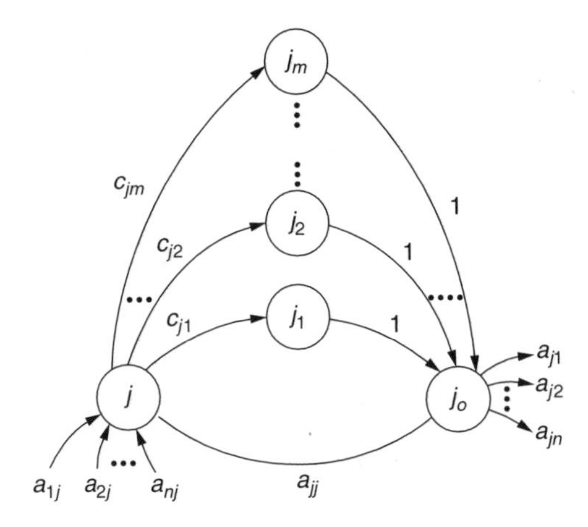

Figure 3.7 Equivalence of the Gaussian mixture HMM and the standard HMM

The transformation can be accomplished in the following way. First we expand the state space of our n-state model as shown in Fig. 3.7, in which we have added states j_0 through j_m for each state j in the original Markov chain. Associated with states $j_1, j_2, \ldots, j_m$ are distinct Gaussian densities corresponding to the m terms of the jth Gaussian mixture in our initial formulation. The transitions exiting state j have probabilities equal to the corresponding mixture weights. State j_0 is a distinguished state that is entered with probability 1 from the other new states, exits to state j with probability a_{jj}, and generates no observation in so doing. The transition matrix for this configuration can be written down by inspection. A large number of the entries in it will be zero or unity. As these are unaltered by (3.112) and (3.113), they need not be reestimated. Using this reconfiguration of the state diagram, Liporace's formulas can be used if $b_j(\mathbf{x})$ is any mixture of elliptically symmetric densities.

A variant on the Gaussian mixture theme results from using $b_j(\mathbf{x})$ of the form of a product of mixtures,

$$b_j(\mathbf{x}) = \prod_{r=1}^{D} \sum_{k=1}^{m} c_{jkr} \mathcal{N}(\mathbf{x}_r, \boldsymbol{\mu}_{jkr}, \mathbf{U}_{jkr}). \tag{3.130}$$

What we have considered so far is the special case of (3.130) for $D = 1$.

From the structure of our derivation it is clear that for hidden Markov chains having densities of the form (3.130), reestimation formulas can be derived as before by solving $\nabla_{\overline{\lambda}} Q(\boldsymbol{\lambda}, \overline{\boldsymbol{\lambda}}) = 0$. Such solutions will yield results quite analogous to (3.112)–(3.116). Note that this case too can be represented as a reconfiguration of the state diagram.

One numerical difficulty which may be manifest in the methods described is the phenomenon noted by Nadas [227] in which one or more of the mean vectors converge to a particular observation while the corresponding covariance matrix approaches a singular matrix. Under these conditions, $\mathcal{L}_{\lambda}(\mathbf{O}) \to \infty$ but the value of $\boldsymbol{\lambda}$ is meaningless. A practical, if unedifying, remedy for this difficulty is to try a different initial $\boldsymbol{\lambda}$. Alternatively, one can drop the offending term from the mixture since it is only contributing at one point of $\boldsymbol{\Lambda}$.

Finally, we call attention to two minor facets of these algorithms. First, for flexibility in modeling, the number of terms in each mixture may vary with state, so that m in (3.109) could as well be m_j. A similar dependence on dimension results if m in (3.130) is replaced by m_{jr}. In either case, the constraints on the mixture weights must be satisfied.

Second, for realistic numbers of observations, for example, $T \geq 5000$, the reestimation formulas will underflow on any existing computer. The basic scaling mechanism described in Section 3.1.1 can be used to alleviate the problem but must be modified to account for the fact that the $\rho_t \beta_t$ product will be missing the tth scale factor. To divide out the product of scale factors, the tth summand in both numerator and denominator of (3.113), (3.114), and (3.115) must be multiplied by the missing coefficient.

3.1.3 *The Autoregressive Observation Case*

Another case which is particularly appropriate for the speech signal is one studied by Poritz [251, 250] in which the stochastic process at each state is Gaussian autoregressive, that is, it assumes the linear prediction model of (2.43) used by Itakura [140]. In the Poritz model, the observation sequence comprises samples of the speech signal itself blocked into frames. The tth frame, designated $\mathbf{O}^{(t)}$ for $1 \leq t \leq T$, is just the sequence of m samples, $\mathbf{O}_1^{(t)}, \mathbf{O}_2^{(t)}, \ldots, \mathbf{O}_m^{(t)}$. The state-dependent stochastic processes are based on an entire frame so that

$$b_j(\mathbf{O}^{(t)}) = \frac{2}{\sqrt{2\pi\sigma_j}} e^{-\mathbf{a}_j \mathbf{R}_j \mathbf{a}_j'/\sigma_j^2} \tag{3.131}$$

where $\mathbf{a}_j$ and $\mathbf{R}_j$ have the same meaning as in (2.48) for a pth-order autocorrelation analysis, and the parameter space for the model is $\mathbf{\Lambda} = \{\mathcal{A}^N \times \{\mathbb{R}^p\}^N \times \{\mathbb{R}^+\}^N\}$ where $\mathcal{R}^+$ denotes the set of strictly positive real numbers.

The $\alpha_t(j)$, $\beta_t(i)$, $\mathcal{L}(\mathbf{O}|\boldsymbol{\lambda})$ and the reestimates $\overline{a}_{ij}$ are calculated directly from (3.110a), (3.110b), (3.111), and (3.112), respectively, using $b_j(\mathbf{O}^{(t)})$ from (3.131) in place of $b_j(\mathbf{O}_t)$. The reestimates of the "gain" $\overline{\sigma}_j^2$ and the autocorrelation matrix $\mathbf{R}_j$ are reminiscent of the usual LPC analysis. Thus

$$\overline{\sigma}_j^2 = \frac{\mathbf{a}_j \mathbf{R}_j \mathbf{a}_j'}{\sum_{t-1}^T \alpha_t(j)\beta_t(j)}, \qquad 1 \leq j \leq N, \tag{3.132}$$

and

$$\mathbf{a}_j = (1, -\mathbf{C}_j \mathbf{D}_j^{-1}), \qquad 1 \leq j \leq n, \tag{3.133}$$

where $\mathbf{C}_j$ and $\mathbf{D}_j$ arise from partitioning the autocorrelation matrix by separating the first row and the first column so that

$$\mathbf{R}_j = \begin{bmatrix} \mathbf{B}_j & \mathbf{C}_j \\ \mathbf{C}_j & \mathbf{D}_j \end{bmatrix}_{(p+1)\times(p+1)}. \tag{3.134}$$

Thus $\mathbf{D}_j$ is a pth-order Toeplitz matrix for which the inverse required by (3.133) can be efficiently computed [211], which method is summarized in (2.50)–(2.54). The $\mathbf{R}_j$ are actually computed from

$$\mathbf{R}_j = \sum_{t=1}^{T} \alpha_t(j)\beta_t(j)\mathbf{Y}_t\mathbf{Y}_t^{'}, \tag{3.135}$$

where $\mathbf{Y}_t$ is the usual data matrix

$$\mathbf{Y}_t = \begin{bmatrix} O_{p+1}^{(t)} & O_{p+2}^{(t)} & \cdots & O_m^{(t)} \\ O_p^{(t)} & O_{p+1}^{(t)} & \cdots & O_{m-1}^{(t)} \\ \vdots & \vdots & & \vdots \\ O_1^{(t)} & O_2^{(t)} & \cdots & O_{m-p-1}^{(t)} \end{bmatrix}_{(p+1)\times(m-p-1)}. \tag{3.136}$$

Poritz has performed a significant experiment based on this method, the implications of which on speech recognition will be examined in Section 3.1.7.

All of the aforementioned types of HMMs are amenable to an interesting modification. The observable process can be made dependent not only on the present state but also on its successor, so that instead of $b_j(O_t)$ we have $b_{ij}(O_t)$, meaning that O_t was generated in the transition from q_i to q_j. This will, of course, increase the size of the parameter space but will simultaneously afford greater representational power.

3.1.4 The Semi-Markov Process and Correlated Observations

Essential to the fidelity of the proposed model is the ability to account for the durations of the symbols of the speech code explicitly and flexibly. The traditional hidden Markov model is inadequate in this respect since the probability of remaining in state q_i for duration τ, Prob$[\tau|q_i]$, is proportional to $a_{ii}^{\tau-1}$, where $a_{ij} =$ Prob$[q_j$ at $t+1|q_i$ at $t]$. This exponential distribution of state durations is inappropriate if the states of the hidden Markov model are to represent linguistically meaningful components of the speech code.

A significant improvement to the standard model results from the introduction of a discrete set of duration probabilities. In the Ferguson [84] model shown in Fig. 3.8a, Prob$[\tau|q_i]$ is specified for $1 \le \tau < \Delta t$ *and* $1 \le i \le n$. At time t, the process enters state q_i for duration τ with probability Prob$[\tau|q_i]$, during which time the observations $\mathbf{O}_{t+1}, \mathbf{O}_{t+2}, \ldots, \mathbf{O}_{t+\tau}$ are generated. It then transits to state q_j with probability a_{ij}. The parameters Prob$[\tau|q_i]$ are estimated from an observation sequence, along with all of the other parameters of the model, by means of a reestimation formula. Unfortunately, $n\Delta t$ additional parameters must be so determined for an n-state model. Since, in general, many observations of the source are required to get a single measurement of duration, the Ferguson method requires an enormous amount of training data. The relation between the Ferguson model and the standard HMM is explored below.

The alternative is to use a parametric family of continuous probability density functions, $d_i(\tau)$, $1 \le i \le n$, $\tau \in \mathbb{R}^+$, to provide the duration probabilities. If only a few parameters are required to completely specify the $d_i(\tau)$, then the complexity of the Ferguson model is greatly reduced. In fact, Ferguson acknowledges the possibility of using continuous densities, and Russell and Moore [286] have studied the case for the Poisson distribution.

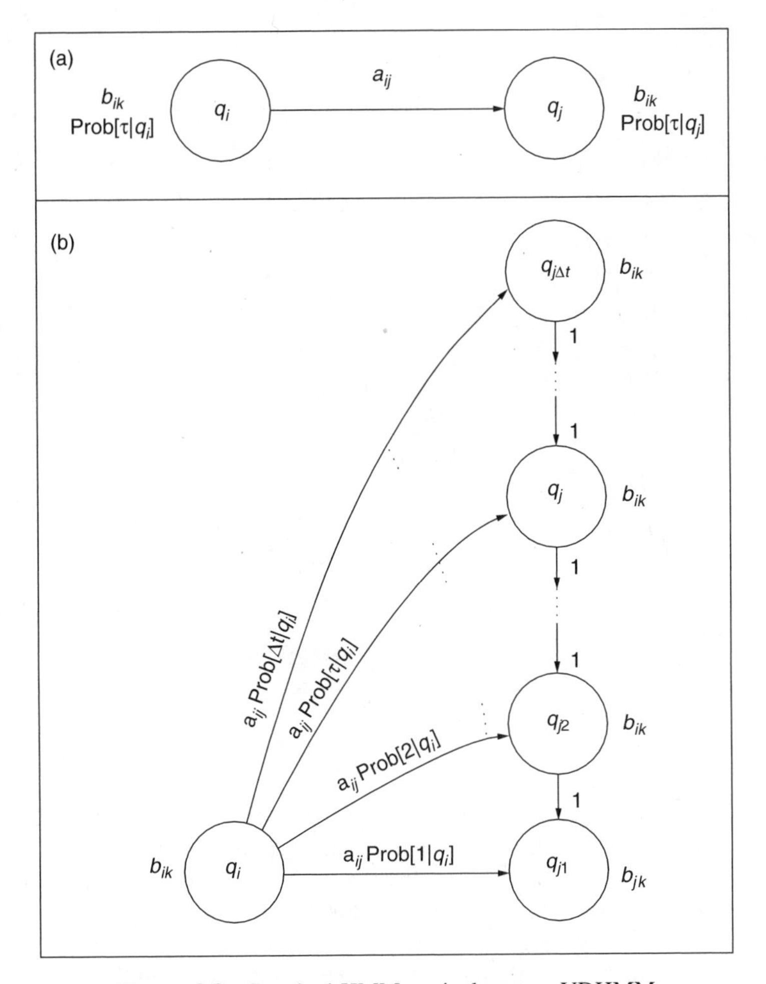

Figure 3.8 Standard HMM equivalent to a VDHMM

It is proposed here that the two-parameter family of gamma distributions given in (3.137) is ideally suited for the purpose [32]:

$$d_i(\tau) = \frac{\eta_i^{v_i}}{\Gamma(v_i)} \tau^{v_i - 1} e^{-\eta_i \tau}, \qquad \tau > 0. \tag{3.137}$$

The mean value of τ is v_i/η_i and its variance is v_i/η_i^2. Several members of the family are shown in Fig. 3.9 displaying $d_i(\tau)$ for varying v_i with fixed η_i and varying η_i with fixed v_i, respectively. The relationship between the variable-duration model and the standard HMM is as follows.

It might first appear that the introduction of duration probabilities destroys the Markovian character of the underlying process. In fact, however, we can construct an ordinary HMM which is identical to the variable duration model. First note that in the Ferguson model $a_{ii} = 0$ for $1 \le i \le n$, which fact allows the construction shown in Fig. 3.8b

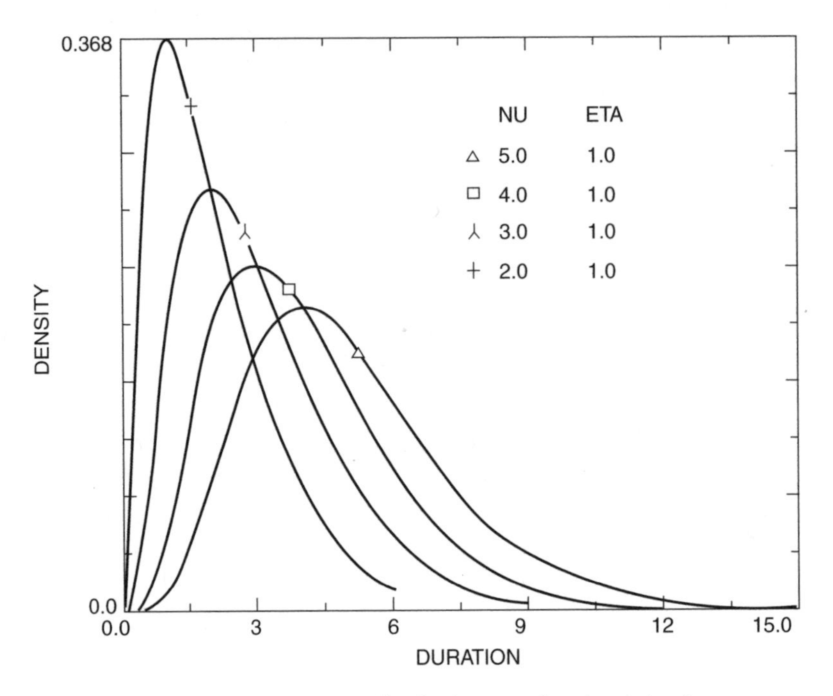

Figure 3.9 The gamma distribution as a durational density

transforming the n-state Ferguson model into an $n\Delta t$-state conventional HMM. The new model has transition probabilities $a_{ij_\tau} = a_{ij}\mathrm{Prob}[\tau|q_i]$ for $1 \le \tau \le \Delta t$, and $a_{j_{\tau+1}j_\tau} = 1$ for $\Delta t - 1 \le \tau \le 1$, for each transition from q_i to q_j of probability a_{ij} in the variable-duration model. The observation probabilities for the new states q_{i_τ}, for $2 \le \tau \le \Delta t$, are b_{ij} for $1 \le k \le m$ when the previous state in the original model was q_i.

The model constructed according to Fig. 3.8b has the desired Markovian property. Clearly $a_{ij_\tau} \ge 0$. Also, for $1 \le i \le n$,

$$\begin{aligned}\sum_{j=1}^{n}\sum_{\tau=1}^{\Delta t} a_{ij_\tau} &= \sum_{j=1}^{n}\sum_{\tau=1}^{\Delta t} a_{ij}\mathrm{Prob}[\tau|q_i] \\ &= \sum_{j=1}^{n} a_{ij}\sum_{\tau=1}^{\Delta t}\mathrm{Prob}[\tau|q_i] \\ &= \sum_{j=1}^{n} a_{ij} = 1,\end{aligned} \tag{3.138}$$

so the transition matrix for the Markov chain is well formed.

It is also true that a given observation sequence will have the same likelihood in either model. To see that this is so, we need only compare one state transition in the variable-duration model with its equivalent state sequence in the new model. Namely,

$$a_{ij}\text{Prob}[\tau|q_i]\prod_{t=1}^{\tau} b_i(\mathbf{O}_t) = a_{ij_\tau} b_i(\mathbf{O}_1)\prod_{t=1}^{\tau-1} a_{j_{\tau-t+1}j_{\tau-t}} b_i(\mathbf{O}_{t+1}) \tag{3.139}$$

since the product of transition probabilities on the right-hand side of (3.139) is unity and, by construction $a_{ij} = a_{ij_\tau}\text{Prob}[\tau|q_i]$. The left-hand side of (3.139) is the likelihood of $O_1, O_2, \ldots, O_\tau$ given q_i at $t = 1$ and q_j at $t = \tau$ in the variable-duration model. The right-hand side of (3.139) is the likelihood of $O_1, O_2, \ldots, O_\tau$ given the unique path of length τ from q_i to q_{j_i} in the constructed model. Since these likelihoods are equal and since the new model is well formed, the two models are equivalent.

The Continuously Variable-Duration (CVD) Hidden Markov Model

In a sense, the CVDHMM is equivalent to the Ferguson model in that, for $1 \le \tau \le \Delta t$,

$$\text{Prob}[\tau|q_i] = \int_{\tau=1}^{\tau} d_i(t)dt \tag{3.140}$$

and if $d_i(t) = 0$ for $t > \Delta t$, then $\sum_{\tau=1}^{\Delta t}\text{Prob}[\tau|q_i] = 1$. Whereas the parameters of the Ferguson model can be obtained by applying the Baum reestimation formulas from Section 3.1.1 to an equivalent standard HMM, parameters of the CVDHMM must be obtained by the methods described in this section.

With duration probabilities derived from (3.137) we are in a position to define our stochastic model (CVDHMM). As in the conventional case, we shall have n states, $q_1, q_2, \ldots, q_n$, in the unobservable process. The state transition matrix $\mathbf{A} = [a_{ij}]_{n\times n}$, however, is subject to the constraint that $a_{ii} = 0$, $1 \le i \le n$.

The observable processes $b_i(\mathbf{x})$, $\mathbf{x} \in \mathbb{R}^m$, $1 \le i \le n$, will be assumed to be Gaussian of mean $\boldsymbol{\mu}_i$ and covariance $\mathbf{U}_i$. The model is illustrated in Fig. 3.10.

The likelihood function, $\mathcal{L}(\mathbf{O}|\boldsymbol{\lambda})$, of the model is defined over observation sequences $\mathbf{O} = \mathbf{O}_1, \mathbf{O}_2, \ldots, \mathbf{O}_\tau$, where $\mathbf{O}_t \in \mathbb{R}^m$ for $1 \le t \le T$, and parameters $\boldsymbol{\lambda}$ in the parameter manifold $\boldsymbol{\Lambda}$. Thus, for $\boldsymbol{\lambda} \in \boldsymbol{\Lambda}$,

$$\boldsymbol{\lambda} = (\mathbf{A}, \{\mu_i\}_{i=1}^n, \{\mathbf{U}_i\}_{i=1}^n, \{\upsilon_i\}_{i=1}^n, \{\eta_i\}_{i=1}^n). \tag{3.141}$$

As in the case of the conventional hidden Markov models, the likelihood function may be recursively evaluated after defining partial forward and backward likelihoods. Let

$$a_i(j) = \mathcal{L}(\mathbf{O}_1, \mathbf{O}_2, \ldots, \mathbf{O}_t \text{ and } q_j \text{ at } t|q_i \text{ at } t+1, i \ne j \text{ and } \boldsymbol{\lambda}) \tag{3.142}$$

and

$$\beta_t(i) = \mathcal{L}(\mathbf{O}_{t+1}, \mathbf{O}_{t+2}, \ldots, \mathbf{O}_\tau \text{ and } q_j \text{ at } t-1, j \ne i|q_i \text{ at } t \text{ and } \boldsymbol{\lambda}). \tag{3.143}$$

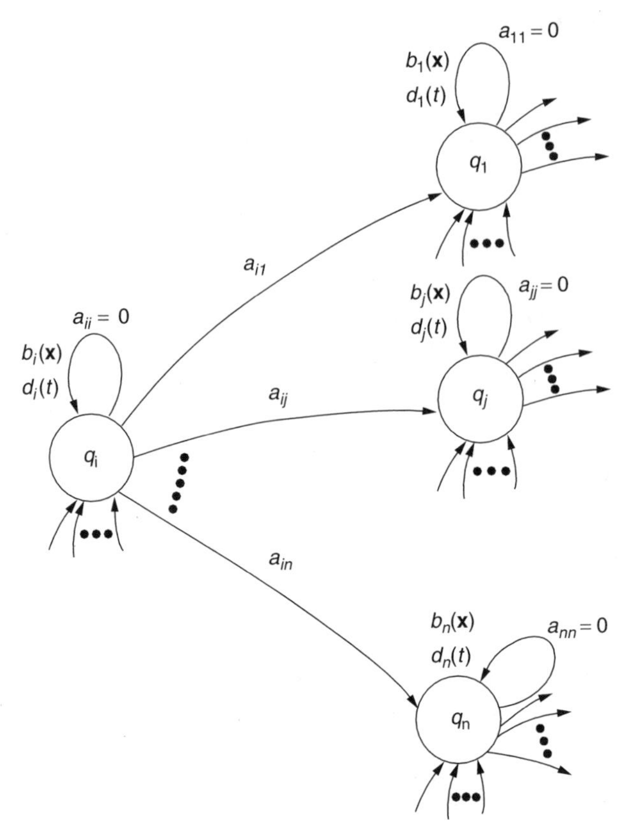

Figure 3.10 The hidden semi-Markov model

Then for $1 \leq j \leq n$ and $1 \leq t \leq T$, the forward likelihoods are computed from

$$\alpha_t(j) = \sum_{\tau \leq t} \sum_{\substack{i=1 \\ i \neq j}}^{n} \alpha_{t-\tau}(i) a_{ij} d_j(\tau) \prod_{\theta=1}^{\tau} b_j(\mathbf{O}_{t-\tau+\theta}). \tag{3.144}$$

The product in (3.144) should, strictly speaking, be replaced by $b_j(\mathbf{O}_{t-\tau+1}, \mathbf{O}_{t-\tau+2}, \ldots \mathbf{O}_t)$, the joint likelihood of the τ observations. If we make the usual assumption of independence we get (3.144).

To begin the recursion (3.144) we set $\alpha_0(i) = 0$ for $i \neq 1$. We assume, without loss of generality, that the process is in state q_1 at $t = 0$ so that $\alpha_0(1) = 1$ and the transition a_{ij} is not included. This accounts for the event of the first τ observations emanating from the initial state prior to the occurrence of any state transitions. In what follows, the same significance of $\alpha_0(t)$ will be assumed wherever it appears.

Similarly, a recursion in reverse temporal order is used to compute the backward likelihoods. For $1 \leq i \leq n$ and $T - 1 \geq t \geq 1$,

$$\beta_t(i) = \sum_{t \le T-t} \sum_{\substack{j=1 \\ j \ne i}}^{n} a_{ij} d_j(\tau) \prod_{\theta=1}^{\tau} b_j(\mathbf{O}_{t+\theta}) \beta_{t+\tau}(j). \quad (3.145)$$

The recursion (3.145) is started by setting $\beta_\tau(i) = 1$ for $1 \le i \le n$. Note that if τ is fixed equal to 1 then (3.144) and (3.145) reduce to the standard forward–backward recursions for ordinary hidden Markov models except that a_{ii} is not necessarily zero and $d_i(1) = 1$; cf. [25].

With the partial likelihoods available, the likelihood function can be evaluated from

$$\mathcal{L}(\mathbf{O}|\boldsymbol{\lambda}) = \sum_{i=1}^{n} \sum_{\substack{j=1 \\ j \ne i}}^{n} \sum_{\tau \le t} \alpha_{t-\tau}(i) a_{ij} d_j(\tau) \prod_{\theta=1}^{\tau} b_j(\mathbf{O}_{t-\tau+\theta}) \beta_t(j) \quad (3.146)$$

for any $t \in [1, T]$. An especially simple form of (3.146) is

$$\mathcal{L}(\mathbf{O}|\boldsymbol{\lambda} = \sum_{i=1}^{n} \alpha_\tau(i). \quad (3.147)$$

Also note that, as in the case of the standard hidden Markov model,

$$\mathcal{L}(q_i \text{ at } t|\mathbf{O}, \boldsymbol{\lambda}) = \alpha_t(i)\beta_t(i). \quad (3.148)$$

Parameter Estimation

Estimation of the parameters of a hidden Markov model is accomplished by maximizing $\mathcal{L}(\mathbf{O}|\boldsymbol{\lambda})$ with respect to $\boldsymbol{\lambda}$ for a given $\mathbf{O}$. Due to the special properties of the likelihood function, it can be locally optimized by an efficient algorithm that does not explicitly require differentiation. This technique, often called reestimation, is described in detail earlier. For the purposes of this discussion, it suffices to say that the solution to

$$\nabla_{\overline{\lambda}} \mathcal{Q}(\boldsymbol{\lambda}, \overline{\boldsymbol{\lambda}}) = 0, \quad (3.149)$$

where

$$\mathcal{Q}(\boldsymbol{\lambda}, \overline{\boldsymbol{\lambda}}) = \mathcal{L}(\mathbf{O}|\boldsymbol{\lambda}) \log \mathcal{L}(\mathbf{O}|\overline{\boldsymbol{\lambda}}), \quad (3.150)$$

is of the form $\overline{\boldsymbol{\lambda}} = \mathcal{T}(\boldsymbol{\lambda})$, where the transformation $\mathcal{T}$ may be shown to have the desired properties that $\mathcal{T} : \boldsymbol{\Lambda} \to \boldsymbol{\Lambda}$ and $\mathcal{L}(\mathbf{O}|\overline{\boldsymbol{\lambda}}) \ge \mathcal{L}(\mathbf{O}|\boldsymbol{\lambda})$ with equality if $\boldsymbol{\lambda}$ is a critical point of $\mathcal{L}$. Thus recursive application of $\mathcal{T}$ to some initial point $\boldsymbol{\lambda}_0 \in \boldsymbol{\Lambda}$ will converge to a local maximum of the likelihood function.

The derivation of a reestimation formula for the transition probabilities for the present model is formally identical to that for the standard hidden Markov model [25]. Without belaboring the point, we simply write:

$$\overline{a}_{ij} = \frac{\sum_{t=1}^{T} \sum_{\tau \le t} a_{t-\tau}(i) a_{ij} d_i(\tau) \prod_{\theta=1}^{\tau} b_j(\mathbf{O}_{t-\tau+\theta}) \beta_t(j)}{\sum_{t=1}^{T-1} \alpha_t(i) \beta_t(i)}. \quad (3.151)$$

In (3.151), as in the formula for the transition probabilities of the standard model, the numerator and denominator are still open to the usual interpretation as the expected number of transitions from state q_i to q_j and the expected number of visits to state q_i, respectively.

The formulas for the parameters of the observable and durational processes are sufficiently different from the standard case to warrant a slightly more detailed treatment. Let us first consider the case of the means of the observable processes. Differentiating (3.184) with respect to $\boldsymbol{\mu}_j$, we get

$$\frac{\partial \mathcal{L}}{\partial \boldsymbol{\mu}_j} = \sum_{T=1}^{T} \sum_{\tau \leq t} \sum_{\substack{i=1 \\ i \neq j}}^{n} \alpha_{t-\tau}(i) a_{ij} d_j(\tau) \left[\frac{\partial}{\partial \boldsymbol{\mu}_j} \prod_{\theta=1}^{\tau} b_j(\mathbf{O}_{t-\tau+\theta}) \right] \beta_t(j), \tag{3.152}$$

where the derivative with respect to a vector is understood to be the vector whose components are the partial derivatives with respect to the components of the parameter vector. In (3.152), the summations on t and τ occur since we are differentiating a product. Recall that $b_j(\mathbf{x}) = \mathcal{N}(\mathbf{x}, \boldsymbol{\mu}_j, \mathbf{U}_j)$, whence the term in brackets on the right-hand side of (3.152) becomes

$$\sum_{r=1}^{\tau} \left[\prod_{\substack{\theta=1 \\ \theta \neq r}}^{\tau} b_j(\mathbf{O}_{t-\tau+\theta}) \right] \frac{\partial \mathcal{N}_j}{\partial \boldsymbol{\mu}_j} | \mathbf{O}_{t-\tau+\theta} \tag{3.153}$$

$$= \sum_{r=1}^{\tau} \left[\prod_{\substack{\tau=1 \\ \theta \neq r}}^{\tau} b_j(\mathbf{O}_{t-\tau+\theta}) \right] b_j(\mathbf{O}_{t-\tau+r}) \mathbf{U}_j^{-1} (\mathbf{O}_{t-\tau+r} - \boldsymbol{\mu}_j)'$$

$$= \left[\prod_{\theta=1}^{\tau} b_j(\mathbf{O}_{t-\tau+\theta}) \right] \cdot \left[\sum_{\theta=1}^{\tau} \mathbf{U}_j^{-1} (\mathbf{O}_{t-\tau+\theta} - \boldsymbol{\mu}_j)' \right],$$

where $'$ denotes the matrix transpose operation. Substituting the rightmost part of (3.153) for the bracketed term in (3.152), setting the result to zero and multiplying by $\mathbf{U}_j^{-1}$ yields

$$\sum_{t=1}^{T} \sum_{\tau \leq t} \sum_{\substack{i=1 \\ i \neq j}}^{n} \alpha_{t-\tau}(i) a_{ij} d_j(\tau) \prod_{\theta=1}^{\tau} b_j(\mathbf{O}_{t-\tau+\theta}) \left[\sum_{\theta=1}^{\tau} \mathbf{O}_{t-\tau+\theta} \right] \beta_t(j) \tag{3.154}$$

$$= \sum_{t=1}^{T} \sum_{\tau \leq t} \sum_{\substack{i=1 \\ i \neq j}}^{n} \alpha_{t-\tau}(i) a_{ij} d_j(\tau) \prod_{\theta=1}^{\tau} b_j(\mathbf{O}_{t-\tau+\theta}) [\tau \boldsymbol{\mu}_j \beta_t(j).$$

Finally, solving (3.154) for $\boldsymbol{\mu}_j$ gives

$$\overline{\boldsymbol{\mu}}_j = \frac{\sum_{t=1}^{T}\sum_{\tau\leq 1}\sum_{i=1}^{n} \alpha_{t-\tau}(i)a_{ij}d_j(\tau)\prod_{\theta=1}^{\tau} b_j(\mathbf{O}_{t-\tau+\theta})\beta_t(j)\left[\sum_{\theta=1}^{\tau}\mathbf{O}_{t-\tau+\theta}\right]}{\sum_{t=1}^{T}\sum_{\tau\leq t}\sum_{\substack{i=1\\ i\neq j}}^{n} \tau\alpha_{t-\tau}(i)a_{ij}d_j(\tau)\prod_{\theta=1}^{\tau} b_j(\mathbf{O}_{t-\tau+\theta})\beta_t(j)}. \tag{3.155}$$

The numerator of the right-hand side of (3.155) is the expected value of the sums of observations due to state q_j, while the denominator is the expected duration of state q_j; hence it is reasonable that their quotient be an estimate of $\boldsymbol{\mu}_j$.

A similar argument leads to a reestimation formula for $\mathbf{U}_j$. First note that:

$$\frac{\partial \mathcal{N}_j}{\partial \mathbf{U}_j}|_{o_t} = b_j(\mathbf{O}_t)\left[-\frac{1}{2}\mathbf{U}_j^{-1} - \frac{1}{2}(\mathbf{O}_t - \boldsymbol{\mu}_j)\mathbf{U}_j^{-1}(\mathbf{U}_j^{-1})'(\mathbf{O}_t - \boldsymbol{\mu}_j)'\right]. \tag{3.156}$$

From (3.156) we obtain an expression for $(\partial\mathcal{L})/(\partial\mathbf{U}_j)$ which can be solved for $\mathbf{U}_j$ by steps analogous to those taken in (3.152)–(3.155). The outcome is a formula for $\overline{\mathbf{U}}_j$ which is identical to (3.156) except that $\sum_{0=1}^{\tau}\mathbf{O}_{t-\tau+\theta}$ is replaced by $\sum_{\theta=1}^{\tau}(\mathbf{O}_{t-\tau+\theta} - \boldsymbol{\mu}_j)(\mathbf{O}_{t-\tau+\theta} - \boldsymbol{\mu}_j)'$. This formula has an intuitive explanation corresponding to that for $\boldsymbol{\mu}_j$ since the numerator is the sample variance of sums of observations.

Turning our attention to the durational parameters, we find that the derivation of a formula for η_j follows the general outline given above. Straightforward differentiation of $d_j(\tau)$ as given in (3.137) produces

$$\frac{\partial d_j}{\partial \eta_j}|_\tau = d_j(\tau)\left[\frac{v_j}{\eta_j} - \tau\right], \tag{3.157}$$

from which an expression for $\partial\mathcal{L}/\partial\eta_j$ is immediately obtained, manipulation of which in the by now familiar way yields

$$\overline{\eta}_j = \frac{v_j\sum_{t=1}^{T}\alpha_t(j)\beta_t(j)}{\sum_{t=1}^{T}\sum_{\tau\leq t}\tau\sum_{\substack{i=1\\ i\neq j}}^{n} \alpha_{t-\tau}(i)a_{ij}d_j(\tau)\prod_{\theta=1}^{\tau} b_j(\mathbf{O}_{t-\tau+\theta})\beta_t(j)}. \tag{3.158}$$

This pleasant pattern does not emerge when we consider the parameter v_j. The problem is immediately apparent as soon as we take the partial derivative of $d_j(\tau)$ with respect to v_j:

$$\frac{\partial d_j}{\partial v_j}|_\tau = \log\eta_j + \log\tau - \frac{d\Gamma}{dv}|_{v_j}/\Gamma(v_j). \tag{3.159}$$

The gamma function and its derivative are inescapable, so that when we carry out the usual algebra on $(\partial\mathcal{L})/(\partial v_j)$ we arrive at:

$$\Psi(v_j) = \frac{\sum_{t=1}^{T}\sum_{\tau\leq t}\log(\eta_j\tau)\sum_{\substack{i=1\\ i\neq j}}^{n} \alpha_{t-\tau}(i)a_{ij}d_j(\tau)\prod_{\theta=1}^{\tau} b_j(\mathbf{O}_{t-\tau+\theta})\beta_t(j)}{\sum_{t=1}^{T}\alpha_t(j)\beta_t(j)}, \tag{3.160}$$

where Ψ is the traditional notation for the digamma function [2]. The Ψ function is continuously differentiable and strictly monotonic increasing on $\mathbb{R}^+$ with a real zero at roughly 1.461. In addition, $\Psi(x)$ is easily obtained from Abromovitz and Stegun [2] for $x \in \mathbb{R}^+$,

$$\Psi(x+n) = \sum_{k=1}^{n=1} \frac{1}{x+k} + \Psi(x+1), \tag{3.161}$$

and for $x \in (-1, 1)$,

$$\Psi(x+1) = -\gamma + \sum_{k=1}^{\infty} (-1)^{k+1} \zeta(k+1) x^k, \tag{3.162}$$

where γ is Euler's constant, 0.557^+, and $\zeta(\cdot)$ is the Riemann zeta function. Twenty terms of (3.162) suffice to compute to full machine precision in 32-bit floating-point format so that (3.160) can be solved numerically for v_j. Applying Newton's method to (3.160), an improving sequence of estimations of v_j is obtained from:

$$v_j^{(k+1)} = v_j^{(k)} - \frac{\Psi(v_j^{(k)}) - C}{\Psi'(v_j^{(k)})}, \tag{3.163}$$

where the superscripts indicate iteration number, the constant C is just the right-hand side of (3.160) and the polygamma function in the denominator is computed from formulas given in Abromovitz and Stegun [2] analogous to (3.161) and (3.162):

$$\Psi'(x+1) = \sum_{n=0}^{\infty} (-1)^n (n+1) \zeta(n+2) x^n \tag{3.164}$$

and

$$\Psi'(x+n) = -\sum_{k=1}^{n-1} \frac{1}{(x+k)^2} + \Psi(x+1). \tag{3.165}$$

Because Ψ is well behaved for positive arguments, the iterates of (3.163) converge very rapidly to a solution of (3.160), making it a useful reestimation formula.

Proofs of the Formulas

As noted at the beginning of the previous section, in order for the reestimation formulas derived therein to be useful, it must be proven that they always provide valid parameter values that increase the likelihood function. Fortunately, it is not difficult to do so in this case since we can rely on powerful theorems proven in connection with standard hidden Markov models.

Since the likelihood function (3.146) is a polynomial in the a_{ij} with positive coefficients, the correctness of (3.151) is ensured by the method of Section 3.1.1. One need only observe that $\mathbf{A}$ is rowwise stochastic and the right-hand side of (3.146) is of the proper form. The former is true by definition, while the latter follows from straightforward differentiation and some rearrangement of terms.

The correctness of the formulas for the other parameters is an immediate consequence of a result due to Baum *et al.* [29] which ensures the correctness of a reestimation formula if $\overline{\boldsymbol{\lambda}}$ is a critical point of $\mathcal{Q}(\boldsymbol{\lambda}, \overline{\boldsymbol{\lambda}})$ and if the probability density function, f, of the observations is log concave in $\boldsymbol{\lambda}$. In the present case, the first criterion is satisfied by construction. The property of log concavity is easily established by differentiation of the relevant density functions with respect to each of the parameters. In all cases except for the v_j, one observes directly that $\frac{\partial^2}{\partial \boldsymbol{\lambda}^2} \log(f(\boldsymbol{\lambda})) < 0$ everywhere on $\boldsymbol{\Lambda}$. For the v_j we find that:

$$\frac{\partial^2}{\partial v_j^2} \log(d_j(\tau)) = -\frac{d\Psi}{dv_j}. \tag{3.166}$$

It is well known [109] that

$$\frac{d\Psi}{dv} = \sum_{k=1}^{\infty} \frac{1}{(v+k)^2}, \tag{3.167}$$

hence the right-hand side of (3.166) is strictly negative for $v_j \in \mathbb{R}^+$, allowing us to conclude that $d_j(\tau)$ is log concave in v_j.

A final technical point should be made regarding the applicability of the result of Baum *et al.* [29]. The auxiliary function considered in their proof has a single summation over time while the present one requires two, one on t and a second on τ. For fixed τ, each summand in our $\mathcal{Q}$ is formally identical to the one treated by Baum *et al.* Thus each term is strictly log concave. The sum of log concave functions is log concave, allowing the result to be extended to the continuously variable-duration case.

Scaling

Although the reestimation formulas given above are theoretically correct, they suffer from the numerical problem that $\alpha_t(i)$ and $\beta_t(j)$ tend rapidly to zero as t increases. To be useful, their values must be kept within the limited dynamic range of a real computer. The same fundamental idea as was described in Section 3.1.1 can be adapted for the variable-duration case, namely,

$$a_t^{'}(i) = w_t \alpha_t(i) \tag{3.168}$$

and

$$\beta_t^{'}(j) = w_t \beta_t(j), \tag{3.169}$$

where the scale factors, $(w_t)_{t=1}^{T}$, are given by:

$$w_t = \left[\sum_{i=1}^{n} \alpha_t(i) \right]^{-1}, \tag{3.170}$$

a fortuitous consequence of which is that:

$$\log(\mathcal{L}(\mathbf{O}|\boldsymbol{\lambda})) = -\sum_{t=1}^{T} \log(w_t). \tag{3.171}$$

Since (3.170) is applied for $t = 1, 2, \ldots$, and since the recursive calculation of α and β now explicitly involves not merely one, but all previous values, the sum on t will be meaningless unless each term is multiplied by the same coefficient. Thus for each τ we must supply the $\tau - 1$ missing scale factors. Doing so, the scaled a and β recursions analogous to (3.144) and (3.145) become

$$\alpha_t(j) = \sum_{\tau \le t} \sum_{\substack{i=1 \\ i \neq j}}^{n} \alpha'_{t-\tau}(i) a_{ij} d_j(\tau) \left[\prod_{\theta=1}^{\tau-1} w_{t-\theta} b_j(\mathbf{O}_{t-\tau+\theta}) \right] b_j(\mathbf{O}_t), \tag{3.172}$$

and

$$\beta_t(i) = \sum_{t \le T-t} \sum_{\substack{i=1 \\ j \neq i}}^{n} a_{ij} d_j(\tau) \left[\prod_{\theta=1}^{\tau-1} w_{t-\tau+\theta} b_j(\mathbf{O}_{t-\theta}) \right] b_j(\mathbf{O}_{t+\tau}) \beta'_{t+\tau}(j), \tag{3.173}$$

respectively. Note that there is still a potential problem since the bracketed terms must fall with the dynamic range of the computer. We have not yet encountered a problem in this respect. It should be clear that as long as this difficulty does not arise, the actual scaling operation (3.170) can be performed at any regular interval, not necessarily at every time step.

When the α and β terms are scaled in this way, the expressions which arise in the reestimation formulas will have a common factor which can be divided out. For example, in determining the state sequence we may observe that

$$\alpha'_t(i)\beta'_t(i)/w_t = \alpha_t(i)\beta_t(i) \prod_{t=1}^{T} w_t \tag{3.174}$$

for any i. Since the scaled α and β terms always include the common factor on the right-hand side of (3.174), the left-hand side can be used in (3.148) and the state sequence can be found by maximizing it over i for each t.

The reestimation formulas can also be evaluated by exploiting the common coefficient formed by the product of scale factors. For example, to scale (3.151) we observe that $w_{t-\tau+1}, w_{t-\tau+2}, \ldots, w_{t-1}$ are absent from each term in the numerator while w_t appears twice in every term in the denominator. Treating the missing and extra coefficients accordingly, we may rewrite (3.151) in terms of the scaled α and β values, namely,

$$\bar{a}_{ij} = \frac{\sum_{t=1}^{T} \sum_{\tau \le t} \alpha'_{t-\tau}(i) a_{ij} d_j(\tau) \left[\prod_{\theta=1}^{\tau-1} w_{t-\tau+\theta} b_j(\mathbf{O}_{t-\tau+\theta}) \right] b_j(\mathbf{O}_t) \beta'_t(j)}{\sum_{t=1}^{T-1} \alpha'_t(i)\beta'_t(i)/w_t}. \tag{3.175}$$

The common term may now be divided out of the right-hand side of (3.175), giving a true value for $\overline{a}_{ij}$ from the scaled α and β terms. Equations (3.155), (3.158), and (3.160) may be treated in just the same manner.

3.1.5 The Non-stationary Observation Case

The stochastic processes studied in Sections 3.1.1–3.1.4 all assume piecewise or quasi-stationarity, that is, the states of the model correspond to stationary regimes in the observed signal within which the statistics are time-invariant. As discussed in Section 2.1, however, the articulatory generation of the speech signal requires that the articulators be in continuous motion, causing the power spectrum of the signal to be a continuous function of time, the behavior of which is governed by mechanical constraints on the articulatory mechanism. We can imagine that as a sequence of sounds is produced by the vocal apparatus, the articulators move along well-determined trajectories between target positions (refer to Figs 2.1 and 2.3) corresponding to the individual sounds. We can account for this co-articulation of sounds by relaxing the assumption of quasi-stationarity to include those non-stationary processes in which the continuous time variation is created by a deterministic function of time. This is accomplished by the non-stationary autoregressive model of Liporace [200] which is a combination of his earlier non-stationary LPC model [199], the autoregressive model discussed in Section 3.1.3, and the semi-Markov model of section 3.1.4.

The general equation for this kind of process is

$$y_t = \sum_{m=1}^{M} c_m^{s_k}(t - \tau_k) y_{t-m} + n_t, \qquad t = \tau_k, \ldots, \tau_k + d_k - 1, \tag{3.176}$$

where τ_k denotes the starting time for the kth segment, d_k denotes the kth segment duration, s_k denotes the state underlying the kth segment, and n_t denotes a Gaussian distributed random noise at time t. The random noise is state-dependent with zero mean and has noise variance $\sigma_{s_k}^2$. The non-stationarity arises from the time variation of the regression coefficients, c_m.

The first segment starts at time $M + 1$; therefore, $\tau_1 = M + 1$. Data points for the first segment, which is generated using (3.176) with $k = 1$, will last from time $t = M + 1$ to $t = \tau_1 + d_1$. Thus, the second segment starts at time $\tau_2 = \tau_1 + d_1$ and ends at $\tau_2 + d_2 - 1$.

After all the data points for a segment have been generated, the next state is selected according to the state transition matrix $\mathbf{A} = \{a_{ij}, i = 1, \ldots, S, \; j = 1, \ldots, S\}$. This matrix gives the probability of going from state i to state j.

After the state for a segment is selected, then the whole process repeats. The duration is selected from the state-dependent duration distribution. Then data are generated using the recursive equation 3.176. This whole process continues for all the segments $k = 1, \ldots, K$, where K is the maximum number of segments in the speech data.

Regression Coefficients

The regression coefficients $\{c_m^j(t), m = 1, \ldots, M, \; j = 1, \ldots, S\}$ are generated as a linear combination of another function $u_n(t)$:

$$c_m^j(t) = \sum_{n=0}^{N} c_{mn}^j u_n(t), \qquad m = 1, \dots, M, \; j = 1, \dots, S. \tag{3.177}$$

From this equation, it can be seen that the c_{mn}^j are just a set of constants not dependent on time. However, the function $u_n(t)$ does depend on time and this makes up the time-varying components of the $c_m^j(t)$. In other words, it is the $u_n(t)$ which makes the data nonstationary. Using either a power series or a Fourier series we get, respectively,

$$u_n(t) = t^n, \tag{3.178}$$

$$u_n(t) = \begin{cases} \cos n\omega t, & n \text{ even}, \\ \sin n\omega t, & n \text{ odd}. \end{cases} \tag{3.179}$$

Parameters

There are five parameters for the model: σ^2, the variance for the first M samples; $\mathbf{A} = \{a_{ij}\}$, the matrix for state transition probabilities, where a_{ij} is the probability from state i to state j; $P(d|j)$, the state duration probabilities, $d = 1, \dots, D$, $j = 1, \dots, S$; $c_m^j(t)$, the parameters of the time-dependent regression coefficients, $m = 1, \dots, M$, $j = 1, \dots, S$; and σ_j^2, the state-dependent noise variance, $j = 1, \dots, S$.

Calculation of Probability

To start the process of finding the parameters given the data, first assume that a set of parameters is known and that the probability density needs to be calculated. It is shown that inductive calculation for finding the probability density is less computationally intensive than direct computation. Then the result from inductive calculations is used in the re-estimation of the parameters.

Direct Calculation

Given the observations sequence $\mathbf{Y}$, but not the original parameters, we want to reestimate the values of the parameters from the observed values. The criterion for doing this is to find the set of parameters that will maximize the probability density of $\mathbf{Y}$.

To do this, suppose that we are already given a set of parameter values known collectively as $\boldsymbol{\lambda}$. Then the probability density of $\mathbf{Y}$ given that the parameter values are $\boldsymbol{\lambda}$ can be written as follows:

$$p_{\boldsymbol{\lambda}}(\mathbf{Y}) = \sum_{\mathbf{x}} p_{\boldsymbol{\lambda}}(\mathbf{Y}, \mathbf{x}) = \sum_{\mathbf{x}} p_{\boldsymbol{\lambda}}(\mathbf{Y}|\mathbf{x}) P_{\boldsymbol{\lambda}}(\mathbf{x}), \tag{3.180}$$

where $\mathbf{x}$ is a possible realization, $p_{\boldsymbol{\lambda}}(\mathbf{Y}, \mathbf{x})$ is the joint probability density of the joint occurrence of $\mathbf{Y}$ and $\mathbf{x}$, $p_{\boldsymbol{\lambda}}(\mathbf{Y}|\mathbf{x})$ is the probability density of $\mathbf{Y}$ given $\mathbf{x}$, and $P_{\boldsymbol{\lambda}}(\mathbf{x})$ is the probability of the sequence $\mathbf{x}$. The state–duration pair sequence for a possible realization, $\mathbf{x}$, is denoted by $(s_1, d_1), (s_2, d_2), \dots, (s_{K_x}, d_{K_x})$, where the number of segments is dependent on the particular sequence $\mathbf{x}$. The term $P_{\boldsymbol{\lambda}}(\mathbf{x})$ denotes the probability of the sequence $\mathbf{x}$ for the set of parameters values $\boldsymbol{\lambda}$. As can be seen from the equation above, this summation is over all possible sequences and will be computationally intensive.

Following is the brief outline from Liporace's paper for the calculation. Since the current active state depends only on the previous state, we have

$$P_\lambda(\mathbf{x}) = a_{s_1} P(s_1, d_1) \prod_{k=2}^{K_x} a_{s_{k-1}} s_k P(d_k | s_k), \tag{3.181}$$

where a_{s_1} denotes the probability of starting from the first state s_1.

Let $G(y; m, \sigma^2)$ denote the univariate Gaussian density on y with mean m and variance σ^2. Then

$$P_\lambda(\mathbf{Y}|\mathbf{x}) = P_\lambda(\mathbf{Y}_{1,M}) \prod_{k=1}^{K_x} P_\lambda(\mathbf{Y}_{\tau_k,\tau_k+d_k-1} | \mathbf{Y}_{1,\tau_k-1}, \mathbf{x}) \tag{3.182}$$

$$= \prod_{t=1}^{M} G(y_t; 0, \sigma^2) \prod_{k=1}^{K_x} \prod_{t=\tau_k}^{\tau_k+d_k-1} G(y_t; E(y_t|\mathbf{Y}_{1,t-1}, \mathbf{x}), \sigma_{s_k}^2). \tag{3.183}$$

The first product term from $t = 1$ to $t = M$ in 3.183 is the probability density function due to the first M samples, which are generated from Gaussian random variables with mean 0 and variance σ^2. The rest of the terms in 3.183 come from the rest of the samples.

For the model, each segment is generated autoregressively; therefore, within each segment:

$$E(y_t|\mathbf{Y}_{1,t-1}, \mathbf{x}) = \sum_{m=1}^{M} \sum_{n=0}^{N} c_{mn}^{s_k} u_n(t\text{-}\tau_K) y_{t-m}, \qquad t \in [\tau_k, \dots, \tau_k + d_k - 1]. \tag{3.184}$$

Substituting equations 3.181, 3.183, and 3.184 into 3.180, we get the probability density function of $\mathbf{Y}$, which can be calculated directly. However, calculating the $p_\lambda(\mathbf{Y})$ for a given set of parameter values $\boldsymbol{\lambda}$ directly using this formula takes a lot of computational power because we need to consider all possible paths, and even the simplest path with the smallest number of possible segments requires considerable computation.

Inductive calculation

In the standard HMM, there is an inductive method of calculating the probability of the observations using forward and backward probabilities. There is also an inductive method to find $p_\lambda(\mathbf{Y})$ in this model. This method is analogous to the forward and backward probabilities for the continuous-duration case of section 3.1.4.

The following notation will be used for the rest of this section:

$p(\cdot)$	Probability density function of an argument
$P(\cdot)$	Probability of an argument
$(j, d)_\tau$	State j and duration d begin at time τ
$(j, \cdot)_\tau$	State j begins at time τ
$(\cdot, d)_\tau$	Duration d begins at time τ

We require the following definitions:

$$\alpha_\tau^*(j) = p_\lambda(\mathbf{Y}_{1,\tau-1}, (j,\cdot)_\tau);$$
$$\beta_\tau^*(j) = p_\lambda(\mathbf{Y}_{\tau,T}|\mathbf{Y}_{1,\tau-1}, (j,\cdot)_\tau);$$
$$\alpha_\tau(j,d) = p_\lambda(\mathbf{Y}_{1,\tau+d-1}, (j,d)_\tau);$$
$$\beta_\tau(j,d) = p_\lambda(\mathbf{Y}_{\tau+d,T}|\mathbf{Y}_{1,\tau+d-1}, (j,d)_\tau);$$
$$u_j(t,\tau) = \sum_{m=1}^{M}\sum_{n=0}^{N} c_{mn}^j u_n(t-\tau)y_{t-m};$$
$$g_j(t,\tau) = G(y_t; u_j(t,\tau), \sigma_j^2)$$

the latter being the univariate Gaussian density on y_t with mean $u_j(t,\tau)$ and variance σ_j^2.

The idea for the inductive calculation is to look at the state which is active at any time $\tau \in [M+1,\dots,T]$. A state j is active at time τ if and only if state j begins at some time $\tau - d' + 1$, $d' \in [1,\dots,D]$ with duration $d \geq d'$.

For $\tau \in [M+1,\dots,T]$,

$$p_\lambda(\mathbf{Y}) = \sum_{j=1}^{S}\sum_{d'=1}^{D}\sum_{d=d'}^{D} p_\lambda(\mathbf{Y}, (j,d)_{\tau-d'+1}) \tag{3.185}$$

and identically in τ. Also

$$p_\lambda(\mathbf{Y}, (j,d)_\tau) = p_\lambda(\mathbf{Y}_{1,\tau+d-1}, (j,d)_\tau)p_\lambda(\mathbf{Y}_{\tau+d,T}|\mathbf{Y}_{1,\tau+d-1}, (j,d)_\tau) \tag{3.186}$$
$$= \alpha_\tau(j,d)\beta_\tau(j,d) \tag{3.187}$$

A comment here is that $\{\alpha_\tau(j,)\}$ is analogous to the forward probability in the standard HMM, while $\{\beta_\tau(j,d)\}$ is analogous to the backward probability in the standard HMM.

Substituting 3.187 into 3.185,

$$p_\lambda(\mathbf{Y}) = \sum_{j=1}^{S}\sum_{d'=1}^{D}\sum_{d=d'}^{D} \alpha_{\tau-d'+1}(j,d)\beta_{\tau-d'+1}(j,d), \qquad \tau \in [M+1,\dots,T], \tag{3.188}$$

where $\{\alpha_\tau(j,d)\}$ and $\{\beta_\tau(j,d)\}$ can be calculated inductively as follows.

The values for $\alpha_\tau^*(j)$ can be computed inductively for all time τ and states j by using the following equations:

$$\alpha_\tau^*(j) = p_\lambda(\mathbf{Y}_{1,\tau-1}, (j,\cdot)_\tau) \tag{3.189}$$
$$= \sum_{i=1}^{S}\sum_{\delta=1}^{D} \alpha_{\tau-\delta}^*(i)a_{ij}P(\delta|i)\prod_{\tau-\delta}^{\tau-1} g_i(t,\tau-\delta). \tag{3.190}$$

Since the first segment always starts from state 1, the initial condition for $\alpha^*_\tau(j)$ is

$$\alpha^*_\tau(j) = 0, \qquad \text{for } j = 1, \ldots, S, \tau = 1, \ldots, M,$$

$$\alpha^*_{M+1}(j) = \begin{cases} 1, & \text{if } j = 1, \\ 0, & \text{otherwise.} \end{cases}$$

Alternatively, we can use the following for each state j:

$$\alpha^*_{M+1}(j) = \frac{1}{M}, \qquad \text{for } j = 1, \ldots, S.$$

After all the values for $\alpha^*_\tau(j)$ are calculated, $\alpha_\tau(j, d)$ can be calculated from $\alpha^*_\tau(j)$ inductively as follows:

$$\alpha_\tau(j, d) = p_\lambda(\mathbf{Y}_{1,\tau+d-1}, (j, d)_\tau) \tag{3.191}$$

$$= p_\lambda(\mathbf{Y}_{\tau,\tau+d-1}|\mathbf{Y}_{1,\tau-1}, (j, d)_\tau)P(d|j)p_\lambda(\mathbf{Y}_{1,\tau-1}, (j, \cdot)_\tau) \tag{3.192}$$

$$= \prod_{t=\tau}^{\tau+d-1} g_j(t, \tau)P(d|j)p_\lambda(\mathbf{Y}_{1,\tau-1}, (j, \cdot)_\tau) \tag{3.193}$$

$$= \prod_{t=\tau}^{\tau+d-1} g_j(t, \tau)P(d|j)\alpha^*_\tau(j), \tag{3.194}$$

where $g_j(t, \tau)$ is defined at the beginning of this section.

Because

$$\beta^*_{\tau+d}(k) = p_\lambda(\mathbf{Y}_{\tau+d,T}|\mathbf{Y}_{1,\tau+d-1}, (k, \cdot)_{\tau+d}) \tag{3.195}$$

it can be concluded that

$$\beta^*_{\tau+d}(k) = \sum_{l=1}^{S}\sum_{\delta=1}^{D} p_\lambda(\mathbf{Y}_{\tau+d,\tau+d+\delta-1}, (l, \cdot)_{\tau+d+\delta}, (\cdot, \delta)_{\tau+d}|\mathbf{Y}_{1,\tau+d-1}, (k, \cdot)_{\tau+d})] \tag{3.196}$$

$$= \sum_{l=1}^{S}\sum_{\delta=1}^{D} \beta^*_{\tau+d+\delta}(l)P(\delta|k)a_{kl} \prod_{t=\tau+d}^{\tau+d+\delta-1} g_k(t, \tau + d). \tag{3.197}$$

The initial conditions used for $\beta^*_t(k)$ are

$$\beta^*_T(k) = 1, \qquad k = 1, \ldots, S.$$

This is because there is the probability of ending in any of the states at the end of the data sequence. All $\beta^*_t(k)$ can be calculated for all time t and states k, starting from time $t = T$ and going back in time.

Then, using $\beta_t^*(k)$, $\beta_\tau(k, d)$ can be calculated as follows:

$$\beta_\tau(j, d) = p_{\boldsymbol{\lambda}}(\mathbf{Y}_{\tau+d,T}|\mathbf{Y}_{1,\tau+d-1}, (j, d)_\tau) \tag{3.198}$$

$$= \sum_{k=1}^{S} a_{jk}\beta_{\tau+d}^*(k). \tag{3.199}$$

From 3.199, it can be seen that

$$\beta_{\tau_m}(j_m, d_m) = \beta_{\tau_n}(j_n, d_n), \qquad \text{if } \tau_m + d_m = \tau_n + d_n \text{ and } j_m = j_n. \tag{3.200}$$

This may save a little computation time. Also, during the implementation, it is found that calculating and saving all $g_j(t, \tau)$ and then calculating and saving all $\prod_{t=\tau}^{\tau+d-1} g_j(t, \tau)$, rather than calling this function every time, saves some computation time.

Parameter Reestimation

There is a set of re-estimation formulas for updating the parameters such that the new parameter values $\overline{\boldsymbol{\lambda}}$ always give a monotonic increase in the likelihood of the observation $\mathbf{Y}$, with respect to the current parameter values $\boldsymbol{\lambda}$, that is, $p_{\overline{\boldsymbol{\lambda}}}(\mathbf{Y}) > p_{\boldsymbol{\lambda}}(\mathbf{Y})$.

As in standard HMM, there is an initial guess for each of the parameters. Then each parameter is updated until the likelihood appears to have converged such that an *ad hoc* criterion has been met (e.g., when the increase in the likelihood value is less than some number ε). For this discussion, a fixed number of iterations is used instead because it can be seen from the plots of the likelihood functions when the *ad hoc* criterion has been reached.

Even though the *ad hoc* criterion or the fixed number of iterations is easier to test for, this is not valid in general. This is because there is a possibility for the likelihood function to reach convergence, but the parameter values may still be changing. The true convergence criterion should depend on the parameter values and not the likelihood values, and this is achieved when the parameter values no longer change. When the true convergence criterion has been met, then the set of values found corresponds to the point where the gradient of the Q function with respect to each of the different parameters is 0. This is explained below.

Auxiliary Function

The reestimation formulas are derived from the auxiliary function

$$Q(\boldsymbol{\lambda}, \overline{\boldsymbol{\lambda}}) = \sum_{\mathbf{x}} p_{\boldsymbol{\lambda}}(\mathbf{Y}, \mathbf{x}) \log p_{\overline{\boldsymbol{\lambda}}}(\mathbf{Y}, \mathbf{x}). \tag{3.201}$$

As noted in Section 3.1.2, we have:

1. $Q(\boldsymbol{\lambda}, \overline{\boldsymbol{\lambda}}) > Q(\boldsymbol{\lambda}, \boldsymbol{\lambda})) \Rightarrow p_{\overline{\boldsymbol{\lambda}}}(\mathbf{Y}) > p_{\boldsymbol{\lambda}}(\mathbf{Y})$.
2. Under mild orthodoxy conditions on $\mathbf{Y}$, $Q(\boldsymbol{\lambda}, \overline{\boldsymbol{\lambda}})$ has a unique global maximum $\overline{\boldsymbol{\lambda}}$. This global maximum is a critical point and coincides with $\boldsymbol{\lambda}$ if and only if $\boldsymbol{\lambda}$ is a critical point of $p_{\boldsymbol{\lambda}}(\mathbf{Y})$.

3. The point $\overline{\boldsymbol{\lambda}}$ at which $Q(\boldsymbol{\lambda}, \overline{\boldsymbol{\lambda}})$ is maximized is expressible in closed form in terms of the inductively computable quantities $\{\alpha_\tau(j, d), \beta_\tau(j, d)\}$.

Statement 1 says that increasing the value of the auxiliary Q function with respect to the new parameters means that the probability density of $\mathbf{Y}$ or likelihood of $\mathbf{Y}$ also increases. Statement 2 says that the likelihood for the new parameter values $\overline{\boldsymbol{\lambda}}$ will always increase the likelihood with respect to current parameters, unless it is a critical point. This is because with every update, the new parameters will always give a better or equal likelihood value. Statement 3 says that it is practicable to determine the new parameter values because an inductive process can be stopped after a certain number of iterations; however, it may be impossible to find the new parameters using the closed-form solution.

Outline of Derivation of Reestimation Formulas

The derivations are very long and will not be given in detail. However, a basic outline is provided. In essence, the gradient of the Q function with respect to each of the five parameters is taken, subject to the stochastic constraints.

First, the auxiliary function $Q(\boldsymbol{\lambda}, \overline{\boldsymbol{\lambda}})$, given earlier in terms of $p_{\boldsymbol{\lambda}}(\mathbf{Y}, \mathbf{x})$ and $\log p_{\overline{\boldsymbol{\lambda}}}(\mathbf{Y}, \mathbf{x})$, is written in terms of the five parameters by substituting 3.181, 3.183, and 3.184 into 3.180, then imposing the following stochastic constraints for some of the parameters in the new parameter set $\overline{\boldsymbol{\lambda}}$:

$$\sum_{j=1}^{S} \overline{a}_{ij} = 1, \qquad i = 1, \ldots, S, \tag{3.202}$$

and

$$\sum_{d=1}^{D} \overline{P}(d|j) = 1, \qquad j = 1, \ldots, S. \tag{3.203}$$

The first constraint 3.202 states that the sum of the probabilities leaving a state should be equal to 1 and the second constraint 3.203 states that the sum of the probability of duration for any state should equal 1.

Next, use Lagrange multipliers to incorporate these constraints into the Q function to give another function $Q^*(\boldsymbol{\lambda}, \overline{\boldsymbol{\lambda}})$. Thus the new auxiliary function, $Q^*(\boldsymbol{\lambda}, \overline{\boldsymbol{\lambda}})$, with the constraints, is

$$Q^*(\boldsymbol{\lambda}, \overline{\boldsymbol{\lambda}}) = Q(\boldsymbol{\lambda}, \overline{\boldsymbol{\lambda}}) - \sum_{i=1}^{S} \theta_i \left(\sum_{j=1}^{S} \overline{a}_{ij} - 1 \right) - \sum_{j=1}^{S} \varepsilon_j \left(\sum_{d=1}^{D} \overline{P}(d|j) - 1 \right) \tag{3.204}$$

where θ_i and ε_j are Langrange multipliers.

Then the gradient of $Q^*(\boldsymbol{\lambda}, \overline{\boldsymbol{\lambda}})$ with respect to each of the five new parameters is set equal to 0 (as in the standard procedure for finding the critical point of a function). For example, $\partial Q^*(\boldsymbol{\lambda}, \overline{\boldsymbol{\lambda}})/\partial \overline{a}_{ij} = 0$ is solved. From the above discussion, the critical point will correspond to a global maximum of Q. Since Q has increased with respect to the new parameters, this means that the likelihood with respect to the new parameter has also increased, that is, $p_{\overline{\boldsymbol{\lambda}}}(\mathbf{Y}) > p_{\boldsymbol{\lambda}}(\mathbf{Y})$.

Reestimation Formulas

The variance for the first M data samples does not require reestimation:

$$\sigma^2 = \frac{1}{M}\sum_{t=1}^{M} y_t^2. \tag{3.205}$$

The rest of the parameters do require re-estimation. The reestimation formulas for $\hat{a}_{ij}$, $\hat{P}(d|j)$, and $\hat{\sigma}_j^2$ are as follows:

$$\hat{a}_{ij} = \frac{\sum_{\tau=M+1}^{T}\sum_{d=1}^{D}\alpha_{\tau-d}(i,d)a_{ij}\beta_\tau^*(j)}{\sum_{\tau=M+1}^{T}\sum_{d=1}^{D}\alpha_{\tau-d}(i,d)\beta_{\tau-d}(i,d)}, \qquad i,\ j = 1,\ldots,S, \tag{3.206}$$

$$\hat{P}(d|j) = \frac{\sum_{\tau=M+1}^{T}\alpha_\tau(j,d)\beta_\tau(j,d)}{\sum_{\tau=M+1}^{T}\sum_{\delta=1}^{D}\alpha_\tau(i,\delta)\beta_\tau(i,\delta)}, \qquad j = 1,\ldots,S, d = 1,\ldots,D, \tag{3.207}$$

$$\hat{\sigma}_j^2 = \frac{\sum_{\tau=M+1}^{T}\sum_{d=1}^{D}\alpha_\tau(i,d)\beta_\tau(i,d)\sum_{t=\tau}^{\tau+d-1}(y_t-\mu_j(t,\tau))^2}{\sum_{\tau=M+1}^{T}\sum_{d=1}^{D}d\alpha_\tau(i,d)\beta_\tau(i,d)}, \qquad j = 1,\ldots,S. \tag{3.208}$$

In the above formulas, $\alpha_\tau(i,d)$, $\beta_\tau(i,d)$, $\beta^*(j)$ and $\mu_j(t,\tau)$ have already been defined.

The reestimates for the autoregressive coefficients were calculated as follows. For each $j = 1,\ldots,S$,

$$\sum_{m=1}^{M}\sum_{n=0}^{N} R_{mqnr}^j \hat{c}_{mn}^j = R_{0q0r} \qquad q = 1,\ldots,M, r = 0,\ldots,N, \tag{3.209}$$

where

$$R_{mqnr}^j = \sum_{\tau=M+1}^{T}\sum_{d=1}^{D}\alpha_\tau(j,d)\beta_\tau(j,d)\sum_{t=\tau}^{t+d-1} y_{t-m}y_{t-q}u_n(t-\tau)u_r(t-\tau). \tag{3.210}$$

Letting $M = 2$, $N = 1$, for $j = 1,\ldots,S$, 3.209 can be written out explicitly as

$$\begin{bmatrix} R_{1100}^j & R_{1110}^j & R_{2100}^j & R_{2110}^j \\ R_{1101}^j & R_{1111}^j & R_{2101}^j & R_{2111}^j \\ R_{1200}^j & R_{2110}^j & R_{2200}^j & R_{2210}^j \\ R_{1201}^j & R_{1211}^j & R_{2201}^j & R_{2211}^j \end{bmatrix} \begin{bmatrix} \hat{c}_{10}^j \\ \hat{c}_{11}^j \\ \hat{c}_{20}^j \\ \hat{c}_{21}^j \end{bmatrix} = \begin{bmatrix} R_{0100}^j \\ R_{0101}^j \\ R_{0200}^j \\ R_{0201}^j \end{bmatrix} \tag{3.211}$$

which can be written compactly as

$$\mathbf{R}^j\hat{\mathbf{c}}^j = \mathbf{r}^j. \tag{3.212}$$

The $\mathbf{R}^j$ matrix is symmetric; this can be seen from 3.210 since it is symmetric about n and r, as well as m and q. Therefore, it is only necessary to compute the values in the upper triangle of this matrix. As always, when using a computer to solve 3.211, one should avoid finding the inverse of $\mathbf{R}^j$ to find $\hat{\mathbf{c}}^j$ because this operation may be unstable numerically and also takes a lot of computation. Rather, other more numerically stable and faster methods for solving the regression coefficients $\hat{\mathbf{c}}^j$ can be used, which do not involve matrix inversion. As in Section 3.1.3, if $\mathbf{R}^j$ is positive definite, since it is already symmetric, Cholesky factorization can be used [265].

The computational complexity of this model renders it difficult to implement and costly to apply. Tan [309] has produced a successful Monte Carlo simulation of the algorithm but, despite its obvious appropriateness, it has yet to be applied to the speech signal.

3.1.6 Parameter Estimation via the EM Algorithm

The parameter estimation problem is analogous to the statistical approach to a seemingly very different problem, that of estimation from incomplete data. Dempster *et al.* [62] indicate that in the 1950s, statisticians were thinking of a random process for which only incomplete data was available as a doubly stochastic process. The output of the "true" but hidden process x was thought of as passing through a second process y which censored the input and produced the observed data. Both processes were considered to be parameterized in $\boldsymbol{\lambda}$ and the problem was to determine $\boldsymbol{\lambda}$ from the observables.

The statistician's solution to this problem has come to be known as the EM algorithm: E for expectation and M for maximization. The algorithm is succinctly described as follows. Let $\tau(x)$ be any sufficient statistic for x, meaning that, in a precise sense, τ contains complete information about x. Suppose that we have an estimate of the parameter λ. Then the so-called E-step of the algorithm calls for the estimation of $\tau(x)$ from

$$\overline{\tau} = E\{\tau(x)|y, \boldsymbol{\lambda}\}, \tag{3.213}$$

where $E\{\cdot|\cdot\}$ is the expectation operator. Thus solving (3.213) gives a new estimate of $\tau(x)$, $\overline{\tau}$, conditioned on the present estimate of the parameter $\boldsymbol{\lambda}$ and the observations y. This is followed by the M-step in which we solve

$$E\{\tau(x)|\overline{\boldsymbol{\lambda}}\} = \overline{\tau} \tag{3.214}$$

to get a new estimate, $\overline{\boldsymbol{\lambda}}$, of the parameter. Iterative applications of (3.213) and (3.214) yield better estimates of $\boldsymbol{\lambda}$. In general, there are no closed-form solutions for (3.213) and (3.214) so the Bayesian solution is used, yielding $\overline{\boldsymbol{\lambda}}$ as the maximum likelihood estimator of the parameter given $\boldsymbol{\lambda}$ and $\overline{\tau}$. Generalizations and proof of convergence of the EM algorithm are given in [62].

Baum and his colleagues [27–29] recognized that the parameter estimation problem for the HMM could be considered an interesting special case of a doubly stochastic process, whereupon they were able to derive a particular solution to (3.213) and (3.214).

3.1.7 The Cave–Neuwirth and Poritz Results

We now turn our attention to two important empirical results based on the theory of hidden Markov models described in Sections 3.1.1 through 3.1.6. The two experiments clearly demonstrate the remarkable power of the HMM to both discover and represent aspects of linguistic structure as manifest in text and speech. This is possible because, like linear prediction described in Section 2.4.2 and time scale normalization discussed in Section 2.6.2, the HMM is a general model of a time series that is particularly appropriate for speech and naturally captures acoustic phonetics, phonology, and phonotactics.

Markov [212] used the stochastic model that now bears his name to analyze the text of *Eugene Onegin*. More recently, Cave and Neuwirth [43] have given a modern interpretation of his experiments in a highly instructive way. Using ordinary text from a newspaper as an observation sequence, they estimated the parameters of a family of HMMs. They then showed how the parameter values can be used to infer some significant linguistic properties of English text.

The observation data for the experiment comprised $T = 30\,000$ letters of ordinary text from the *New York Times*. The text was preprocessed to remove all symbols except the 26 letters of the English alphabet and a delimiter (blank or space) used to separate words. Thus, M was fixed at 27. The observation sequence, $\mathbf{O}$, was then the entire text.

A family of models corresponding to $N = 1, 2, 3, \ldots, 12$ was generated by applying the algorithm of (3.11)–(3.13) to the entire observation sequence until convergence was reached. The state sequence for each model was determined from (3.7) and put into correspondence with the observation sequence. Many of the entries in the $\mathbf{A}$ and $\mathbf{B}$ matrices converged to zero, indicating impossible state transitions and impossible observations for a given state, respectively. These four quantities make possible the formulation of two kinds of rules governing English text. First are those rules that identify a particular state with a linguistic property of an observation. Conversely, we have rules that determine which state generated a particular observation based on the linguistic properties of the observation.

The case for $N = 6$ provides a good example. In this case, it was discovered that only vowels were produced by state 2. This is an example of the first type of rule. Whenever a blank was present in the observation sequence, the model was in state 6. This is an example of the second type of rule. In addition, it was found that state 3 was associated exclusively with consonants, state 1 was a vowel successor, and state 5 was a consonant successor. Some more complicated rules were also discovered; for example, word-final letters could come only from states 1, 2, or 4 whereas word-initial letters could only arise in states 2, 3, or 5. Of course, the state transition matrix determines the allowable sequences of letters having specific properties.

It is remarkable that the HMM discovers these important properties of text (for which linguistic categories had already been named) without any information other than that which is implicit but camouflaged in the text.

The Poritz [250] result comes from an experiment which was, no doubt, inspired by the Cave–Neuwirth result. The experiment is identical in spirit, with speech as the observable process instead of text. In this case, the autoregressive model of Section 3.1.3 was derived from readings of short paragraphs of about 40 seconds in duration. A five-state model was estimated and then the state sequence corresponding to the speech signal was determined. By listening to those intervals of the signal that correspond to each of the five states, their identities were easily determined. State 1 produced only vowels. State 2 indicated silence. State 3 was reserved for nasals; state 4, plosives; and state 5, fricatives. Moreover, the state transition matrix determined the phonotactic rules of these five broad phonetic categories. The spectra corresponding to the state-dependent autoregressive processes were exactly those that would be predicted by solutions to the Webster equation for the vocal tract geometry appropriate to the phonetic category. In a manner analogous to that of the Cave–Neuwirth result, the Poritz result shows how broad phonetic categories and phonotactics can be discovered in the speech signal without recourse to known linguistic categories.

These two results are the best available demonstration of the appropriateness of the HMM formalism for extraction and representation of linguistic structure. These results form the empirical basis for the speech recognition systems which we will study in Chapter 7.

3.2 Formal Grammars and Abstract Automata

We may think of a language as a (possibly infinite) list of sentences, each composed of a sequence of words. Allowable sequences of words are produced according to a finite set of grammatical rules. We shall call such a set of rules a formal grammar G, and the list generated by it, the language, $\mathcal{L}(G)$.

The grammar G is a mathematical object composed of four parts and is designated by $G(V_N, V_T, S, R)$. V_N signifies a finite set of non-terminal symbols disjoint with respect to V_T, a finite set of terminal symbols. The terms "terminal" and "non-terminal" refer to whether or not, respectively, the symbols may appear in a sentence. Non-terminals are traditionally designated by upper-case letters, terminals by lower-case. Thus a sentence will contain only lower-case letters. The distinguished non-terminal S is called the start symbol since all sentences in the language are derived from it. The kernel of the grammar is the set R of rewriting or production rules, a member of which is customarily indicated by $\alpha \rightarrow \beta$, where α and β stand for elements of the set $\{V_N \cup V_T\}^*$, and the notation $\mathcal{L}^*$ is taken to mean the set of all subsets of $\mathcal{L}$. A rewriting rule allows the replacement of the arbitrary string appearing on the left-hand side of the rule by the arbitrary string on the right wherever the left-hand member appears in any other string. A special case is that for which $\beta = \phi$, the null symbol. The effect of this rule is to cause the string α to vanish.

A grammar generates sentences in a language by the following operations. A string γ is said to derive the string δ, written $\gamma \Rightarrow \delta$, if and only if $\gamma = \eta_1 \alpha \eta_2, \delta = \eta_1 \beta \eta_2$, and $\exists \alpha \rightarrow \beta \in R$ where η_1 and η_2 are arbitrary strings. The transitive closure of the derivation operator is $\gamma \overset{*}{\Rightarrow} \delta$, read γ ultimately derives δ, meaning that $\gamma = \zeta_0, \delta = \zeta_T$, and $\zeta_{t-1} \Rightarrow \zeta_t$ for $1 \leq t \leq T$. Then

$$\mathcal{L}(G) = \{W \in V_T^* \mid S \overset{*}{\Rightarrow} W\}, \tag{3.215}$$

meaning that a language comprises all those strings of terminal symbols that S ultimately derives under the set R. From (3.215) one may thus infer that if G imposes any constraint on word order then $\mathcal{L}(G) \subset V_T^*$.

A simple example will serve to clarify the notation. Let $V_N = \{S, A\}$, $V_T = \{a, b\}$, and R be defined by

$$\begin{aligned} S &\rightarrow aS, \\ S &\rightarrow bA, \\ A &\rightarrow bA, \\ A &\rightarrow \phi. \end{aligned} \tag{3.216}$$

According to (3.216), the start symbol can be transformed into a sequence of arbitrarily many as followed by another sequence of any number of bs. Thus we may write

$$\mathcal{L}(G) = \{a^m b^n \mid m, n > 0\}. \tag{3.217}$$

3.2.1 The Chomsky Hierarchy

The Chomsky hierarchy [46] is a particularly significant taxonomy of formal grammars according to their complexity and representational powers. Depending upon the form of R, grammars are classified as either regular, context-free, context-sensitive, or phrase-structure, in increasing order of complexity. Regular grammars, of which (3.216) is an example, have rules of the form $A \to a$ or $A \to aB$.

Context-free grammars, so called because their production rules may be applied independent of the symbols surrounding the non-terminal on the left-hand side, have rules of the form $A \to \alpha$. Naturally, context-sensitive grammars have rules of the form $\alpha A \beta \to \alpha\gamma\beta$ which map the non-terminal A onto the string γ only in the left and right context of α and β, respectively. Finally, the phrase-structure grammars have the unrestricted rules $\alpha \to \beta$. Each class in the hierarchy is properly included in the one above it.

That this differentiation among formal grammars is deeply meaningful requires a proof which goes too far afield for the purposes of this exposition. The reader interested in pursuing these ideas should consult [112], [120], and [133]. Since, however, that fact has implications for the algorithms we are about to discuss, the following examples should lend it some credence. Note that in (3.217) there is no relationship between m and n. If we require, for example, $m = n$ for any m, then no set of regular rules will suffice. However, the context-free rules $S \to aSb$, $S \to \phi$ will generate exactly the language $\mathcal{L}(G) = \{a^n b^n \mid n > 0\}$. In other words, regular grammars cannot count while context-free grammars can count the elements of one set, that is, the length of the string of bs. Similarly, if we wish to append another string of as, then no set of context-free rules is powerful enough. The context-sensitive rules $S \to ABSa$, $BA \to AB$, $S \to \phi$, $Aa \to aa$, $Ab \to ab$, $Bb \to bb$, $Ba \to ba$ do, in fact, generate the language, $\mathcal{L}(G) = \{a^n b^n a^n \mid n > 0\}$. Thus whereas context-free grammars can count one set, context-sensitive grammars can count two. Finally, we note, without offering any justification, that, in a certain sense, any computational process can be expressed as a phrase-structure grammar [319].

In (3.219) we derive the sentence $a^3b^3c^3$ for the context-sensitive case: according to the rewrite rules of (3.218). The rules are numbered and the rule used for a particular step in the derivation is indicated by the number of the rule in parentheses.

$$\begin{array}{ll} (1) & S \to ABSc \\ (2) & S \to \lambda \\ (3) & BA \to AB \\ (4) & Ab \to ab \\ (5) & Aa \to aa \\ (6) & Bb \to bb \\ (7) & Bc \to bc \\ (8) & A \to a \end{array} \tag{3.218}$$

$$
\begin{array}{ll}
 & S \\
(1) & ABSc \\
(1) & ABABScc \\
(1) & ABABABSccc \\
(2) & ABABABccc \\
(7) & ABABAbccc \\
(3) & ABAABbccc \\
(6) & ABAAbbccc \\
(3) & AABAbbccc \\
(3) & AAABbbccc \\
(6) & AAAbbbccc \\
(5) & AAabbbccc \\
(5) & Aaabbbccc \\
(8) & aaabbbccc
\end{array}
\tag{3.219}
$$

Properties and Representations

The languages generated by formal grammars have some useful computational properties and representations. A particularly interesting case is the finite language for which

$$|\mathcal{L}(G)| = \mathcal{N} < \infty. \tag{3.220}$$

Any finite language can be generated by a regular grammar from which we can count the sentences in the language. Let the matrix **C** have elements c_{ij} defined by

$$c_{ij} = |\{A_i \rightarrow vA_j\}| \tag{3.221}$$

for any $v \in V_T$ and $1 \leq i, j \leq |V_N|$.

Powers of **C** have a particular significance, namely the number, N_k, of sentences of length k is given by

$$N_k = \mathbf{e}_s \mathbf{C}^k \mathbf{e}_f, \tag{3.222}$$

where $\mathbf{e}_s$ is the vector $(1, 0, 0, \ldots, 1)$ and $\mathbf{e}_f$ is the vector $(0, 0, 0, \ldots, 1)$. These vectors correspond to an arrangement of the grammar such that the start symbol is the first non-terminal and the null symbol is the last. The generating function

$$P(Z) = \sum_{k=1}^{K} N_k Z^k \tag{3.223}$$

will be important in the discussion of grammatical constraint in Section 6.2.4. In (3.223) K is the length of the longest sentence, which must be finite since

$$\sum_{k=1}^{K} N_k = \mathcal{N} \tag{3.224}$$

which, from (3.220), is assumed finite.

Any context-free grammar can be written in a canonical form called Chomsky normal form (CNF) in which all rules are either of the form $A \rightarrow a$ or $A \rightarrow BC$. The CNF is

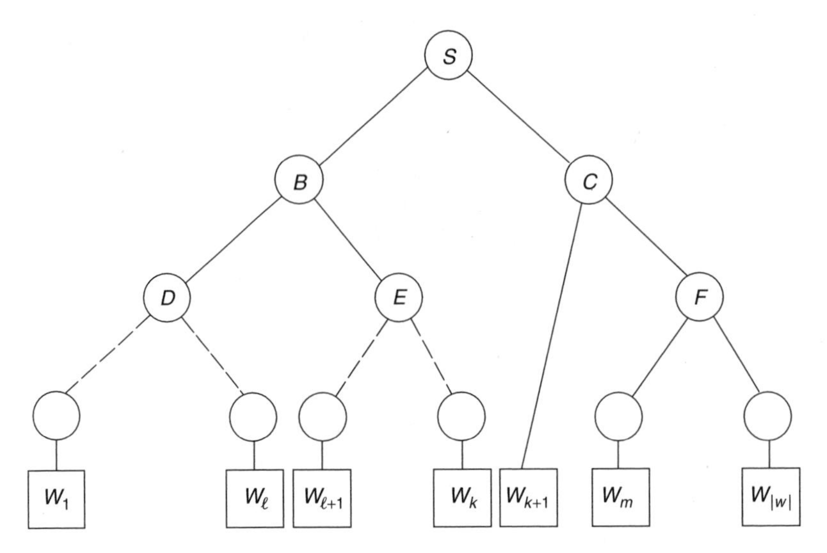

Figure 3.11 Parse tree for a sentence in a context-free language

naturally related to a binary tree as shown in Fig. 3.11. The root of the tree is the start symbol. All other nodes of the tree are labeled with non-terminals. Each node, except for the leaves of the tree and their predecessors, have two successors corresponding to the left and right non-terminals on the right-hand side of $A \rightarrow BC$, respectively. The CNF of a context-free grammar gives rise to the binary tree structure of Fig. 3.11, which is analogous to the directed graph structure for regular grammars shown in Fig. 3.12.

$$A \rightarrow \alpha = B_1 B_2 \ldots B_m, \qquad B_i \in \{V_N \cup V_T\}, \tag{3.225}$$

becomes

$$\begin{cases} A \rightarrow \alpha C_1 C_2 \ldots C_m \\ C_i \rightarrow B_i \Leftrightarrow B_i \in V_T \end{cases} \tag{3.226}$$

then

$$\begin{aligned} A &\rightarrow C_1 D_1, \\ D_1 &\rightarrow C_2 D_2, \\ D_2 &\rightarrow C_3 D_3, \\ &\vdots \\ D_{m-3} &\rightarrow C_{m-2} D_{m-2}, \\ D_{m-2} &\rightarrow C_{m-1} C_m. \end{aligned} \tag{3.227}$$

An example of the construction of (3.225)–(3.227) is given below. Consider the CF production rules

$$\begin{array}{llll} \text{a)} & S \rightarrow bA, & \text{e)} & S \rightarrow aB, \\ \text{b)} & A \rightarrow a, & \text{f)} & B \rightarrow b, \\ \text{c)} & A \rightarrow aS, & \text{g)} & B \rightarrow bS, \\ \text{d)} & A \rightarrow bAA, & \text{h)} & B \rightarrow aBB. \end{array} \quad (3.228)$$

The rules (3.228) generate the language

$$\mathcal{L}(G) = \left\{ a^{n_1} b^{m_1} a^{n_2} b^{m_2} \ldots a^{n_t} b^{m_t} \ldots a^{n_T} b^{m_T} \mid \sum_{t=1}^{T} n_t = \sum_{t=1}^{T} m_t \right\}. \quad (3.229)$$

Note that (3.228b) and (3.228f) are already in CNF. Then (3.228a) and (3.228e) become

$$\begin{aligned} S &\longrightarrow C_1 A, \\ C_1 &\longrightarrow b, \end{aligned} \quad (3.230)$$

and

$$\begin{aligned} S &\longrightarrow C_2 B, \\ C_2 &\longrightarrow a, \end{aligned} \quad (3.231)$$

respectively. Similarly, (3.228c) and (3.228g) are transformed into

$$\begin{aligned} A &\longrightarrow C_3 S, \\ C_3 &\longrightarrow a, \end{aligned} \quad (3.232)$$

and

$$\begin{aligned} B &\longrightarrow C_4 S, \\ C_4 &\longrightarrow b, \end{aligned} \quad (3.233)$$

respectively. Finally, (3.228d) and (3.228h) result in the CNF rules

$$\begin{aligned} A &\longrightarrow C_5 AA, \\ C_5 &\longrightarrow b, \\ A &\longrightarrow C_5 C_6, \end{aligned} \quad (3.234)$$

and similarly,

$$\begin{aligned} B &\longrightarrow C_7 BB, \\ C_7 &\longrightarrow a, \\ B &\longrightarrow C_7 C_8. \end{aligned} \quad (3.235)$$

3.2.2 Stochastic Grammars

Another approach to the use of formal syntax in speech recognition is one based upon stochastic grammars by means of which we shall be able to uncover some interesting interrelationships among the algorithms we have discussed thus far. Stochastic grammars are similar to the deterministic ones we have been examining except that their production rules have associated probabilities. The stochastic grammar $G_s(V_N, V_T S, R_s, \theta)$ has non-terminal, terminal, and start symbols over which its stochastic productions R_s are defined to have the form

$$\alpha \overset{p_{\alpha\beta}}{\rightarrow} \beta, \tag{3.236}$$

where $p_{\alpha\beta} > 0$ is understood to be the probability of applying the rule $\alpha \rightarrow \beta$. The characteristic grammar $\overline{G_s}$ of the stochastic grammar G_s is just the deterministic grammar formed by removing the probabilities from all rules in R_s. Thus stochastic grammars are assigned to classes of the Chomsky hierarchy according to the classes of their respective characteristic grammars. If $W \in \mathcal{L}(\overline{G_s})$, then $W \in \mathcal{L}(G_s)$ and W has the probability $P(W)$. If G_s is unambiguous, that is, there exists exactly one derivation for each $W \in \mathcal{L}(G)$, then $P(W)$ is just the product of the probabilities associated with the rules from which $S \overset{*}{\Rightarrow} W$. That is,

$$P(W) = \sum_{S \overset{*}{\Rightarrow} W} \prod_{t} p_{\alpha_{t-1}\alpha_t}. \tag{3.237}$$

If

$$\sum_{W \in \mathcal{L}(G_s)} P(W) = 1, \tag{3.238}$$

the grammar G_s is said to be consistent. All regular stochastic grammars are consistent. The conditions under which stochastic context-free grammars are consistent are known [34]. Consistency conditions for more complex grammars are not known. A summary of the theory of stochastic grammars is available in [94].

In the case of stochastic languages, $W \in \mathcal{L}(G_s)$ if and only if $P(W) > \theta$. This determination can be made for the regular and context-free classes by the algorithms discussed in Chapter 4.

3.2.3 Equivalence of Regular Stochastic Grammars and Discrete HMMs

There is an important relationship between doubly stochastic processes and stochastic grammars. First, discrete-symbol HMMs are equivalent to regular stochastic grammars. Production rules of the form $A_i \overset{p_{ij}}{\rightarrow} vA_j$ account for the hidden Markov chain in that the non-terminal symbols A_i, A_j correspond to states q_i, q_j. Productions of the form $A_j \overset{p_{jk}}{\rightarrow} v_k$ correspond to the observable process, so that p_{jk} is equivalent to $b_{jk} = b_j(O_t)$ when $O_t = v_k \in V_T$.

Note that the likelihood function of the discrete symbol HMM for the observation sequence $\mathbf{O} = O_1 O_2 \ldots O_T$ can be written as

$$\mathcal{L}(\mathbf{O}|\lambda) = \sum_{\mathbf{q} \in Q^T} \prod_{t=1}^{T} a_{q_{t-1}q_t} b_{q_t}(O_t). \tag{3.239}$$

If we identify the sentence $W \in \mathcal{L}(G)$ with the observation sequence, $\mathbf{O}$, meaning that $O_t \in V_T$, then we can compare (3.226) and (3.229) to see that

$$p_{\alpha_t \alpha_{t+1}} = a_{q_t q_{t+1}} b_{q_{t+1}}(\theta_{t+1}). \tag{3.240}$$

Going in the other direction has no unique solution. If the $p_{\alpha\beta}$ are known then the a_{ij} and b_{jk} can be assigned any values so that their product is $p_{\alpha\beta}$.

A stochastic context-free grammar is not equivalent to any HMM but it has a related structure. Rules of the form $A_i \stackrel{p_{ijk}}{\rightarrow} A_j A_k$ constitute a hidden stochastic process, and rules of the form $A_i \stackrel{A_k}{\rightarrow} U_k$ an observable one.

3.2.4 Recognition of Well-Formed Strings

Given any sequence of terminal symbols, $W \in V^*$, we would like to determine whether or not $W \in \mathcal{L}(G)$. In the cases of regular and context-free grammars, this question is answered by abstract automata called finite-state automata (FSA) and push-down automata (PDA), respectively.

An FSA is a labeled, directed graph in which the vertices (states) are labeled with non-terminal symbols, and the edges or state transitions are labeled with terminal symbols. For every rule of the form $A \rightarrow aB$, the FSA contains a state transition labeled a joining the two states labeled A and B. This is depicted in Fig. 3.12. A state, labeled A, may be designated as a final state if there is a rule of the form $A \rightarrow a$. Then $W \in \mathcal{L}(G)$ if and only if there is a path from state S to a final state whose edges are labeled, in order, with the symbols of W.

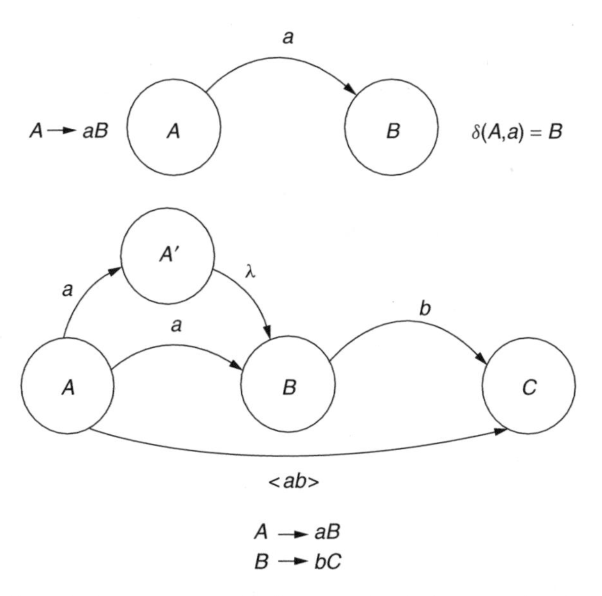

Figure 3.12 The relation between right-linear production rates and finite-state automata

For the context-free case, the PDA can construct a tree such as the one shown in Fig. 3.11 whenever the string, W, is well formed.

For the regular and context-free cases, the FSA and PDA, respectively, are minimal automata in the sense that they are the least complex machines that can answer the membership question. As such they are rather inefficient for determining the structure (e.g. derivation) of $W \in \mathcal{L}(G)$. For that reason, we will not consider them further. In Chapter 4 we will again address these questions in computationally efficient algorithms not based on either the FSA or PDA.

3.2.5 Representation of Phonology and Syntax

Formal grammars were developed, at least in part, to provide an abstract representation of the grammar of natural language. The definitions given above can be made more intuitively appealing by emphasizing that motivation.

Context-sensitive grammars are ideally suited for describing phonological phenomena such as the way the pronunciation of a given sound changes because of its phonetic environment. Rules of the form $\alpha A \beta \longrightarrow \alpha \gamma \beta$ exactly capture this effect since they may be interpreted to mean that the phones derived from the non-terminal A are rendered as the

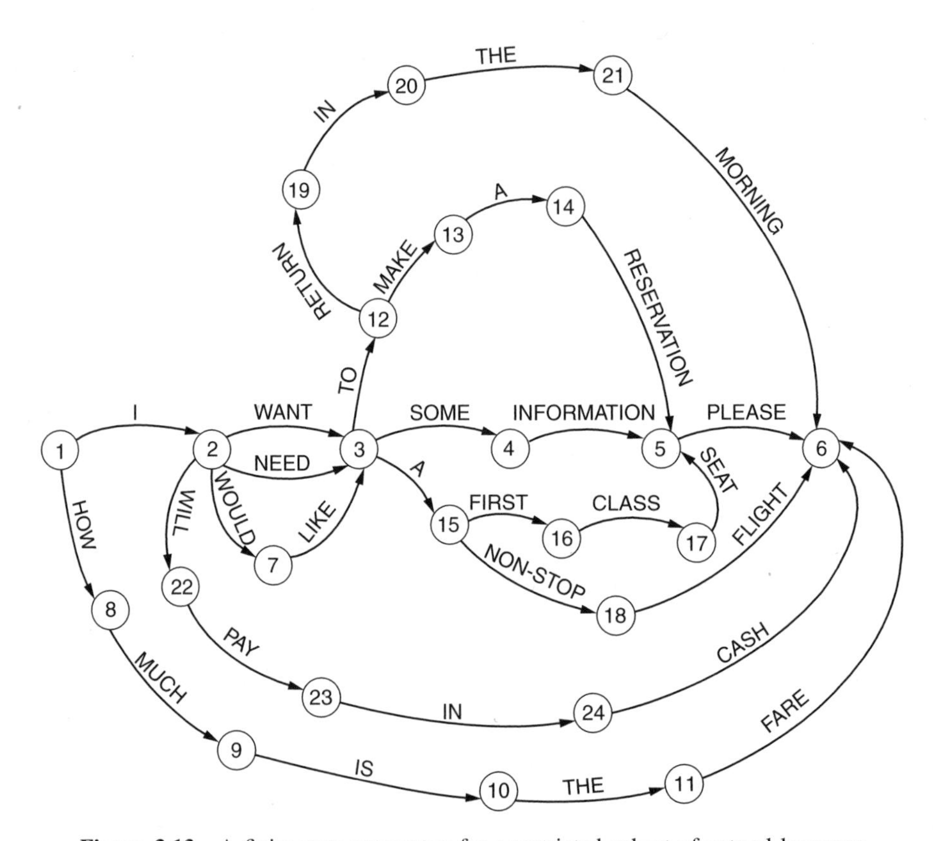

Figure 3.13 A finite state automaton for a restricted subset of natural language

specific phones derived from the string γ (thereby allowing for insertions and deletions) when the preceeding and succeeding sounds derive from α and β, respectively. Cohen and Mercer [50] have compiled a comprehensive grammar for American English.

Regular languages are appropriate for the representation of carefully circumscribed subsets of natural language applicable to a particular limited domain of discourse. In Fig. 3.13 we have used ordinary vocabulary words for the terminal symbols. We can thus generate an English sentence by starting at node 1 and following any path to either node 5 or node 6, reading the associated labels as we traverse each edge. For obvious reasons, such a directed graph is often called a state-transition diagram of the grammar.

For context-free rules, an approximation of a natural language may be obtained. For example, non-terminal symbols can be thought of as generalized parts of speech. In particular, the start symbol S represents the part of speech ⟨sentence⟩. (The angle brackets are

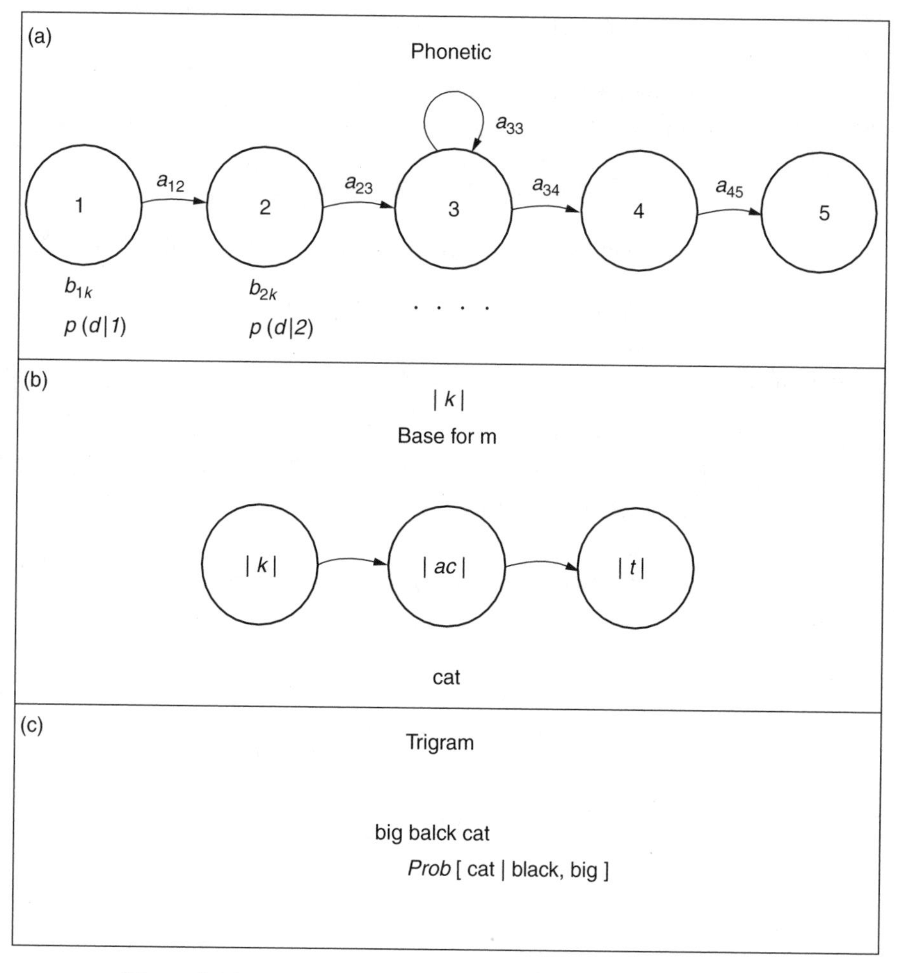

Figure 3.14 The non-ergodic HMM used for speech recognition

used to distinguish the word they enclose from the same vocabulary word.) The production rule ⟨sentence⟩ → ⟨subject⟩ ⟨predicate⟩ may be interpreted to mean that ⟨sentence⟩ may have two components called ⟨subject⟩ and ⟨predicate⟩ which appear in that order. A grammar of English might also contain the rules ⟨subject⟩ → ⟨adjective⟩ ⟨noun⟩ and ⟨predicate⟩ → ⟨verb⟩ ⟨adverb⟩. If we also include the rules ⟨adjective⟩ → white, ⟨noun⟩ → horses, ⟨verb⟩ → run, and ⟨adverb⟩ → fast, then we can produce the sentence "white horses run fast". Note that the absence of the angle brackets signifies terminal symbols which are, in this case ordinary members of the English lexicon. If we now add further rules which invite the replacement of the names of the parts of speech by specific vocabulary words of that type, then the rules given above can be used to produce many sentences having the same grammatical structure; namely, ⟨adjective⟩ ⟨noun⟩ ⟨verb⟩ ⟨adverb⟩.

The context-free rule discussed in Section 3.2.1 of the form $S \longrightarrow aSb$ is, for obvious reasons, called a "center embedding" rule. This rule exists in ordinary English. Start with the sentence

The rat ate the cheese,

which is easily derived from the context-free rules above.

Now use the center embedding rule to add a modifier of "rat" giving the sentence

The rat the cat chased ate the cheese.

Similarly, adding a modifier for cat we get an unusual but well-formed sentence,

The rat the cat the dog bit chased ate the cheese.

Even more unwieldy sentences can be created by recursive application of center embedding.

Stochastic grammars can also be used to represent ordinary syntax. This is usually done in the form of a Markov chain as shown in Fig. 3.14c. The word order probability $P(W_n|W_{n-1}, W_{n-2})$ is equivalent to a set of production rules of the form $A_1 \stackrel{p_1}{\rightarrow} W_{n-2}A_2$, $A_2 \stackrel{p_2}{\rightarrow} W_{n-1}A_3$, and $A_3 \stackrel{p_3}{\rightarrow} W_nA_4$.

We will return to a more detailed consideration of these ideas in Section 4.3.

4

Syntactic Analysis

4.1 Deterministic Parsing Algorithms

A particular problem addressed by formal language theory that is directly relevant to our discussion is that of parsing sentences in a language. Specifically, given G and $W \in V_T^*$, we wish to know whether or not $W \in \mathcal{L}(G)$ and, if so, by what sequences of rules $S \overset{*}{\Rightarrow} W$.

Recall from Section 3.2.4 that these questions can be answered by different kinds of automata for languages in the different complexity classes of the Chomsky hierarchy. In particular, right-linear or regular languages can be analyzed by finite-state automata, and context-free languages by push-down automata.

Unfortunately, the conceptual simplicity of these machines makes them inefficient analyzers of their respective languages in the most general cases. There are, however, optimally efficient parsers that can be implemented on more complex machines such as general-purpose computers or other Turing-equivalent models of computation. We shall consider the significance of this fact in Chapter 9.

4.1.1 The Dijkstra Algorithm for Regular Languages

The optimal general parser for regular languages is obtained by casting the parsing problem as an optimization problem. Suppose $W \in V_T^*$, $W = w_1 w_2 \ldots w_n$. Each $w_j \in W$ corresponds to the interval from t_{j-1} to t_j in the speech signal and has the cost $C(w_j)$, and the sentence W has total cost $C(W)$ given by

$$C[W] = \min_{V \in V_T^*} \left\{ \sum_{j=1}^{|W|} \mathcal{C}[v_j \mid t_{j-1}, t_j] \right\} \tag{4.1}$$

(see Fig. 4.1). Parsing symbol strings has no explicit notion of time. Only word order is required. Hence, the interval (t_{j-1}, t_j) can be replaced by the word order index, j. If we let $C(v_j) = 0$ if and only if $v_j = w_j$ and 1 otherwise, then $C(W) = 0$ if and only if $W \in \mathcal{L}(G)$. In the process of computing $C(W)$, we will get $S \overset{*}{\Rightarrow} W$ as a by-product.

Mathematical Models for Speech Technology. Stephen Levinson
 ISBN: 0-470-84407-8

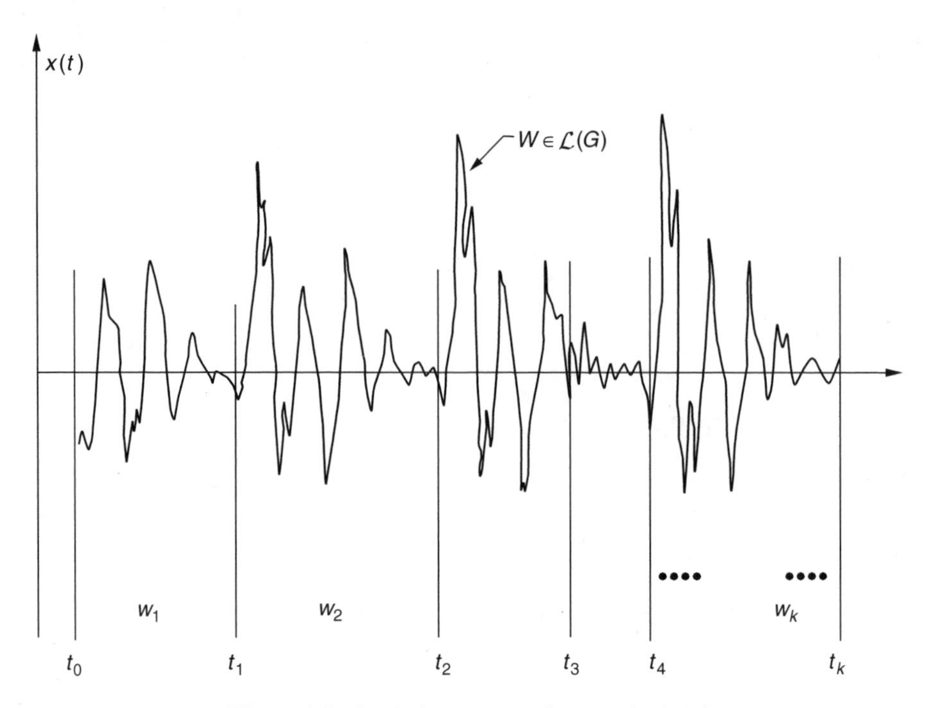

Figure 4.1 Lexical segments of a speech signal

The solution to this problem can be obtained by a dynamic programming algorithm due to Dijkstra [65]. In the case of regular grammars, let $\psi_j(B)$ be the prefix of length j of some $W \in \mathcal{L}(G)$ having minimum cost, and denote by $\phi_j(B)$ its cost. Initially $\psi_0(B) = \lambda \forall B \in V_N$ and $\phi_0(B) = 0$ if and only if $B = S$, and ∞ otherwise. Then for $1 \leq j \leq |W|$ and $\forall B \in V_N$,

$$\phi_j(B) = \min_{\{A \to aB\}} \{\phi_{j-1}(A) + \mathcal{C}[a \mid t_{j-1}, t_j]\} \tag{4.2}$$

and

$$\psi_j(B) = \psi_{j-1}(A) \otimes \overline{a}, \tag{4.3}$$

where

$$\overline{a} = \arg \min_{\{A \to aB\}} \{\phi_j(B)\}. \tag{4.4}$$

Then $\hat{W} = \psi_{|W|}(\lambda)$ and $\mathcal{C}[\hat{W}] = \phi_{|W|}(\lambda)$. For notational convenience we have assumed that $\lambda \in V_N$. Note that (4.2) and (4.3) are similar to the Viterbi algorithm (3.7) and (3.8) except that the set over which the optimization occurs is different. Note also that in both algorithms the required number of operations is proportional to $|W| \cdot |V_N|$.

4.1.2 *The Cocke–Kasami–Younger Algorithm for Context-free Languages*

The context-free case is based on the general context-free parsing algorithm of Younger [338]; cf. [101]. There is an added complication in this case since the grammar must first be transformed into Chomsky normal form [133, pp. 51ff.] so that the rules are of either the form $A \rightarrow BC$ or $A \rightarrow a$ (see Section 3.2.1).

Let $\psi_{ij}(A)$ be the string, α, spanning the ith to the jth word position in W of minimum cost such that $S \stackrel{*}{\Rightarrow} \alpha$, and let $\phi_{ij}(A)$ be its cost. Initially,

$$\phi_{ii}(A) = \min_{\{A \rightarrow a\}} \{\mathcal{C}[a \mid t_{i-1,i}]\} \tag{4.5}$$

and

$$\psi_{ii}(A) = \overline{a} = \arg \min_{\{A \rightarrow a\}} \{\phi_{ij}(A)\} \tag{4.6}$$

for $1 \leq i \leq |W|$. All other values of ϕ and ψ are undefined at this point. Then for $1 \leq i, j \leq |W|$,

$$\phi_{ij}(A) = \min_{\{A \rightarrow BC\}} \left\{ \min_{i \leq l < j} \{\phi_{il}(B) \cdot \phi_{l+1,J}(C)\} \right\} \tag{4.7}$$

and

$$\psi_{ij}(A) = \psi_{il} \otimes \psi_{l+1,j}(\overline{C}), \tag{4.8}$$

where

$$(\overline{l}, \overline{B}, \overline{C}) = \arg \min_{\substack{\{A \rightarrow BC\} \\ i \leq l < j}} \{\phi_{ij}(A)\}. \tag{4.9}$$

The derivation $S \stackrel{*}{\Rightarrow} W$ can be reconstructed from (4.9) according to

$$(l, B, C) = \psi_{1N}(s), \tag{4.10}$$

where $N = |W|$. Then for the left subtrees

$$(l, B, C) = \psi_{1l}(B), \tag{4.11}$$

and for the corresponding right subtrees

$$(l, B, C) = \psi_{l+1,N}(l, B, C) \tag{4.12}$$

Finally, $\hat{W} = \phi_{1|W|}(S)$ and $\mathcal{C}[\hat{W}] = \psi_{1|W|}(S)$. The operation of this algorithm is shown in Fig. 4.2. Note that the number of operations required by (4.7)–(4.9) is proportional to $|W|^3$ and in fact the algorithm itself is a form of matrix multiplication [120, pp. 442 ff.]. An important application of this algorithm will appear in Section 5.4.

As we move further up the Chomsky hierarchy, the computational complexity of parsing algorithms increases drastically. An algorithm similar to (4.7)–(4.9) for context-sensitive grammars was given by Tanaka and Fu [310] but has an operation count that is exponential in $|W|$.

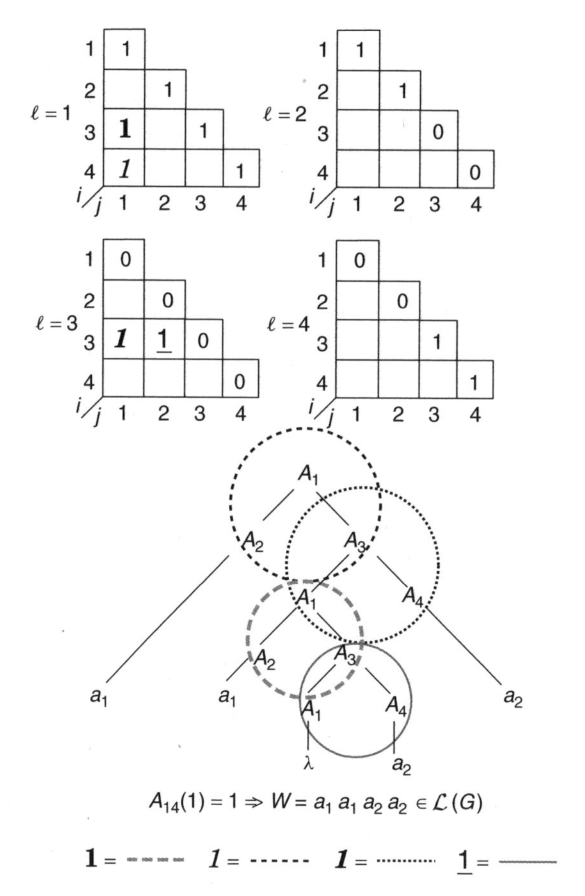

Figure 4.2 An example of the Cooke–Kasami–Younger algorithm

4.2 Probabilistic Parsing Algorithms

In the following cases we shall assume that word boundaries t_j are known so that only word order, j, is important. However, the words themselves are only known probabilistically.

4.2.1 Using the Baum Algorithm to Parse Regular Languages

Recall from Section 3.2.3 that an HMM is equivalent to a regular stochastic grammar. Recall also from Section 3.2.4 that regular grammars have an equivalent FSA the transitions of which correspond to the production rules of the equivalent grammar. From these observations we conclude that finding the state transitions in an HMM is tantamount to parsing a sentence.

From Section 3.1.1 we know that the state, q_t, at time t is found from

$$q_t = \arg \max_{1 \leq i \leq N} \{\alpha_t(i)\beta_t(i)\}. \tag{4.13}$$

Determination of the state sequence from (4.13) is equivalent to parsing since each q_t corresponds to some $A_i \in V_N$. A state transition from q_i to q_j at time t corresponds to the rule $A_i \rightarrow w_t A_j$. The a_{ij} are assumed known and $b_j(O_t)$ is the probability that w_t is the word v_k given q_j at t.

4.2.2 Dynamic Programming Methods

The type of parser system we are concerned with is shown in Fig. 4.3, and its operation is formally described below.

Let the language, $\mathcal{L}$, be the subset of English used in a particular speech recognition task. Sentences in $\mathcal{L}$ are composed from the vocabulary, V, consisting of the M words $v_1, v_2, \ldots, v_M$. Let $\mathbf{W}$ be an arbitrary sentence in the language. Then we write $\mathbf{W} \in \mathcal{L}$ and

$$\mathbf{W} = w_1 w_2 \cdots w_k, \tag{4.14}$$

where each w_i is a vocabulary word which we signify by writing $w_i \in V$ for $1 \leq i \leq k$. Clearly $\mathbf{W}$ contains k words, and we will often denote this by writing $|\mathbf{W}| = k$. Similarly, the number of sentences in $\mathcal{L}$ will be denoted by $|\mathcal{L}|$.

The sentence $\mathbf{W}$ of (4.14) is encoded in the speech signal $x(t)$ and input to the acoustic recognizer from which is obtained the probably corrupted string

$$\tilde{\mathbf{W}} = \tilde{w}_1 \tilde{w}_2 \cdots \tilde{w}_k, \tag{4.15}$$

where $\tilde{w}_i \in V$ for $1 \leq i \leq k$ but $\tilde{\mathbf{W}}$ is not, in general, a sentence in $\mathcal{L}$.

The acoustic recognizer also produces the matrix $[d_{ij}]$ whose ijth entry, d_{ij}, is the distance, as measured by some metric in an appropriate pattern space, from the ith word, $\tilde{w}_i$, to the prototype for the jth vocabulary word, v_j, for $1 \leq i \leq k$ and $1 \leq j \leq \mathbf{M}$.

The syntax analyzer then produces the string

$$\hat{\mathbf{W}} = \hat{w}_1 \hat{w}_2 \cdots \hat{w}_k, \tag{4.16}$$

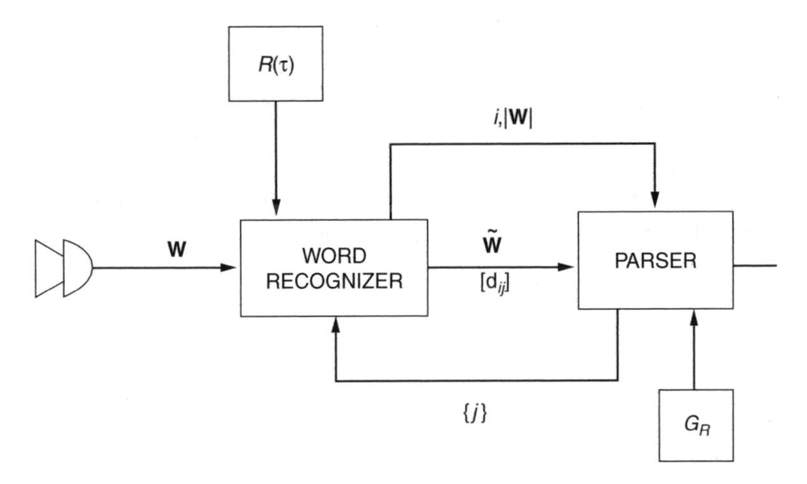

Figure 4.3 Maximum likelihood parsing system

for which the total distance, $D(\hat{\mathbf{W}})$, given by

$$D(\hat{\mathbf{W}}) = \sum_{i=1}^{k} d_{ij_i}, \qquad 1 \leq j_i \leq M, \tag{4.17}$$

is minimized subject to the constraint that $\hat{\mathbf{W}} \in \mathcal{L}$. Thus the syntax analysis is optimal in the sense of minimum distance.

Since, in general, $\tilde{\mathbf{W}} \notin \mathcal{L}$, whereas $\mathbf{W}$ was assumed to be grammatically well-formed (i.e. $\mathbf{W} \in \mathcal{L}$), the process should correct word recognition errors.

In principle one could minimize the objective function of (4.17) by computing $D(\mathbf{W})$ for all $\mathbf{W} \in \mathcal{L}$ and choosing the smallest value. In practice, when $|\mathcal{L}|$ is large, this is impossible. One must perform the optimization efficiently. It has been shown by Lipton and Snyder [201] that for a particular class of languages one can minimize $D(\mathbf{W})$ in time proportional to $|W|$. In fact one can optimize any reasonable objective function in time linear in the length of the input.

The particular class of languages for which the efficiency can be attained is called the class of regular languages. For the purposes of this discussion we shall define the class of regular languages as that class for which each member language can be represented by an abstract graph called a state transition diagram.

A state transition diagram consists of a finite set of vertices or states, Q, and a set of edges or transitions connecting the states. Each such edge is labeled with some $v_i \in V$. The exact manner of the interconnection of states is specified symbolically by a transition function, δ, where

$$\delta\colon (Q \times V) \rightarrow Q. \tag{4.18}$$

That is, if a state $q_i \in Q$ is connected to another state $q_j \in Q$ by an edge labeled $v_m \in V$, then

$$\delta(q_i, v_m) = q_j. \tag{4.19}$$

We also define a set of accepting states, $Z \subset Q$, which has the significance that a string $\mathbf{W} = w_1 w_2 \cdots w_k$, where $w_i \in V$ for $1 \leq i \leq k$, is a well-formed sentence in the language, $\mathcal{L}$, represented by the state transition diagram if and only if there is a path starting at q_1 and terminating in some $q_j \in Z$ whose edges are labeled, in order, $w_1, w_2, \ldots, w_k$.

Alternatively, we may write $\mathbf{W} \in \mathcal{L}$ if and only if

$$\begin{aligned} \delta_1(q_1, w_1) &= q_{j1}, \\ \delta_2(q_j, w_2) &= q_{j2}, \\ &\vdots \\ \delta_k(q_{jk-1}, w_k) &= q_{jk} \in Z. \end{aligned} \tag{4.20}$$

We may then define the language, $\mathcal{L}$, as the set of all $\mathbf{W}$ satisfying (4.17). An example of these concepts is shown in Fig. 4.4. The accepting states are marked by asterisks.

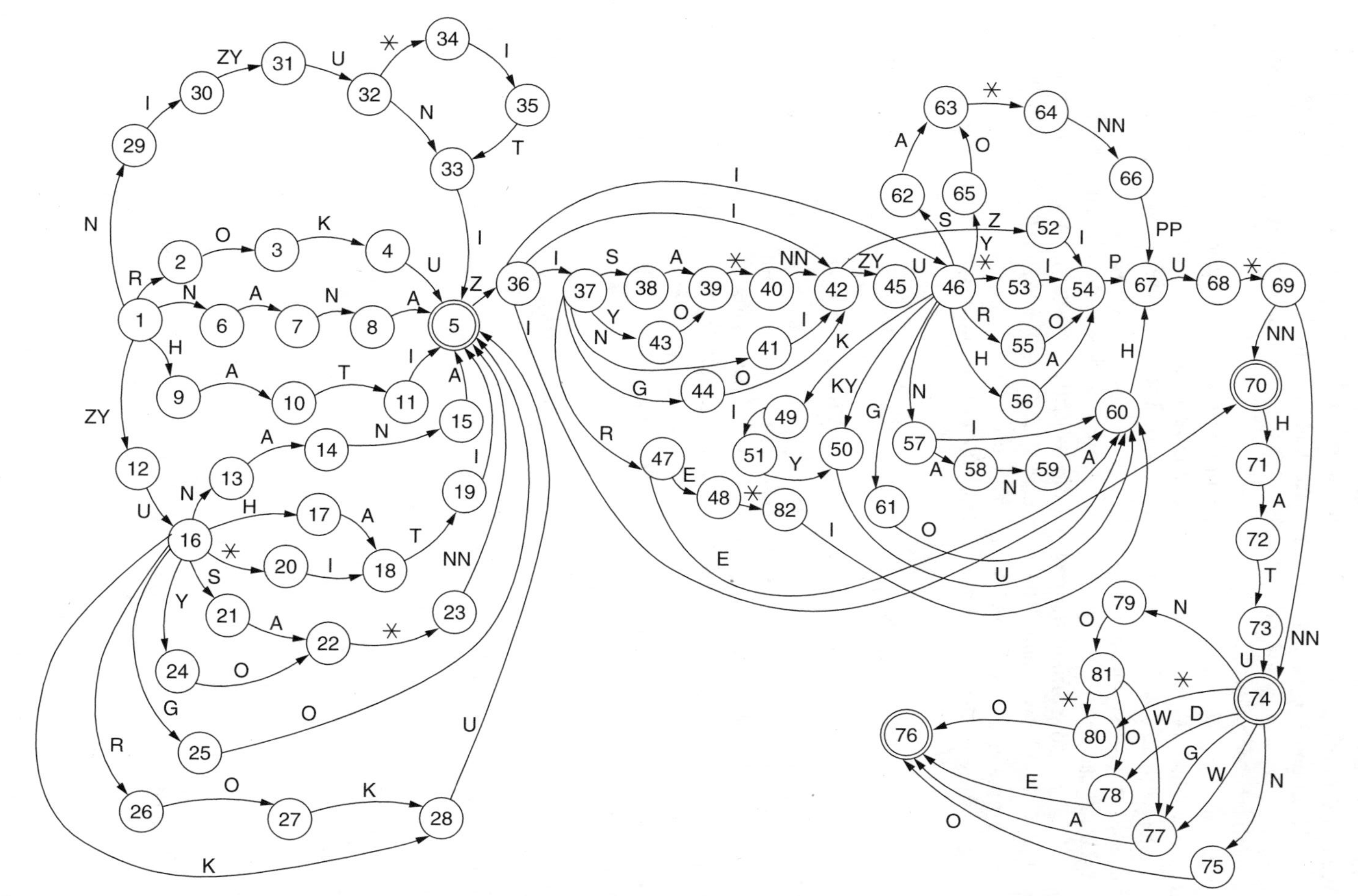

Figure 4.4 A finite state automaton for the phonotactics of a fragment of Japanese

While the definition given above of a regular language is mathematically rigorous, it is not the standard one used in the literature on formal language theory but rather has been specifically tailored to the notational requirements of this section. The interested reader is urged to refer to Hopcroft and Ullman [133] for a standard and complete introduction to formal language theory.

In the following discussion we shall restrict ourselves to finite regular languages, that is, those for which $|\mathcal{L}|$ is finite. This restriction in no way alters the theory but its practical importance will become obvious in what follows. The finiteness of the language implies that its state transition diagram has no circuits, that is, no paths of any length starting and ending at the same state. Thus there is some maximum sentence length which we shall denote, $l_{\max}$. We now turn to the problem of efficiently solving the minimization problem of (4.17). To do this we shall define two data structures, Φ and Ψ, which will be used to store the estimates of $D(\hat{\mathbf{W}})$ and $\hat{\mathbf{W}}$, respectively.

The first stage of the algorithm is the initialization procedure in which we set

$$\Phi_i(q) = \begin{cases} 0, & \text{for } q = q_0 \text{ and } i = 0, \\ \infty & \text{otherwise,} \end{cases} \tag{4.21}$$

$$\Psi_i(q) = 0, \qquad 1 \le i \le |\mathbf{W}| = k, \forall q \in Q. \tag{4.22}$$

The data structures have two indices. The subscript is the position of the word in the sentence and the argument in parentheses refers to the state so that the storage required for each array is, at most, the product of $l_{\max} + 1$ and $|Q|$, the number of states in the set Q.

After initialization we utilize a dynamic programming technique defined by the following recursion relations:

$$\Phi_i(q) = \min_{\Delta}\{\Phi_{i-1}(q_p) + d_{ij}\} \tag{4.23}$$

where the set Δ is given by

$$\Delta = \{\delta(q_p, v_j) = q\}. \tag{4.24}$$

Then

$$\Psi_i(q) = \Psi_{i-1}(q_p)\hat{w}_i, \tag{4.25}$$

where $\hat{w}_i$ is just the v_j which minimizes $\Phi_i(q)$. Equation (4.25) is understood to mean that the word $\hat{w}_i$ is simply appended to the string $\Psi_{i-1}(q_p)$.

Unfortunately, the concatenation operation is not easily implemented on general-purpose computers so we change the recursion of (4.25) by making Ψ into a linked list structure of the form

$$\Psi_{1i}(q) = q_p, \tag{4.26}$$

$$\Psi_{2i}(q) = \hat{w}_i. \tag{4.27}$$

Then when $i = k$ we can trace back through the linked list of (4.27) and construct the sentence $\hat{\mathbf{W}}$ as follows: First, find $q_f \in Z$ such that

$$\phi_k(q_f) = \min_{q \in Z}\{\Phi_k(q)\}, \tag{4.28}$$

set $q = q_f$ and then, for $i = k, k-1, k-2, \ldots, 1$,

$$\hat{w}_i = \Psi_{2i}(q), \tag{4.29}$$

$$q = \Psi_{1i}(q). \tag{4.30}$$

Thus the sentence $\hat{\mathbf{W}}$ is computed from right to left.

The operation of the above algorithm is illustrated in Figs 4.5–4.7. Figure 4.5 shows the vocabulary words of the language diagrammed in Fig. 4.8 along with numerical codes and a sample $[d_{ij}]$ matrix. Figure 4.6 shows the details of the operation of the algorithm for $i = 0, 1, 2$. Figure 4.7 shows the results after the sentence has been completely analyzed.

By locating the smallest entry in each column of the sample $[d_{ij}]$ matrix of Fig. 4.7, it can be seen that the acoustic transcription of the sentence from which this matrix was produced is WOULD SOME IS TO FARE. Clearly this is not a valid English sentence nor is there any path through the state transition diagram of Fig. 4.8 whose edges are so labeled.

Following Fig. 4.6 the reader can trace the operation of the algorithm as it computes the valid sentence having the smallest total distance. First the Φ and Ψ arrays are initialized according to (4.21). To make the figure easier to read, this has been shown only for $i = 0$.

Note that there are two transitions from state 1: one to state 2 labeled I, and the other to state 8 labeled HOW. Accordingly, $\Phi_1(2)$ is set to 7, the metric for I; $\Psi_{11}(2)$ is set to 1, the state at the beginning of the transition and Ψ_{21} is set to 3, the code for the transition label I. Similarly, $\Phi_1(8)$ is set to 2, the metric for HOW; $\Psi_{11}(8)$ is set to 1 as before and $\Psi_{21}(8)$ is set to 26, the code for HOW. All other entries remain unchanged.

In the next stage more transitions become possible. Note, in particular, that there are two possible transitions from state 2 to state 3. In accordance with (4.23), the one labeled NEED is chosen since it results in the smallest total distance, 13, which is entered in $\Phi_2(3)$; $\Psi_{12}(3)$ is set to 2 the previous state, and $\Psi_{22}(3)$ is set to 28, the code for NEED. Transitions to states 7, 9, and 22 are also permissible and thus these columns are filled in according to the same procedure.

The completion of this phase of the algorithm results in Φ and Ψ as shown in Fig. 4.7. We can now trace back according to (4.29) and (4.30) to find $\hat{\mathbf{W}}$. From Fig. 3.13 we see that there are two accepting states, 5 and 6. $\Phi_5(6)$ is 8 which is less than 30, the value of $\Phi_5(5)$, so we start tracing back from state 6. The optimal state sequence is, in reverse order, 6, 11, 10, 9, 8, 1. The corresponding word codes which, when reversed, decode to the sentence: HOW MUCH IS THE FARE.

One final note: from the operation of the algorithm it should be clear that it is not necessary to retain $\Phi_i(q)$ for $0 \le i \le |\mathbf{W}|$. At the ith stage one needs only $\Phi_{i-1}(q)$ to compute $\Phi_i(q)$. Thus the storage requirements are nearly halved in the actual implementation.

In closing, we should note that this scheme is formally the same as (though conceptually different from) the Viterbi [323] algorithm and similar to methods used by Baker [19] and

(a)

CODE	VOCABULARY WORD	$I = 1$	$I = 2$	$I = 3$	$I = 4$	$i = 5$
1	IS	9	9	1	8	4
2	FARE	2	2	5	3	1
3	I	7	3	2	3	2
4	WANT	2	9	4	7	3
5	WOULD	1	5	4	8	2
6	LIKE	2	5	2	6	5
7	SOME	2	1	9	8	3
8	INFORMATION	7	7	4	7	8
9	PLEASE	2	3	2	4	9
10	TO	4	5	8	1	7
11	MAKE	6	3	9	8	5
12	A	4	7	6	9	8
13	RESERVATION	3	6	7	8	9
14	RETURN	9	7	6	4	8
15	THE	8	6	5	2	3
16	MORNING	3	4	5	6	7
17	FIRST	8	6	8	7	5
18	CLASS	5	5	4	3	9
19	SEAT	9	9	8	7	3
20	NON-STOP	3	3	4	5	8
21	FLIGHT	9	8	3	5	6
22	WILL	6	7	7	6	5
23	PAY	4	4	4	4	3
24	IN	3	3	3	6	9
25	CASH	5	4	3	7	6
26	HOW	2	9	8	7	5
27	MUCH	6	2	8	4	9
28	NEED	7	6	5	4	3

(b)

		POSITION	
CODE	WORD	1	2
3	I	7	3
4	WANT	2	9
28	NEED	7	6
5	WOULD	1	5
22	WILL	6	7
26	HOW	2	9
27	MUCH	6	2

Figure 4.5 (a) The full distance matrix for a five-word sentence. (b) Distance matrix for the first two words of a five word sentence

i \ q	1	2	3	4	5	6	7	8	9	10	11	12	13	14	15	16	17	18	19	20	21	22	23	24
ϕ_0	0	∞	∞	∞	∞	∞	∞	∞	∞	∞	∞	∞	∞	∞	∞	∞	∞	∞	∞	∞	∞	∞	∞	∞
ψ_{10}	0	0	0	0	0	0	0	0	0	0	0	0	0	0	0	0	0	0	0	0	0	0	0	0
ψ_{20}	0	0	0	0	0	0	0	0	0	0	0	0	0	0	0	0	0	0	0	0	0	0	0	0
ϕ_1		7						2																
ψ_{11}		1						1																
ψ_{21}		3						26																
ϕ_2			13				12		4													14		
ψ_{12}			2				2		8													2		
ψ_{22}			28				5		27													22		

Figure 4.6 The recognition matrix Φ for the first two words of a sentence

i \ q	1	2	3	4	5	6	7	8	9	10	11	12	13	14	15	16	17	18	19	20	22	23	24
ϕ_0	0	∞	∞	∞	∞	∞	∞	∞	∞	∞	∞	∞	∞	∞	∞	∞	∞	∞	∞	∞	∞	∞	∞
ψ_{10}	0	0	0	0	0	0	0	0	0	0	0	0	0	0	0	0	0	0	0	0	0	0	0
ψ_{20}	0	0	0	0	0	0	0	0	0	0	0	0	0	0	0	0	0	0	0	0	0	0	0
ϕ_1		7						2															
ψ_{11}		1						1															
ψ_{21}		3						26															
ϕ_2			13				12		4												14		
ψ_{12}			2				2		8												2		
ψ_{22}			28				5		27												22		
ϕ_3			14	22						5		21			19							18	
ψ_{13}			7	3						9		3			3							22	
ψ_{23}			6	7						1		10			12							23	
ϕ_4				22	29						7	15	29		23	26		24	25				24
ψ_{14}				3	4						10	3	12		3	15		15	12				23
ψ_{24}				7	8						15	10	11		12	17		20	14				24
ϕ_5					30	8							20	37		28	35	31	23	34			
ψ_{15}					4	11							12	13		15	16	15	12	19			
ψ_{25}					8	2							11	12		17	18	20	14	24			

Figure 4.7 The complete recognition matrix Φ for a five-word sentence

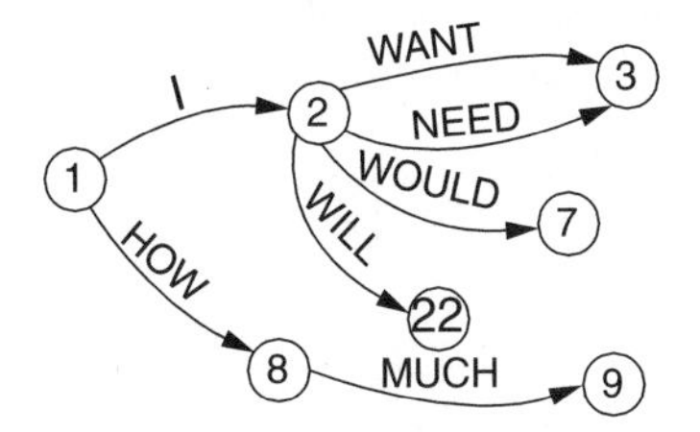

Figure 4.8 FSA for the first two words of the English fragment

stochastic parsing techniques discussed in Fu [96] and Paz [243]. The crucial difference is that in the cited references estimates of transitions probabilities are used, whereas in this method the transitions are deterministic and the probabilities used are only those conditioned on the input $x(t)$.

4.2.3 Probabilistic Cocke–Kasami–Younger Methods

The Cocke–Kasami–Younger algorithm of (4.5)–(4.12) can be adapted to the case in which the words are known only probabilistically. The only modification required is that the initialization procedure, (4.5) becomes

$$\phi_{ii}(A) = \min_{\{A \to a\}} \{-\log P(w_i = a|x(t))\}, \tag{4.31}$$

where $P(w_i = a|x(t))$ is the probability that the ith word in the sentence is the word $a \in V_T$ given the signal, $x(t)$. Once this initialization has been accomplished, the rest of the algorithm is exactly as described in Section 4.1.2.

4.2.4 Asynchronous Methods

The most difficult case of probabilistic parsing is the one in which there is uncertainty with respect to both the words in the sentence and the word boundaries in the speech signal. Unknown word boundaries force us to abandon the simple notion of word order and replace word indicies with actual time intervals in the signal. This greatly increases the computation complexity.

In this case, as in Section 4.2, the sentence W is assumed to be the word string $v_1 v_2 \ldots v_n$ but the words are only known probabilistically. Now, however, we will assume that W is encoded in the speech signal $x(t)$ which is represented by a sequence of measured feature vectors $\mathbf{O} = \mathbf{O}_1, \mathbf{O}_2, \ldots, \mathbf{O}_t, \ldots, \mathbf{O}_T$. Each $\mathbf{O}_t$ occurs at real time t and $\mathbf{O}_t \in \mathbb{R}$. Thus the word v_j corresponds to some subsequence of $\mathbf{O}, \mathbf{O}_{t_1}, \mathbf{O}_{t_2}, \ldots, \mathbf{O}_{t_m}$. The problem is asynchronous in the sense that words can be of different durations and the hypothetical word v_j can end at a different time from an alternative, w_j.

Then the parsing problem becomes finding

$$\hat{W} = \arg \max_{W \in \mathcal{L}(G)} \left\{ \arg\max_{t, \Delta t} \left\{ \sum_j C(w_j | t, t + \Delta t) \right\} \right\}. \tag{4.32}$$

For regular languages the maximization problem of (4.32) can be solved by modifying (4.23) to account for time instead of word index. Thus

$$\phi_t(j) = \min_{\{A_i \overset{p_{ij}}{\to} v A_j\}} \{ \min_{\tau < t} \{ \phi_{t-\tau}(i) - \log(\mathcal{L}(\mathbf{O}_{t-\tau}, \mathbf{O}_{t-\tau+1}, \ldots, \mathbf{O}_t | \boldsymbol{\lambda}_v)) - \log(p_{ij}) \} \},$$
$$t = 1, 2, \ldots T. \tag{4.33}$$

In (4.33) $\log(\mathcal{L}(\mathbf{O}_{t-\tau}, \mathbf{O}_{t-\tau+1}, \ldots \mathbf{O}_t | \boldsymbol{\lambda}_v))$ is the cost, $C(v, t - \tau, t)$, of classifying the interval from $t - \tau$ to t as word v. The likelihood function $\mathcal{L}(\mathbf{O}_{t-\tau}, \mathbf{O}_{t-\tau+1}, \ldots \mathbf{O}_t | \boldsymbol{\lambda}_v)$ is the likelihood function of an HMM for word v and having parameter vector $\boldsymbol{\lambda}_v$.

The solution to the context-free case is a variant on (4.7). Initially we set

$$\phi_{t_I t_F}(A) = \min_{\substack{p_{Av} \\ \{A \to v\}}} \{C(v, t_I, t_F) - \log p_{Av}\}, \qquad \text{for } 1 \leq t_I < t_F \leq T. \tag{4.34}$$

Then (4.34) becomes

$$\phi_{t_I t_F}(A) = \min_{\substack{p_{ABC} \\ \{A \to BC\}}} \{ \min_{t_I < \tau < t_F} \{\phi_{t_I \tau}(B) + \phi_{\tau+1 t_F}(C) - \log P_{ABC}\}\}, \text{ for } 1 \leq t_I < t_F \leq T. \tag{4.35}$$

The optimal $\hat{W}$ is recovered as in (4.8)–(4.12).

In both the regular and context-free cases, we may carry out the computations exhaustively, allowing any word v to be associated with any interval of $x(t)$–i.e. any subsequence of $\mathbf{O}$–as long as the intervals corresponding to successive words are contiguous.

Alternatively, we can reduce the computation by searching over a word lattice of the most likely words and intervals. Such a word lattice is shown in two equivalent graphical forms in Fig. 4.9.

4.3 Parsing Natural Language

The algorithms described in Sections 4.1 and 4.2 can be applied to natural language to determine linguistic structure of different types. Since the algorithms are independent of specific grammars, all that is missing is a formal expression of the grammatical structure

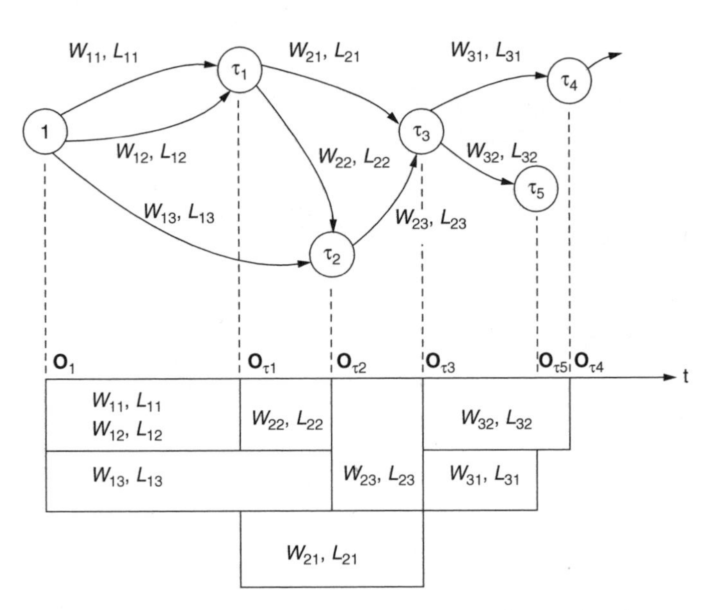

Figure 4.9 Two graphical representations of a word lattice

under consideration. The following are examples of grammars used in natural language applications.

4.3.1 The Right-Linear Case

Regular languages are best suited to describing the phonotactics of natural language although, as was illustrated in Section 4.2.2, they may also be used to analyze subsets of natural language syntax. In fact, a complete and moderately complex example appears in Chapter 8.

An example of Japanese phonotactics is given in Fig. 4.4 which shows the finite-state automaton corresponding to the regular grammar for generating phrases and short sentences used in requesting time table information. The terminal alphabet comprises five vowels and twelve consonants. State 1 is the start state, and states 70, 74 and 76 are final states. Figure 4.10 shows two forms of the same phone lattice typical of those generated by an acoustic-phonetic pattern recognition scheme. The actual likelihood values are not shown, but the phones are arranged in order of decreasing likelihood from top to bottom. Time is ordered from left to right in units of fixed duration.

An example of the operation of the parsing algorithm is shown in Fig. 4.11. The top of the figure shows a phone lattice like the one in Fig. 4.10. The correct phonetic classification is indicated by circles around the best transcription of each segment. Notice that several correct phones appear far down the list.

After parsing the phone lattice with the algorithm of (4.33) the correct transcription of the utterance is obtained as indicated at the bottom of the figure. The indicated word boundaries are found by another procedure not included in the parser. The correct transcription "Roku zi ni hun hatsu no" means "one (a train) leaving at 6:02 AM".

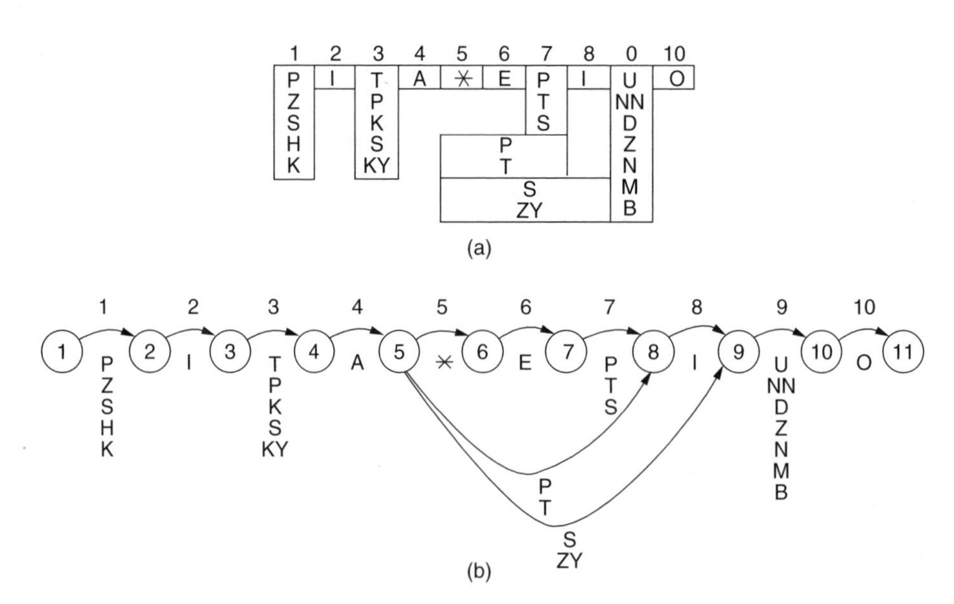

Figure 4.10 Two representations of a phone lattice in Japanese

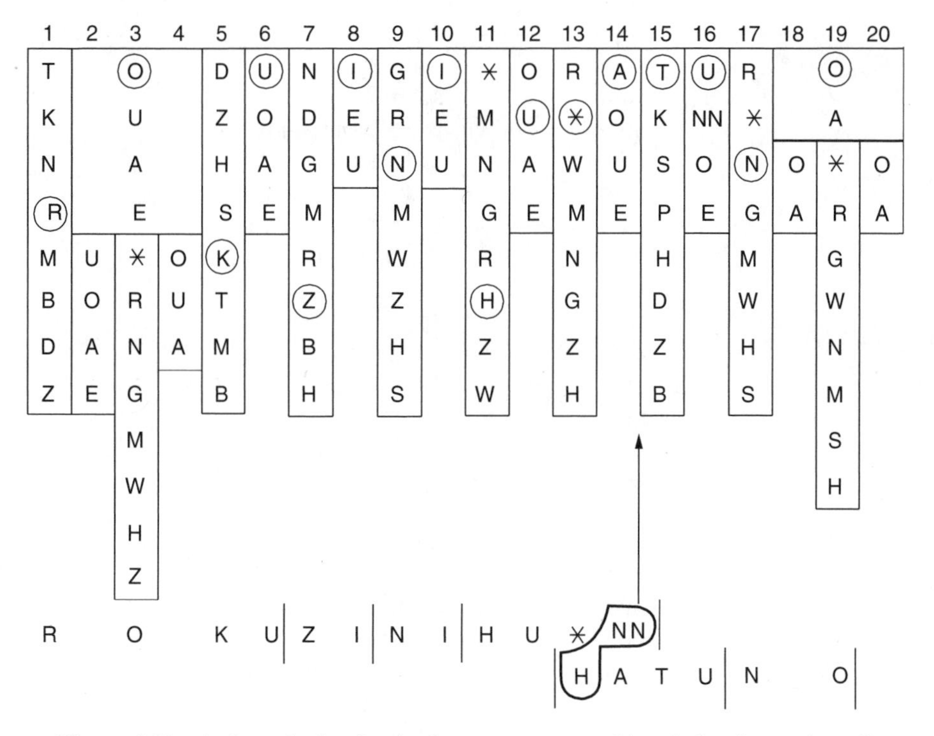

Figure 4.11 A phone lattice for the Japanese sentence "A train leaving at 6 a.m."

4.3.2 The Markovian Case

Present-day commercial speech recognizers represent natural language syntax as a Markov n-chain. That is, word order is governed by the n-gram probabilities, $p(w_n|w_{n-1}, w_{n-2}, \ldots, w_1)$, the probability that the word w_n will be preceeded in order by the words $w_{n-1}, w_{n-2}, \ldots, w_1$. Tables of n-grams are compiled from large corpora of text containing more than 10^9 words. Even from such large databases, it is difficult to get reliable statistics for $n > 3$. In some cases of frequently occurring n-grams, fifth- or sixth-order statistics can be estimated.

Parsing, in this case, reduces to application of a dynamic programming algorithm similar to (4.22) except that states represent $n - 1$ grams, and the transitions the n-gram probabilities. Even in cases where $n \geq 5$, this algorithm enforces word order but gives little indication of syntactic structure. We will see the significance of this deficiency in Chapter 8.

4.3.3 The Context-Free Case

Unlike the Markovian models of syntax, context-free models both respect word order and provide syntactic structure. Natural language is not strictly context-free but is well approximated by such grammars. Tomita [316] gives the following skeletal grammar of English containing only 40 rules but capturing a surprising degree of important structure. Word strings can be parsed with respect to this grammar in a highly efficient way using the algorithms of Sections 4.1.2 and 4.2.3.

$$
\begin{array}{lll}
\text{adjm} & \longrightarrow & \text{*adj} \\
\text{adjm} & \longrightarrow & \text{*adj adjm} \\
\text{adjm} & \longrightarrow & \text{advm *adj} \\
\text{adjm} & \longrightarrow & \text{adjm *conj adjm} \\
\text{advm} & \longrightarrow & \text{*adv advm} \\
\text{advm} & \longrightarrow & \text{*adv} \\
\text{advm} & \longrightarrow & \text{advm *conj advm} \\
\text{dir} & \longrightarrow & \text{dir *conj dir} \\
\text{dir} & \longrightarrow & \text{pp vp} \\
\text{dir} & \longrightarrow & \text{vp} \\
\text{dir} & \longrightarrow & \text{vp pp} \\
\text{nm} & \longrightarrow & \text{*n} \\
\text{nm} & \longrightarrow & \text{*n nm} \\
\text{np} & \longrightarrow & \text{np *conj np} \\
\text{np} & \longrightarrow & \text{np1 *that s} \\
\text{np} & \longrightarrow & \text{np1 s} \\
\text{np} & \longrightarrow & \text{np1} \\
\text{np0} & \longrightarrow & \text{nm} \\
\text{np0} & \longrightarrow & \text{adjm nm} \\
\text{np0} & \longrightarrow & \text{*det nm} \\
\text{np0} & \longrightarrow & \text{*det adjm nm} \\
\text{np1} & \longrightarrow & \text{adjm np0 pp pp} \\
\text{np1} & \longrightarrow & \text{adjm np0 pp} \\
\text{np1} & \longrightarrow & \text{adjm np0} \\
\text{np1} & \longrightarrow & \text{np0 pp} \\
\text{np1} & \longrightarrow & \text{np0} \\
\text{np1} & \longrightarrow & \text{np0 pp pp} \\
\text{pp} & \longrightarrow & \text{pp *conj pp} \\
\text{pp} & \longrightarrow & \text{*prep np} \\
\text{s} & \longrightarrow & \text{np vp pp pp} \\
\text{s} & \longrightarrow & \text{np vp pp} \\
\text{s} & \longrightarrow & \text{pp np vp} \\
\text{s} & \longrightarrow & \text{np vp} \\
\text{s} & \longrightarrow & \text{s *conj s} \\
\text{start} & \longrightarrow & \text{dir} \\
\text{start} & \longrightarrow & \text{np} \\
\text{start} & \longrightarrow & \text{s} \\
\text{vc} & \longrightarrow & \text{*aux *v} \\
\text{vc} & \longrightarrow & \text{*v} \\
\text{vp} & \longrightarrow & \text{vc np} \\
\text{vp} & \longrightarrow & \text{vp *conj vp} \\
\text{vp} & \longrightarrow & \text{vc}
\end{array}
$$

These rules are only the structure-building rules expressed in general context-free form. That is, all the symbols used in these rules are non-terminals. Symbols beginning with * are preterminals which they appear on the left-hand side of lexical assignment rules whose right-hand side is a single terminal symbol signifying an ordinary vocabulary word. For example,

*conj	⟶	AND
*conj	⟶	BUT
*conj	⟶	OR
*n	⟶	BOY
*prep	⟶	TO
*prep	⟶	IN
*prep	⟶	ON
*prep	⟶	WITH
*v	⟶	RUN
*v	⟶	BE

The number of such rules is limited only by the size of the vocabulary and could include all entries in a standard dictionary.

The non-terminal symbols refer to specific parts of speech as follows:

adj	adjective
aux	auxiliary verb
adv	adverb
conj	conjunction
det	determiner
dir	directive
n	noun
np	noun phrase
pp	prepositional phrase
prep	preposition
s	sentence
that	indicator
v	verb
vc	compound verb
vp	verb phrase

A far better grammar for English is given in [316, pp. 177ff.]. This one has about 400 rules and is still practical with modern computers.

5

Grammatical Inference

5.1 Exact Inference and Gold's Theorem

The term "grammatical inference" refers to that aspect of language acquisition in which the rules of grammar are learned. Here, grammar is construed in the broad definition used in Section 2.7 to include phonology, phonotactics, morphology, and syntax. Grammatical inference also connotes a formal logical method for acquiring the rules of grammar. Humans appear to learn language reliably and effortlessly, and this ability is regarded by Chomsky [45] and his followers to be an extraordinary feat inexplicable by any theory except that it is inherent in the architecture of our brains.

An early result that appears to justify this conclusion is due to Gold [105], who proved that the rules of right-linear or context-free grammar cannot be correctly inferred from any number of well-formed examples even if they were actually generated by such a formal mechanism. Gold also showed that if, as shown in Fig. 5.1, there were available an oracle that could indicate whether or not a string was well formed, and if both types of examples were given, then, asymptotically, the exact grammar can be correctly inferred. This result is often cited to demonstrate the futility of formal methods of language acquisition. We will return to this matter in Chapter 10. This chapter, however, will treat grammatical inference as a problem in estimating the parameters of the stochastic models introduced in Chapters 3 and 4. When the problem is cast in this way, we find that, in a well-defined sense, it is quite tractable.

5.2 Baum's Algorithm for Regular Grammars

Section 3.2.3 and 3.2.4 provide the basis for treating grammatical inference as a parameter estimation problem for regular grammars. Equations (3.239) and (3.240) relate the parameters of an HMM to those of a regular stochastic grammar. From this it follows that the Baum algorithm of (3.11)–(3.13) is a method of grammatical in inference. There is, however, a more direct method. Figure 3.12 shows that the syntactic structure of a sentence in a regular grammar is represented by a state sequence in the FSA that recognizes the corresponding regular language.

Combining these ideas, we can devise an algorithm for the inference of regular grammars by counting states in the state sequence. The desired grammar has production rules

Mathematical Models for Speech Technology. Stephen Levinson
 ISBN: 0-470-84407-8

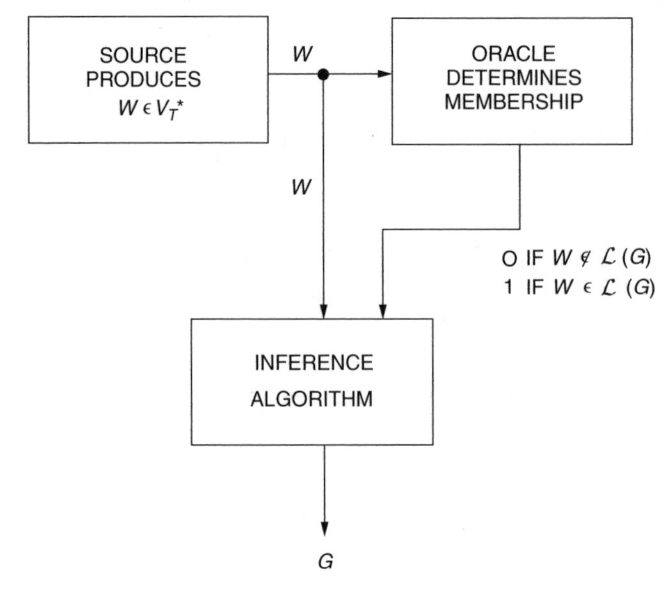

Figure 5.1 A block diagram of grammatical inference

of the form

$$A_i \overset{a_{ij}}{\rightarrow} vA_j, \qquad \forall v \in V_T, \tag{5.1}$$

and

$$A_j \overset{b_{jk}}{\rightarrow} v_k \qquad \text{for some } v_k \in V_T. \tag{5.2}$$

Recalling the EM algorithm of Section 3.1.6, we can estimate a_{ij} in (5.1) according to

$$\overline{a}_{ij} = \frac{\mathcal{N}_{ij}}{\mathcal{N}_i}, \tag{5.3}$$

where $\mathcal{N}_{ij}$ is the number of transitions from state q_i to state q_j in the state sequence obtained by parsing a set of sentences, $\{W_i\}$. Similarly, b_{jk} is estimated by

$$\overline{b}_{jk} = \frac{\mathcal{N}_k}{\mathcal{N}_j}, \tag{5.4}$$

where $\mathcal{N}_k$ is the number of occurrences of v_k while in state q_j and, as in (5.3), $\mathcal{N}_j$ is the number of visits to state q_j.

According to the EM algorithm, the estimates of (5.3) and (5.4) are refined by recursive application of the parsing process followed by recounting the events (i.e. state pairs and state–word pairs) that are the sufficient statistics for the parameters. When this process converges, we have our desired grammar in the form of (5.1) and (5.2). If one wishes to have a characteristic grammar instead of the stochastic one, all production rules whose probabilities are zero are dropped and the non-zero probabilities of the remaining rules are ignored.

5.3 Event Counting in Parse Trees

The observed relationship between discrete symbol HMMs and regular stochastic grammars can, in fact, be generalized in the sense that stochastic grammars may be thought of as doubly stochastic processes. The hidden process is that which rewrites the non-terminal symbols while the observed process produces the terminal symbols from them. The former are called structure-building rules, and the latter lexical assignment rules. Thus in the context-free case, the hidden process comprises all the production rules of the form $A \to BC$ while the observable process results from rules of the form $A \to a$. Although the separability of the processes vanishes if we move further up the Chomsky hierarchy, the concept is still valid and we can imagine stochastic models in which the underlying process is of any desired complexity.

At least at the context-free level, and perhaps even beyond it, algorithms corresponding to those we examined for HMMs exist. We have already seen that probabilistic parsing algorithms provide a way to evaluate a likelihood function $P(W)$. They also provide a means to determine derivations, the analog of state sequences, and to estimate parameters. The key to these algorithms lies in replacing (4.8) by

$$\psi_{ij}(A) = (\bar{l}, \overline{B}, \overline{C}), \tag{5.5}$$

where $\bar{l}$, $\overline{B}$, and $\overline{C}$ are as defined in (4.9). At the termination of the recursion (4.7)–(4.8) we can recover the derivation of W from ψ. The process is one of constructing the "parse tree" illustrated in Fig. 3.11. The root of the tree is S so we begin by examining $\psi_{1|W|}(S) = (B, C, k)$. This means that there are two subtrees whose roots are B and C and which account for w_1 through w_k and w_{k+1} through $w_{|W|}$, respectively. Now we examine $\psi_{1k}(B)$ and $\psi_{k+1|W|}(C)$ each of which indicates the formation of two subtrees. The two branching from B, D, and E, span symbols 1 to l and $l+1$ to k, respectively, while those branching from C account for the single terminal symbol w_{k+1} and F the rest of the sentence, that is, $m = k+2$. Recursion over ψ allows the construction of the entire tree from which we may simply read the derivation. A node B with two successors, C and D, indicates the production rule $B \to CD$. A node C with the single successor signals the application of the rule $C \to c$. This process is described by (4.10)–(4.12).

The "parse tree" also offers the solution to the parameter estimation problem, in which we are given $W = w_1 w_2 \ldots w_T$ and from which we must determine $\{P_{ABD} \mid A, B, C \in V_N\}$ and $\{P_{Aa} \mid A \in V_N, a \in V_T\}$. The solution to this problem advocated by Fujisaki [100] follows the EM algorithm and the Viterbi algorithm. Given an estimate of the parameters, produce a parse tree for W. Next, count events. Let $N(A, B, C)$ be the number of occurrences of the rule $A \to BC$ and let $N(A, X, Y)$ be the number of occurrences of rules $A \to XY$ where $X, Y \in V_N$. Then a new estimate of P_{ABC} is

$$\overline{P}_{ABC} = \frac{N(A, B, C)}{N(A, X, Y)}. \tag{5.6}$$

Similarly

$$\overline{P}_{Aa} = \frac{N(A, a)}{N(A, x)}, \tag{5.7}$$

where $N(A, a)$ is the number of applications in the derivation of W of $A \to a$ and $N(A, x)$ of $A \to x$ for any $x \in V_T$. Iteration of this procedure will give improved estimates of the parameter values.

5.4 Baker's Algorithm for Context-Free Grammars

Baker [21] has devised another version of the algorithm of Section 5.2 based on Baum's idea of a likelihood function. For reasons that will become clear, this method is sometimes called the inside-outside algorithm. The algorithm is used to estimate the parameters, $\boldsymbol{\lambda}$, of a CNF grammar with structure-building rules of the form

$$A_i \overset{a_{ijk}}{\to} A_j a_k \tag{5.8}$$

and lexial assignment rules of the form

$$A_j \overset{b_{jk}}{\to} v_k. \tag{5.9}$$

Thus $\boldsymbol{\lambda} = (\ldots, a_{ijk} \ldots, b_{jk}, \ldots)$.

Given a sentence $W = \mathcal{L}(G_s)$, where $W = w_1 w_2 \ldots w_T$ with $w_t \in V_T$ and G_s having rules given by (5.8) and (5.9), we can compute the likelihood, $P(W|\boldsymbol{\lambda})$, that $W \in \mathcal{L}(G)$. From Figure 5.2 we see that in any parse tree of W, we have the condition that the subtree that spans the string $w_s w_{s+1} \ldots w_{t-1} w_t$ has the root A_i. We denote this condition by the term $A_i(s, t)$. In particular, $A_1 = S$ must span the sentence W.

We denote the "inside probability" that A_i spans $w_s \ldots w_t$, for $1 \le s < t \le T$, by

$$\alpha_{st}(i) = P[w_s, w_{s+1}, \ldots w_{t-1} w_t | A_i(s, t), \boldsymbol{\lambda}], \tag{5.10}$$

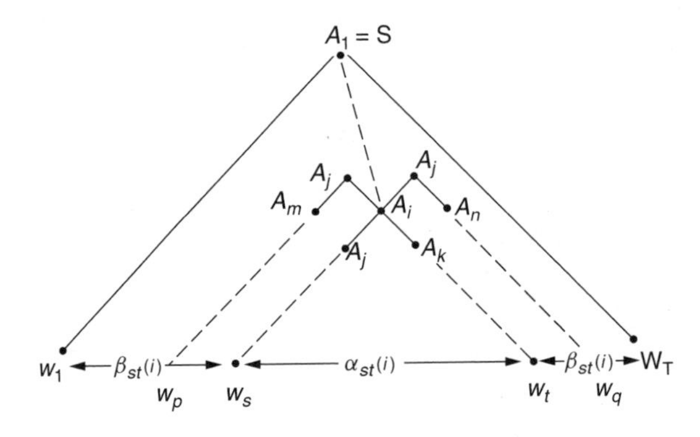

Figure 5.2 Operation of the Baker algorithm

with the special case that

$$\alpha_{tt}(i) = b_{ik} \qquad \text{if } w_t = v_k. \tag{5.11}$$

The $\alpha_t(i)$ can be computed recursively according to

$$\alpha_{st}(i) = \sum_{r=s}^{t-1} \sum_{j=1}^{|V_N|} \sum_{k=1}^{|V_N|} \alpha_{sr}(j) a_{ijk} \alpha_{r+1t}(k), \tag{5.12}$$

for $1 \leq s < t \leq T$. In particular, the likelihood function

$$P(w|\boldsymbol{\lambda}) = \alpha_{1T}(s), \tag{5.13}$$

and it is this value that we wish to maximize with respect to $\boldsymbol{\lambda}$.

Analogous to (5.7) we have the "outside probability" of the phrases to the left and right of $w_s \ldots w_t$ on condition $A_i(s, t)$, namely,

$$\beta_{st}(i) = P[w_1 w_2 \ldots w_{s-1}, w_{t+1} w_{t+2} \ldots w_T | A_i(s, t), \boldsymbol{\lambda}], \tag{5.14}$$

with $\beta_{1T}(i) = 1$ if and only if $i = 1$, and 0 otherwise. The $\beta_{st}(i)$ can also be computed recursively according to

$$\beta_{st}(i) = \sum_{p=1}^{s-1} \sum_{m=1}^{|V_N|} \sum_{l=1}^{|V_N|} \beta_{pt}(l) a_{lmi} \alpha_{ps-1}(m) + \sum_{q=t+1}^{T} \sum_{n=1}^{|V_N|} \sum_{l=1}^{|V_N|} \beta_{sq}(l) a_{lin} \alpha_{t+1q}(n). \tag{5.15}$$

Note that the first term on the right hand side of (5.15) accounts for the left complement of $w_s \ldots w_t$ and the second term, the right.

Baker now invokes the EM procedure obtaining the analog of (5.3) for the Hidden process

$$\bar{a}_{ijk} = \frac{\displaystyle\sum_{s=1}^{T} \sum_{t=1}^{T} \sum_{r=s}^{t-1} \alpha_{sr}(j) a_{ijk} \alpha_{r+1t}(k) \beta_{st}(i)}{\displaystyle\sum_{s=1}^{T} \sum_{t=1}^{T} \sum_{r=s}^{t-1} \sum_{m=1}^{|V_T|} \sum_{n=1}^{|V_T|} \alpha_{sr}(m) a_{imn} \alpha_{r+1t}(n) \beta_{st}(i)} \tag{5.16}$$

and for the observable process

$$\bar{b}_{jk} = \frac{\displaystyle\sum_{t \ni w_t = v_k} \alpha_{tt}(j) \beta_{tt}(j)}{\displaystyle\sum_{t=1}^{T} \alpha_{tt}(j) \beta_{tt}(j)}. \tag{5.17}$$

To obtain reliable results, (5.16) and (5.17) must be carried out for a large set of $W \in \mathcal{L}(G_s)$ and combined as the parameters were in the case of the non-ergodic HMM; refer to (3.92) and (3.93). Note also that the parameter values obtained from (5.16) and (5.17) define the grammars G_s and G_c in the same way as they did in the event counting procedure.

6

Information-Theoretic Analysis of Speech Communication

6.1 The Miller *et al.* Experiments

Chapters 2 through 5 have provided us with a mathematically well-defined and computationally tractable means of discovering and representing several different aspects of linguistic structure. One naturally asks what purpose is served by this structure. This chapter gives the first of two different but related answers to that question.

Experience with the expanding nationwide telephone network led to the intuition that "simple" speech was more likely correctly understood than "complicated". Unfortunately, this observation leaves undefined measures of the complexity and reliability of a message. In 1948, Claude Shannon clarified these issues with his mathematical theory of communication [295] which we now call information theory.

In the early 1950s psychologists and linguists began to be influenced by Shannon's work and, in particular Miller *et al.* [220] devised experiments to quantify the reliability of natural language communication in an information-theoretic sense.

6.2 Entropy of an Information Source

Information theory defines entropy and explains its role in the recovery of messages that have been corrupted by noise. These ideas are usually developed for abstract codes based on certain algebraic structures. We humans communicate in a special code called natural language which has a very complex structure quite different from that of its artificial cousins. Despite the structural differences, natural language performs many of the same functions as other codes and, in principle, admits of the same information-theoretic treatment. The code of natural language is its grammar, which comprises phonology, phonotactics, morphology and syntax. These are the categories of rules which govern the formation of permissible sentences. This aspect of linguistic structure is quite well understood and well modeled mathematically. For the purposes of computing entropy it is most convenient to treat grammar as a high-order Markov chain. Unfortunately, this ignores some important properties of grammar that are not captured by the Markov model.

Mathematical Models for Speech Technology. Stephen Levinson
 ISBN: 0-470-84407-8

However, even when formal models of grammar have no inherent probabilistic properties, it is possible to compute the entropy of a language directly from its grammatical rules without appeal to some putative Markovian principle. Once the entropy is calculated, we can use it to construct a bound on the probability of decoding error for a special optimal decoder used in conjunction with an acoustic classifier characterized by its confusion matrix and represented as a noisy channel with an appropriate equivocation.

We now consider algorithms for computing various statistical properties of finite languages. Relative redundancy is of major interest since it is a measure of grammatical constraint which is, in turn, indicative of the acoustic error correction ability provided by the language when used in a speech recognition system.

6.2.1 *Entropy of Deterministic Formal Languages*

For the purposes of this discussion we shall regard a finite language as one which has a state transition diagram with no circuits in it. Readers interested in a complete and rigorous introduction to formal language and automata theory may consult Hopcroft and Ullman [133].

The state transition diagram consists of a finite set of states, Q, a finite vocabulary, V, and a transition function, δ, which imposes the grammatical constraints. This function is defined as the mapping

$$\delta\colon (Q \times V) \rightarrow Q, \tag{6.1}$$

which means that each transition in the diagram is of the form

$$\delta(q_i, v) = q_j, \tag{6.2}$$

where (6.2) is understood to mean that from state $q_i \in Q$ one transverses a branch labeled by some $v \in V$ to get to the state $q_j \in Q$. There may be many such branches connecting any pair of states although each branch connecting the same pair of states must have a distinct label.

The state transition diagram defines sentences in the following way. There is an initial state, q_i, and a set of accepting states $Z \subset Q$. A valid sentence w consists of the labels of all the branches traversed (in order) along any path starting at q_1 and terminating at some accepting state $q_i \in Z$. We shall call the set of all sentences in the language $\mathcal{L}\ (G)$, and the set of state sequences $\mathbf{S}$.

We require that the grammar, G, be unambiguous. That is, given any sentence $W \in \mathcal{L}(G)$, there is exactly one state sequence in $\mathbf{S}$ associated with it. Note, however, that because there may be many transitions between a pair of states, there are, in general, many sentences which are generated by the same state sequence.

We define an important quantity characterizing the state transition diagram, its connectivity matrix, $\mathbf{C}$, whose ijth entry, c_{ij}, is the number of branches from q_i to q_j. Formally,

$$c_{ij} = |\{\delta(q_i, v) = q_j : v \in V; q_i, q_j \in Q\}|. \tag{6.3}$$

Because $\mathcal{L}(G)$ is finite, a state, q_i may appear at most once in any state sequence in $\mathbf{S}$. This implies that $q_i \neq q_j$ in equation (6.2), making the diagonal entries of $\mathbf{C}$ all zero.

Efficient Enumeration Algorithms

In this section we develop efficient algorithms for computing:

(a) the probability, $P(v)$, of the occurrence of the vocabulary word v in the set of sentences, $\mathcal{L}(G)$;
(b) the probability, $P(q_i)$ of the occurrence of the state q_i in the set of state sequences, $\mathbf{S}$;
(c) the probability of transition from state q_i, to state q_j, denoted by $P(q_j|q_i)$, $q_i, q_j \in Q$.

These algorithms are efficient counting techniques and the probabilities are based on relative frequencies of events assuming the sentences of the language, $\mathcal{L}$ (G), are equiprobable. In a later section we will compute the entropy for another set of sentence probabilities – the one that maximizes entropy.

We begin, for convenience, by adding one absorbing state, q_a, and the transitions

$$\delta(q_j, \cdot) = q_a \qquad \text{whenever } q_j \in Z \tag{6.4}$$

to the state transition diagram. The period is a sentence termination symbol which may or may not be required to locate sentence boundaries depending upon whether or not the language is self-punctuating.

We redefine the connectivity matrix, $\mathbf{C}$, of the state transition diagram by

$$\mathbf{C} = [c_{ij}], \qquad 1 \leq i, j \leq |Q| + 1, \tag{6.5}$$

where the ijth entry c_{ij} is the number of transitions from state q_i to state q_j. Similarly, we define a connectivity matrix $\mathbf{C}(v)$ for all $v \in V$. This is just the connectivity matrix of the state diagram obtained by removing all branches whose label is different from v. Thus the ijth entry of $\mathbf{C}(v)$, $c_{ij}(v)$, is either zero or one according to

$$c_{ij}(v) = \begin{cases} 1 & \text{if } \delta(q_j, v) = q_j, \\ 0 & \text{otherwise.} \end{cases} \tag{6.6}$$

Clearly, then,

$$\mathbf{C} = \sum_{v \in V} \mathbf{C}(v). \tag{6.7}$$

From $\mathbf{C}$ we shall compute the following quantities:

(a) the vector $\mathbf{s}$ whose ith element, s_i, is the total number of occurrences of the state q_i in the set of all permissible state sequences, $\mathbf{S}$;
(b) the matrix, $\mathbf{T}$, whose ijth entry, t_{ij}, is the number of transitions from state q_i to state q_j in the set of all permissible state sequences, $\mathbf{S}$;
(c) the number, $\mathcal{N}(v)$, of occurrences of the vocabulary word v for all $v \in V$ in all sentences in the language, $\mathcal{L}(G)$.

The required counting procedures follow from the following lemma, stated here without proof.

Lemma: Let

$$\mathbf{B} = (\mathbf{I} - \mathbf{C})^{-1}. \tag{6.8}$$

Then for $i \neq j$ the ijth entry of $\mathbf{B}$, b_{ij}, is exactly the number of paths from q_i to q_j.

Now let $\mathbf{r}'$ and $\mathbf{u}$ be the first row and last column of $\mathbf{B}$, respectively. Then from the lemma, r_i, the ith element of $\mathbf{r}'$, is the number of paths from q_1 to q_i; similarly, u_i, the ith element of $\mathbf{u}$, is exactly the number of paths from q_i to q_a.

The number of paths from q_1 to q_a passing through q_i is just $r_i u_i$. We have assumed that $\mathcal{L}(G)$ is finite, hence q_j occurs at most once in any state sequence. Thus $r_j u_j$ is exactly the number of occurrences of q_j in $\mathbf{S}$. We have thus shown that

$$s_j = r_i u_j. \tag{6.9}$$

Similarly, $c_{ij} u_j$ is the number of paths from q_1 to q_a having q_i as the second state. Therefore $r_i c_{ij} u_j$ is the number of paths from q_1 to q_a in which a transition from q_i to q_j occurs. This shows that

$$t_{ij} = r_i c_{ij} u_j. \tag{6.10}$$

Finally, let

$$\mathbf{m}(v) = \mathbf{C}(v)\mathbf{u}, \qquad \text{for all } v \in \mathbf{V}. \tag{6.11}$$

The ith component of $\mathbf{m}(v)$, $m_i(v)$, is the number of paths from q_i to q_a whose first transition is $\delta(q_i, v) = q_j$ for some $q_j \in Q$. Thus $r_i m_i(v)$ is the number of sentences in which the state q_i is succeeded by the word v. The total number of occurrences, $\mathcal{N}(v)$, of the word v in all sentences in the language is therefore

$$\mathcal{N}(v) = \sum_{i=1}^{|Q|+1} r_i m_i(v) = \mathbf{r}'\mathbf{C}(v)\mathbf{u}. \tag{6.12}$$

Let $\mathcal{Q}$ be the total number of state occurrences in $\mathbf{S}$. Then

$$\mathcal{Q} = \sum_{i=1}^{|Q|+1} s_i = \sum_{i=1}^{|Q|+1} r_i u_i = \mathbf{r}'\mathbf{u}. \tag{6.13}$$

From (6.12), the total number of words, $\mathcal{N}$ used in all sentences of $\mathcal{L}(G)$ is just

$$\mathcal{N} = \sum_{v \in V} \mathcal{N}(v) = \mathbf{r}'\mathbf{C}\mathbf{u} \tag{6.14}$$

or

$$\mathcal{N} = \mathbf{r}'\mathbf{u} - \mathbf{r}'(\mathbf{I} - \mathbf{C})\mathbf{u} = \mathcal{Q} - u_1; \tag{6.15}$$

the last relationship in (6.15) follows directly from the definitions of $\mathbf{r}'$ and $\mathbf{u}$.

One last quantity will be required. The total number of sentences in the language is just

$$|\mathcal{L}(G)| = u_1 = r_{|Q|+1}. \tag{6.16}$$

The desired probabilities are now easily computed according to:

$$P(v) = \frac{\mathcal{N}(v)}{\mathcal{N}} = \frac{\mathbf{r}'\mathbf{C}(v)\mathbf{u}}{\mathbf{r}'\mathbf{u} - u_1}, \tag{6.17}$$

$$P(q_1) = \frac{s_1}{\mathcal{Q}} = \frac{r_1 u_1}{\mathbf{r}'\mathbf{u}}, \tag{6.18}$$

and

$$P(q_j|q_i) = \frac{r_i c_{ij} u_j}{r_i u_i} = \frac{c_{ij} u_j}{u_i}. \tag{6.19}$$

From these quantities, some useful, well-known, information-theoretic properties of $\mathcal{L}(G)$ can be computed. The zeroth-order entropy per word of $\mathcal{L}(G)$ is

$$H_0(\mathcal{L}(G)) = \log_2(|\mathbf{V}|), \tag{6.20}$$

while the first-order entropy is just

$$H_1(\mathcal{L}(G)) = -\sum_{v \in 1} P(v) \log_2(P(v)). \tag{6.21}$$

The uncertainty associated with the state q_i can be shown to be

$$H(q_i) = -\sum_{i=1}^{|Q|+1} P(q_j|q_i) \log_2 \left[\frac{P(q_i|q_j)}{c_{ij}} \right]. \tag{6.22}$$

Substituting the value of $P(q_i|q_j)$ from (6.19), we get

$$H(q_i) = \log_2(u_i) - \sum_{j=1}^{|Q|+1} \frac{c_{ij} u_j}{u_i} \log_2(u_j). \tag{6.23}$$

Finally, the entropy of the language can be computed by substituting (6.18) and (6.23) into the well-known expression for the entropy of a Markov process:

$$H(\mathcal{L}(G)) = \sum_{i=1}^{|Q|+1} H(q_i) P(q_i). \tag{6.24}$$

In the above discussion, $\mathcal{N}$ includes occurrences of ".", the terminating symbol, and $\mathcal{Q}$ includes occurrences of q_a, the absorbing state. If these are to be ignored because the language is self-punctuating then u_1 must be subtracted from $\mathcal{N}$ and $\mathcal{Q}$ since "." is the final word in each sentence and q_a is the last state in the corresponding state sequence.

A Direct Method for Finding the Entropy

If one is not interested in the state, symbol or transition probabilities then $H(\mathcal{L}(G))$ can be computed directly.

Since the selection of a sentence uniquely determines the words in it, it is intuitively obvious that the entropy of a language is the quotient of the uncertainty associated with the selection of a sentence from the language and the average sentence length. We omit the formal proof of this observation and simply write

$$H(\mathcal{L}(G)) = \frac{-\sum_{w\in\mathcal{L}(G)} P(w)\log_2(P(w))}{E\{|w|\}} \tag{6.25}$$

where $P(w)$ is the probability of the sentence $w \in \mathcal{L}(G)$ and $E\{|w|\}$ is the average sentence length over all $w \in \mathcal{L}(G)$. Then assuming all sentences to be equally likely, we finally get:

$$H(\mathcal{L}(G)) = \frac{\log_2(|\mathcal{L}(G)|)}{E\{|w|\}} = \frac{u_1 \log_2(u_1)}{(\mathcal{N})}. \tag{6.26}$$

Computation of Maximum Entropy

In general, in speech communication tasks, all sentences are not equally likely as semantic and pragmatic knowledge bias the listener's expectations of what the speaker is about to say. If one had an accurate estimate of the probability of each sentence in the language, then the entropy could be computed from

$$H(\mathcal{L}(G)) = \frac{-\sum_{w\in\mathcal{L}(G)} P(w)\log(P(w))}{\sum_{w\in\mathcal{L}(G)} P(w)|w|} = \frac{H}{E}. \tag{6.27}$$

In (6.27) the numerator, H, is the entropy per sentence and the denominator, E is simply the average sentence length.

For any language which could be used for practical communication the $P(w)$ are too difficult to estimate. An alternative is to find the maximum value of $H(\mathcal{L}(G))$ attained by any set of sentence probabilities. This would then be a measure of the minimum amount of grammatical constraint imposed by the language.

This can be accomplished by maximizing the right-hand side of (6.27), subject to the constraint that

$$\sum_{w\in\mathcal{L}(G)} P(w) = 1. \tag{6.28}$$

We shall use the method of Lagrange multipliers in which the desired extremum can be found by solving

$$\frac{\partial}{\partial P(w)}\left[\frac{H}{E}+\lambda\sum_{w\in\mathcal{L}(G)}P(w)\right]=0,\quad w\in\mathcal{L}(G) \tag{6.29}$$

where λ is the Lagrange multiplier. Carrying out the differentiation gives

$$-\frac{E(1+\log(P(w)))+|w|H}{E^2}+\lambda=0,\qquad \forall w\in\mathcal{L}(G). \tag{6.30}$$

Let $\hat{P}(w), \forall w \in \mathcal{L}(G)$ be the solution of these equations, and let $\hat{E}$ and $\hat{H}$ be the corresponding values of E and H. For convenience, we let

$$\alpha=\frac{\hat{H}}{\hat{E}} \tag{6.31}$$

Then rearranging (6.30) gives

$$\log(\hat{P}(w))=\alpha|w|+\lambda(\hat{E}-1). \tag{6.32}$$

Multiplying (6.32) by $-\hat{P}(w)$ and summing over $w \in \mathcal{L}(G)$, we have

$$\hat{H}=\hat{H}-[\lambda(\hat{E}-1)], \tag{6.33}$$

which shows that the term in brackets is identically zero. Thus

$$\log(\hat{P}(w))=\alpha|w|, \tag{6.34}$$

where α is chosen so as to satisfy the constraint of (6.28).

In order to determine α, and ultimately $H(\mathcal{L}(G))$, we proceed as follows. Substituting the values of $\hat{P}(w)$ from (6.34) into the constraint of (6.28) yields

$$\sum_{w\in\mathcal{L}(G)}e^{-\alpha|w|}=1. \tag{6.35}$$

Let $\mathcal{N}_k$ be the number of sentences of length k. That is,

$$\mathcal{N}_k=|\{w\in\mathcal{L}(G):|w|=k\}|. \tag{6.36}$$

We can then rewrite (6.35) as

$$\sum_{k=1}^{l_{\max}}\mathcal{N}_k e^{-\alpha k}-1=0, \tag{6.37}$$

where $l_{\max}$ is the length of the longest sentence. Clearly, (6.37) is a polynomial in $e^{-\alpha}$ and has a real root, x_0, with $0 \le x_0 \le 1$ since $\mathcal{N}_k > 0$ for all k. This root is easily and accurately computed numerically.

Substituting the values of $\hat{P}(w)$ from (6.34) into the expression for the entropy of (6.27) and simplifying gives the result

$$H(\mathcal{L}(G)) = \log_2(x_0) = \alpha, \tag{6.38}$$

by recalling that $x_0 = e^{-\alpha}$.

Similarly, using the values of $\hat{P}(w)$ in the expression

$$E\{|w|\} = \sum_{w \in \mathcal{L}(G)} \hat{P}(w)|w|, \tag{6.39}$$

we finally compute the average sentence length which corresponds to the maximum entropy of (6.38) as

$$E\{|w|\} = \sum_{k-1}^{l_{\max}} \mathcal{N}_k k e^{-\alpha k}. \tag{6.40}$$

For eqs. (6.37) and (6.40) the values of $\mathcal{N}_k$ are easily computed

$$\mathcal{N}_k = c_{1a}^{(k)}. \tag{6.41}$$

That is, $\mathcal{N}_k$ is simply the last entry in the first row of the kth power of the connectivity matrix, **C**.

The relative redundancy, $R(\mathcal{L}(G))$, which measures grammatical constraint is just

$$R(\mathcal{L}(G)) = 1 - \frac{H(\mathcal{L}(G))}{\log_2(|V|)}. \tag{6.42}$$

6.2.2 *Entropy of Languages Generated by Stochastic Grammars*

The most direct way to obtain the entropy of a source is to assume that it is Markovian, generating the infinite sequence of words $W = w_1 w_2 w_3 \ldots$ with $w_i \in V_T$. Then

$$\begin{aligned} H(W) &= H(\mathcal{L}(G)) \\ &= \lim_{n \to \infty} \left\{ -\sum_{i=1}^{n} p(w_i | w_{i-1}, w_{i-2}, \ldots, w_1) \log p(w_i | w_{i-1}, w_{i-2}, \ldots, w_1) \right\}. \end{aligned} \tag{6.43}$$

The stochastic grammar, G, according to which the source generates words, w_i, is implicit in the n-gram probabilities but can be expressed explicitly. The set, Q, of states of the underlying stochastic right-linear grammar is

$$Q = \left\{ q_{k_{i-1}, k_{i-2}, \ldots, k_1} | w_l = v_{k_l}, l = 1, 2, \ldots i - 1 \right\} \tag{6.44}$$

and the production rules are

$$q_{k_{i-1},k_{i-2},\ldots,k_1} \xrightarrow{p(w_i|w_{i-1},w_{i-2},\ldots,w_1)} w_i q_{k_i,k_{i-1},\ldots,k_2}. \tag{6.45}$$

Notice that

$$|Q| = |\underbrace{V_T \times V_T \times \ldots \times V_T}_{n-1 \text{ times}}|^{=|V_T|^{n-1}=N}. \tag{6.46}$$

Then the state transition matrix, **A**, of the Markov chain is

$$\mathbf{A} = [a_{ij}]_{N\times N} = \frac{\mathcal{N}(q_i, q_j)}{\sum_k \mathcal{N}(q_i, q_k)}, \tag{6.47}$$

where $\mathcal{N}(q_i, q_j)$ is the number of occurrences of the state sequence (q_i, q_j) in the generalization of W.

Using the well-known formula for the entropy of a Markov chain, we can compute $H(\mathcal{L}(G))$ from

$$H(\mathcal{L}(G)) = \sum_{i=1}^{N} p(q_i)H(q_i), \tag{6.48}$$

where the $p(q_I)$ are the stationary probabilities of the Markov chain and the $H(q_i)$ are the uncertainties associated with the individual states, q_i, given by

$$H(q_i) = \sum_{v 6 q_i} p(v)\log p(v), \tag{6.49}$$

where the $p(v)$ are the probabilities that the word, v, is generated in the state, q_i. Thus $p(v)$ is just the n-gram probability

$$p(v) = p(\mathbf{v}|w_{i-1}, w_{i-2}, \ldots, w_1) \tag{6.50}$$

and the stationary probabilities are the eigenvectors of **A** found by solving

$$\mathbf{Ap} = \mathbf{p}, \tag{6.51}$$

where

$$\mathbf{p} = (p(q_1), p(q_2), \ldots, p(q_N)). \tag{6.52}$$

Note that (6.51) relies on the fact that the eigenvalues of **A** all have unit magnitude.

Now, instead of making the Markovian assumption, let us suppose that W is generated by G_s, a known stochastic context-free grammar. We define the generating function of this grammar according to

$$g_j(z_1, z_2, \ldots, z_K) = \sum_{\Gamma(A_j)} p_{A_j\beta} z_1^{r_1} z_2^{r_2} \ldots z_K^{r_K}, \tag{6.53}$$

where $\Gamma(A_j)$ is the set of all production rules of the form $A_j \xrightarrow{p_{A_j\beta}} \beta$ and $A_j \in V_T$, $\beta \in \{V_N \cup V_T\}^*$. The variable z_j accounts for β derived from A_j. The generating function (6.53) is thus a function of K variables, where $K = |V_N|$. The exponents, r_l, are the numbers of occurrences of $A_l \in V_N$ in the string β.

The generating function (6.53) allows us to define the first moment matrix, $\mathbf{E}$, whose ijth entry, e_{ij}, is the expected number of occurrences of $A_j \in \beta$ for the production rule $A_i \xrightarrow{p} \beta$. The expectation is taken with respect to all sentences, $W \in \mathcal{L}(G_s)$. The significance of the generating function is that

$$e_{ij} = \frac{\partial g_j}{\partial z_i}|_{z_1=z_2=z_3\ldots=z_K=1} = \sum_{\Gamma(A_j)} r_i p_{A_j\beta}. \tag{6.54}$$

Note that if G_s is transformed into Chomsky normal form then

$$r_l = \begin{cases} 0 & \text{if } A_j \xrightarrow{p} A_m A_n,\ m, n \neq l, \\ 1 & \text{if } A_j \xrightarrow{p} A_m A_l,\ m \neq l, \\ 2 & \text{if } A_j \xrightarrow{p} A_l A_l. \end{cases} \tag{6.55}$$

This simplifies the evaluation of (6.54).

We also need to know the expected sentence length over all $W \in \mathcal{L}(G_s)$. Using the formula for the sum of a geometric series, we have (cf. (6.8))

$$\sum_{k=0}^{\infty} \mathbf{E}^k = (\mathbf{I} - \mathbf{E})^{-1}. \tag{6.56}$$

As in (6.39), the average sentence length is

$$|\overline{W}| = \sum_{W \in \mathcal{L}(G_s)} |W| P(W) \tag{6.57}$$

$$= (1, 0, 0, \ldots 0)[\mathbf{I} - \mathbf{E}]^{-1}\mathbf{x}, \tag{6.58}$$

where $\mathbf{x} = (x_1, x_2, \ldots, x_K)$ and x_k is the expected number of terminal symbols appearing on the right-hand side of $A_j \xrightarrow{p_{jk}} \beta_{jk}$, thus

$$x_j = \sum_k p_{jk}, \tag{6.59}$$

and the sum is indexed over all productions of the form $A_j \overset{p_{jk}}{\rightarrow} V_k$ or $A_j \overset{p_k}{\rightarrow} \rho_{jk}$ and $V_k \in \beta_{jk}$.

Finally, we get the entropy of the language from

$$H(\mathcal{L}(G_s)) = (1, 0, 0, \ldots, 0)[\mathbf{I} - \mathbf{E}]^{-1}\mathbf{y}/|\overline{W}|, \tag{6.60}$$

in which $|\overline{W}|$ comes from (6.57) and $\mathbf{y} = (y_1, y_2, \ldots, y_K)$, and each y_j comes from

$$y = -\sum_k p_{jk} \log(p_{jk}). \tag{6.61}$$

In (6.61) the index set and the p_{jk} are the same as those of (6.59).

6.2.3 *Epsilon Representations of Deterministic Languages*

The discussion in Section 6.2.1 applies to finite deterministic languages, whereas that of Section 6.2.2 addresses infinite stochastic languages. There is a surprising connection between the two. Recall that a stochastic grammar is consistent if and only if (cf. (3.238))

$$\sum_{w \in \mathcal{L}(G_s)} P(w) = 1. \tag{6.62}$$

We generalize (6.62) to define an ε-representation, $\mathcal{L}_\varepsilon$, of $\mathcal{L}(G_s)$ whenever $\mathcal{L}_\varepsilon \subset \mathcal{L}(G_s)$ such that

$$\sum_{w \in \mathcal{L}_\epsilon} p(w) = 1 - \varepsilon. \tag{6.63}$$

If G_s is a consistent context-free stochastic grammar then for any $\varepsilon > 0$ there is a finite N_ε such that

$$\mathcal{L}_\varepsilon = \{W \in \mathcal{L}(G_s) | \; |W| \leq N_\varepsilon\} \tag{6.64}$$

is an ε-representation of $\mathcal{L}(G_s)$ and thus satisfies (6.63); see [34].

The result (6.64) is remarkable because in an infinite language it is intuitive that sentences of all lengths would be needed to satisfy (6.63). However, the context-free structure of the grammar provides for an arbitrarily good approximation to $\mathcal{L}(G_s)$ using sentences of finite length. The implication of this result is that the methods of Section 6.2.1 are good approximations to those of Section 6.2.2.

6.3 Recognition Error Rates and Entropy

We are now in a position to formalize and quantify the Miller *et al.* results of Fig. 6.1. In the classification system shown in Fig. 6.2, a message, W encoded on $x(t)$, from an information source is assumed to be well formed with respect to the grammar, G. The first stage of the classification process is a word-by-word identification of W without regard

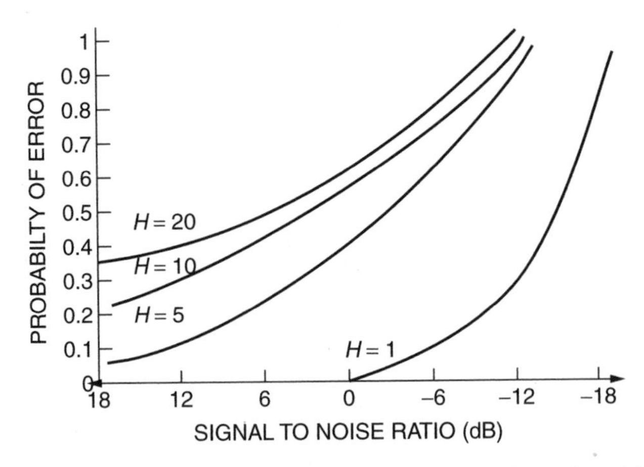

Figure 6.1 Error probability as a function of signal to noise ratio for codes of different entropies (after Miller *et al.*)

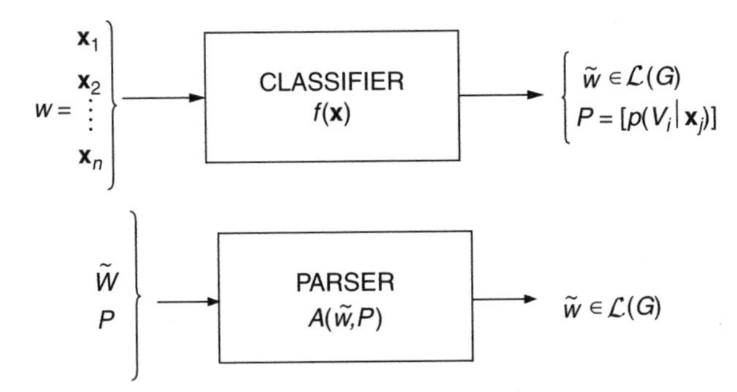

Figure 6.2 Block diagram of a speech recognizer using a probabilistic parser

for G. This process is error-prone and results in the corrupted interpretation, $\tilde{W}$, of W. In general, $\tilde{W}$ is not well formed. The second stage of the process uses the methods of Chapter 4 to find $\hat{W} \in \mathcal{L}(G)$ such that $p(\hat{W}|x(t))$ is maximized. The following analysis will show the relationship among the complexity of the source defined by its entropy, $H(\mathcal{L}(G))$, the accuracy of the classifier as characterized by its equivocation, $H(W|\tilde{W})$, and p_e, the average probability of word error in $\hat{W}$.

6.3.1 Analytic Results Derived from the Fano Bound

A classic result in information theory is the Fano bound [81] relating source entropy, channel equivocation, and decoding error probability. It states that

$$H(v|\tilde{v}) \leq H(p_e) + p_e \log_2(|V_T| - 1), \tag{6.65}$$

where the symbol $v \in V_T$ is transmitted, $\tilde{v} \in V_T$ is received and p_e is the probability that $\tilde{v} \neq v$. The equivocation of the channel, $H(v|\tilde{v})$ is the average uncertainty in bits that v

was transmitted given that $\tilde{v}$ was received:

$$H(v|\tilde{v}) = -\sum_{v,\tilde{v}\in V_T} p(v,\tilde{v}) \log p(v|\tilde{v}) \tag{6.66}$$

We define

$$H(p_e) = p_e \log_2 p_e + (1-p_e)\log_2(1-p_e), \tag{6.67}$$

which may be interpreted to mean the uncertainty in bits of just whether or not $v = \tilde{v}$.

Inequality (6.65) may be considered to state that the equivocation is bounded above by the sum of two uncertainties. The first is the uncertainty in the correctness of the decision and the second is the uncertainty of the remaining $|V_T| - 1$ symbols if $\tilde{v}$ is incorrect. This term has a factor of p_e accounting for the fact that an error occurs with probability p_e.

The effective vocabulary size, $|V_T|_{\text{eff}}$, of the source is

$$|V_T|_{\text{eff}} = 2^{H(\mathcal{L}(G))}, \tag{6.68}$$

and the efficiency, η, of the source is

$$\eta = \frac{H(\mathcal{L}(G))}{\log_2|V_T|} = \frac{\log_2|V_T|_{\text{eff}}}{\log_2|V_T|} \le 1, \tag{6.69}$$

with equality if and only if $\mathcal{L}(G) = V_T^*$, which is equivalent to $|V_T|_{\text{eff}} = |V_T|$. We may thus rewrite (6.65) as

$$\begin{aligned} H(v|\tilde{v})\eta &\le H(p_e) + p_e \log_2(2^{H(\mathcal{L}(G))} - 1) \\ &\le H(p_e) + p_e \log_2(2^{H(\mathcal{L}(G))}) \\ &= H(p_e) + p_e H(\mathcal{L}(G)). \end{aligned} \tag{6.70}$$

Substituting η from (6.69), we get

$$\frac{H(v|\tilde{v})}{\log_2|V_T|} - \frac{1}{H(\mathcal{L}(G))} \le p_e. \tag{6.71}$$

The first term on the left-hand side of (6.71) is fixed for a given language and classifier, in which case increasing the entropy of the source increases the lower bound on the error probability.

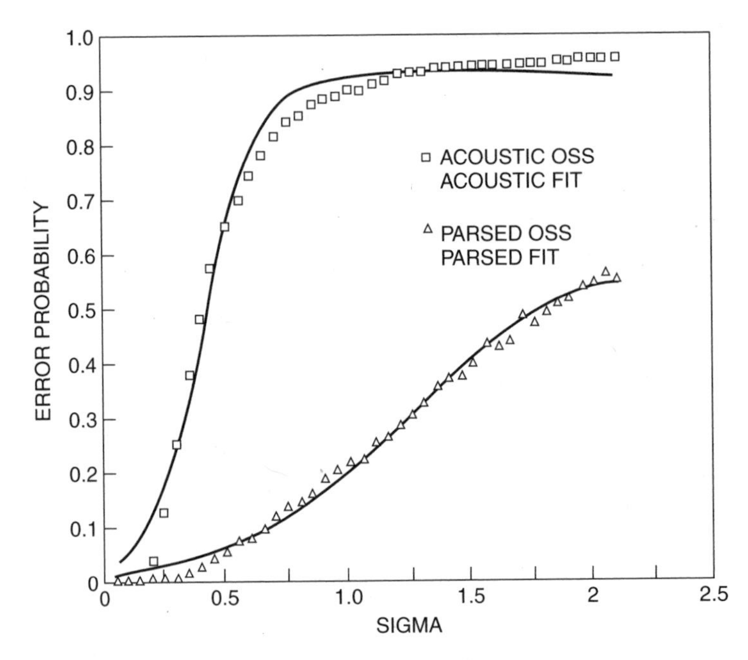

Figure 6.3 Error rate as a function of the negative of signal to noise ratio for the recognition system of Levinson [183]

6.3.2 Experimental Results

The four different conditions used in the Miller *et al.* experiment as shown in Figure 6.3 actually correspond to four sources of different entropies: digits, words, nonsense syllables and sentences. For any particular source the error probability increases with decreasing signal to noise ratio. The important result, however, is that the degradation decreases monotonically with decreasing source entropy at any fixed signal to noise ratio. This is exactly the behavior predicted by (6.71).

The Miller *et al.* experiments are based on the performance of human listeners. However, the same behavior is obtained for automatic speech recognizers. The results of an early experiment by Levinson *et al.* [183] are shown in Figure 6.3. The two curves are, respectively, the probability of error in $\tilde{W}$ and $\hat{W}$. The uncertainty in $\tilde{W}$ is $\log_2 |V_T| > H(\mathcal{L}(G))$, the uncertainty in $\hat{W}$. Thus the two curves are ordered as predicted by (6.71) at any fixed value of σ which is inversely related to signal to noise ratio.

7

Automatic Speech Recognition and Constructive Theories of Language

The theories and methods described in Chapters 3, 4, and 5 can be combined and used to form a constructive theory of language and a technology for automatic speech recognition. We will consider two approaches to the "language engine", one integrated and the other modular. We will then briefly describe the way that our mathematical models can be applied to the problems of speech synthesis, language translation, language identification, and a low-bit-rate speech communication.

7.1 Integrated Architectures

The first approach to the "language engine" is one in which several levels of linguistic structure are captured and compiled into a single monolithic model that can then be used to automatically transcribe speech into text. This model was introduced by Jelinek *et al.* [146] and has been refined over the past three decades. It is the state of the art in automatic speech recognition and the basis for most commercial speech recognition machines.

The basic model is shown in Fig. 7.1 in which the dashed lines indicate that all representations of linguistic structure are analyzed by a single process. Acoustic phonetics and phonology are represented by an inventory of sub-word models. Typically there are several allophonic variants of each of the phones listed in Table 2.2. Each phone is represented by a three-state HMM of the type illustrated in Fig. 3.5 in which the Markov chain is non-ergodic and the observable processes are Gaussian mixtures.

Allophonic variation is described by triphone models in which the acoustic properties of a given phone are a function of both the preceeding and following phones. That is, each phone appears in as many different forms as are required to account for all of the phonetic contexts in which it occurs.

The lexicon is simply a pronouncing dictionary in which each word is "spelled" in terms of the triphones of which it is made in citation form. Thus the word v is the sequence of

Mathematical Models for Speech Technology. Stephen Levinson
 ISBN: 0-470-84407-8

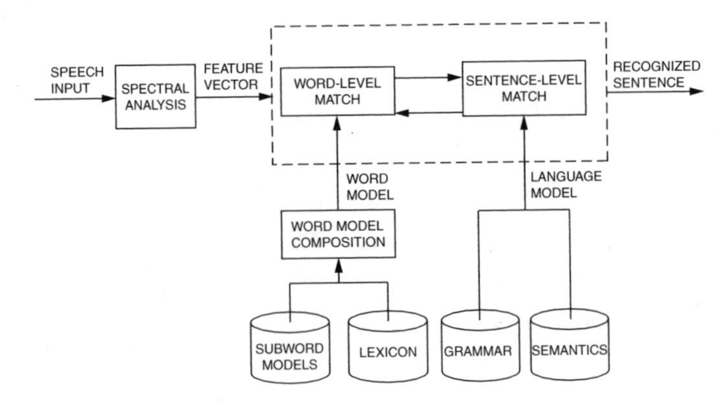

Figure 7.1 The integrated architecture for automatic speech recognition (from Rabiner [257])

phones

$$v = f_1 f_2 \dots f_k, \tag{7.1}$$

where the ith phone, f_i, is the particular allophone (triphone)

$$f_i = (f_{i-1}, f_i, f_{i+1}). \tag{7.2}$$

The intuition behind the triphone model is that each phonetic unit has a target articulatory position characterized by the second state and transitions into and out of it characterized by the first and third states, respectively. The initial transition of phone f_i is modified by the preceeding phone, f_{i-1}. Similarly, the final transition is affected by f_{i+1}. The idea is illustrated in Fig. 7.2.

In the model of Fig. 7.1, the only notion of syntax is that of word order as specified by an n-gram probability. If the sentence, W, is the word sequence

$$W = v_1 v_2 \dots v_K, \tag{7.3}$$

where v_k is some entry (word) in the lexicon of the form of (7.1), then word order is specified by $p(v_k | v_{k-1}, v_{k-2}, \dots, v_{k-n})$. If an n-gram has probability 0, then the corresponding word sequence is not allowed.

All of the information described above may be compiled into a single, large lattice of the form shown in Fig. 7.3 and 7.4. The large oval states represent words, and the small

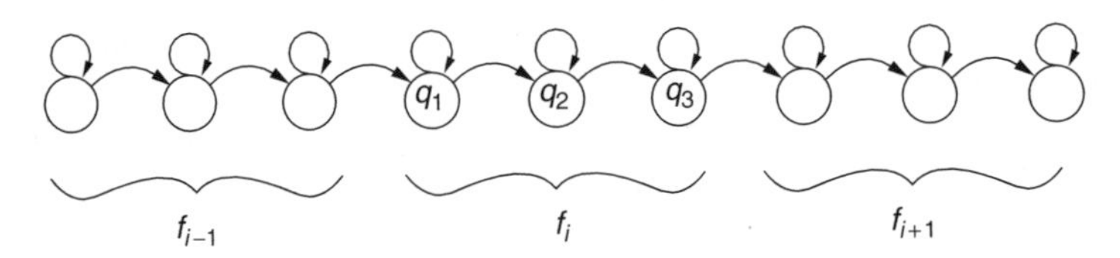

Figure 7.2 The triphone model for f_i

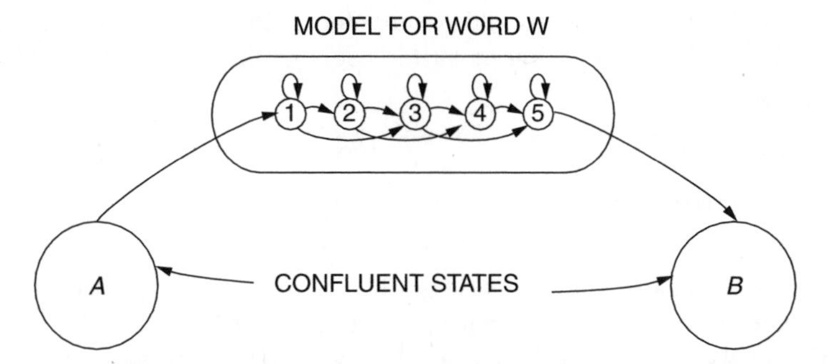

Figure 7.3 The HMM representation of the phonology of the production rule $A \rightarrow WB$

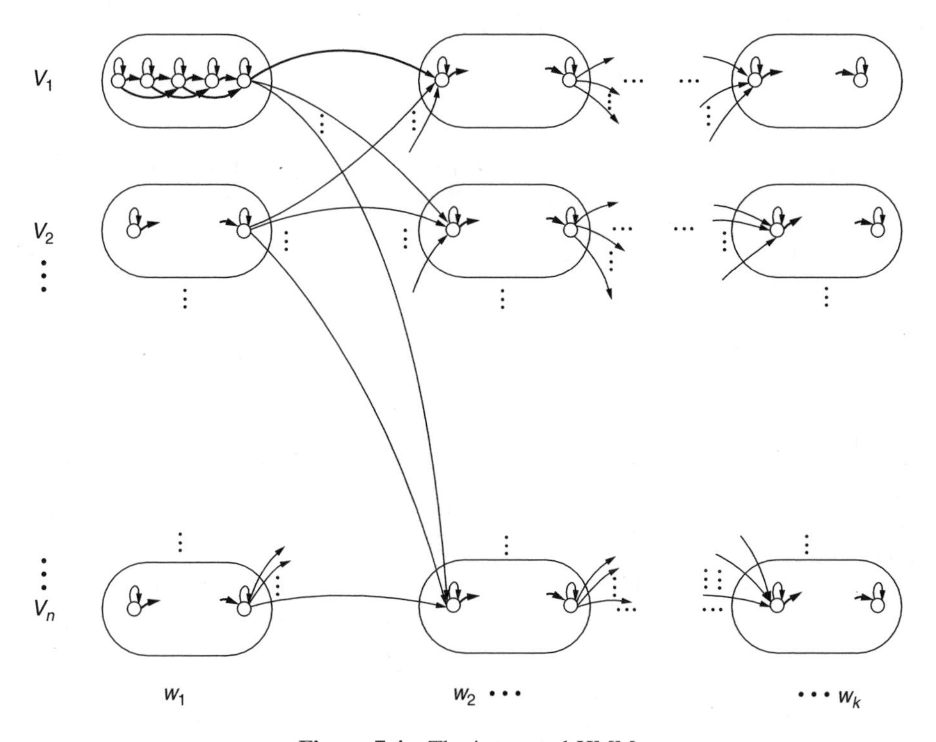

Figure 7.4 The integrated HMM

circular states the quasi-stationary regions of the phones. Decoding the sentence, W, is the process of finding the path through the lattice of maximum likelihood.

The dynamic programming algorithm for this process is just

$$\phi_t(j, v) = \max_{\{j-1 \leq i \leq j\}} \{\phi_{t-1}(i, v) a_{ij}^{(v)} b_j^{(v)}(\mathbf{O}_t)\}, \tag{7.4}$$

which accounts for state transitions and observations within the word v. The model for the word v is just the concatenation of all the triphone models listed for the word v in the lexicon.

The transitions between words are evaluated according to

$$\phi_t(1, w) = \max_{\{p(w|v)\}} \{\phi_{t-1}(1, w), \phi_{t-1}(N, v)p(w|v)\}. \tag{7.5}$$

Note that, for convenience, we have written (7.5) to maximize only over bigram probabilities. The algorithm can easily be generalized to account for n-grams.

The trellis can also be searched by means of a "best-first" algorithm according to which

$$\phi_t(q) = \phi_{t-1}(p)a_{pq}^{(v)}b_q^{(v)}(\mathbf{O}_t)\hat{L}, \tag{7.6}$$

where the priority queue is initialized according to

$$\phi_1(1) = b_1^{(v)}(\mathbf{O}_1)\hat{L}^{T-1}. \tag{7.7}$$

$\hat{L}$ is the heuristic function, p and q are any two states in the trellis that are connected by a transition, and the priority queue, ϕ, is arranged so that

$$\phi_{t_{i_1}}(p_{i_1}) > \phi_{t_{i_2}}(p_{i_2}) > \cdots > \phi_{t_{i_N}}(p_{i_N}). \tag{7.8}$$

The value of $\hat{L}$ is usually taken to be

$$\hat{L}_t = E\{\mathcal{L}(\mathbf{O}_t, \mathbf{O}_{t+1}, \ldots, \mathbf{O}_T|\boldsymbol{\lambda})\}. \tag{7.9}$$

In light of (7.8), (7.6) is interpreted to mean that p is the first entry of the queue and is extended to state 1, the value of which is then inserted into the queue in the proper position to maintain the ordering of (7.8). This algorithm will yield the same result as that of (7.4).

In actual practice the trellis is unmanageably large. If there are 30 000 words in the lexicon each containing five phones thus requiring 15 states, then for each frame, $\mathbf{O}_t$, there are of the order of 10^5 nodes in the lattice. If a typical sentence is 5 seconds in duration, then, at 100 frames per second, there are 10^7 nodes in the lattice each one of which requires the computation of either (7.4) or (7.6). This cannot be accomplished within the constraints of real time.

The solution to the real-time problem is to heuristically prune the lattice. There are two methods for doing so. The first method is usually called "beam search", according to which (7.4) is modified to allow only "likely" states by thresholding $\phi_t(j, v)$ according to

$$\phi_t(j, v) = \begin{cases} \phi_t(j, v), & \text{if } \phi_t(j, v) < \Theta\phi_{\max}, \\ 0, & \text{otherwise.} \end{cases} \tag{7.10}$$

According to (7.10) only those nodes of the lattice are evaluated whose likelihood is some small factor, Θ, times the maximum value.

The best first search of (7.6) is pruned by limiting the size, N, of the priority queue of (7.8). As a result only the N best nodes will be explored and less likely ones will be dropped from the queue.

These heuristics may induce search errors. That is, at some time t, the heuristic may terminate the globally optimal state sequence because it is locally poor. Empirical studies are required to set the values of either Θ or N to achieve the desired balance between computation time and rate of search errors.

7.2 Modular Architectures

The modular architecture is based on a completely different model of the "language engine" than is used in the integrated architecture. The modular design uses a separate representation of each level of linguistic structure and analyzes each level independently but in sequence. This design is shown in Fig. 7.5, from which we see that the phonetic and phonological structure is analyzed first with respect to the hidden semi-Markov model of Fig. 7.6, yielding a phonetic transcription. The phone sequence is then analyzed to determine the identities and locations of the words, which process is called lexical access and which produces a word lattice. Finally, the syntactic analysis is performed by means of an asynchronous parsing algorithm.

7.2.1 Acoustic-phonetic Transcription

Since each state of the acoustic-phonetic model corresponds to a single phone, a phonetic transcription may be obtained by finding the optimal state sequence using the dynamic programming algorithm

$$\phi_t(j) = \max_{1\leq i\leq N} \left\{ \max_{\tau<t} \left\{ \phi_{t-\tau}(i) a_{ij} d_{ij}(\tau) \prod_{\theta=0}^{\tau-1} b_{ij}(\mathbf{O}_{t-\theta}) \right\} \right\}. \tag{7.11}$$

Recall that (7.11) maximizes the joint likelihood of the state and observation sequences, $\mathbf{q}$ and $\mathbf{O}$, respectively. Also note that (7.11) differs from (3.7) in that the durational densities and the observation densities are indexed by state transition rather than present state. This means that there are more parameters to estimate but a better representation of phonology is obtained.

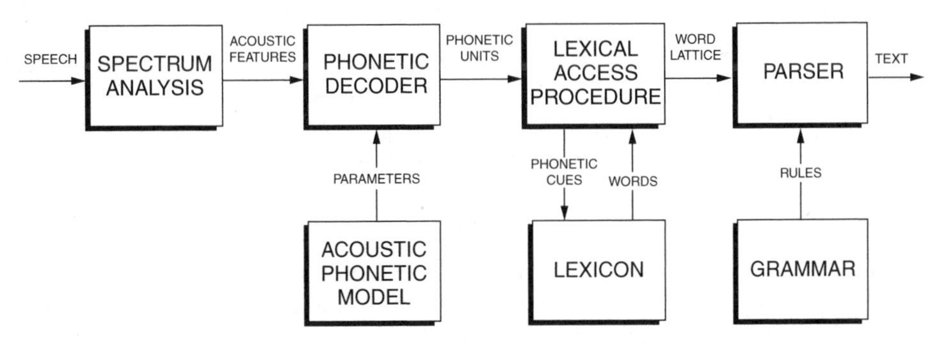

Figure 7.5 Block diagram of a modular automatic speech recognition system (from [188])

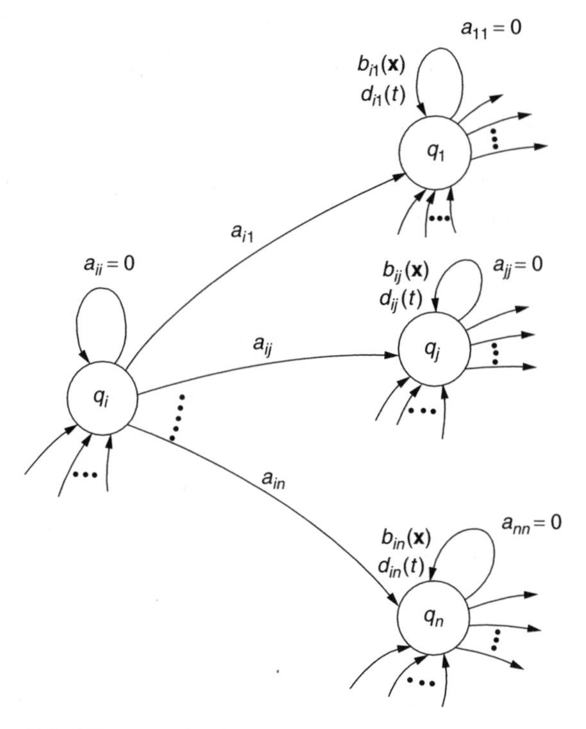

Figure 7.6 The acoustic-phonetic model of the modular architecture

The state sequence is recovered from the lattice of $\phi_t(i)$ by setting

$$\gamma_t(j) = i^*, \qquad \delta_t(j) = \tau^*, \tag{7.12}$$

where i^* and τ^* maximize the right-hand side of (7.11). Then set

$$q_T = \arg\max_i \{\phi_T(i)\} \tag{7.13}$$

and

$$q_{t-\delta(q_t)} = \gamma_t(q_t). \tag{7.14}$$

The q_t are the desired state sequence, and the duration of each state is $\hat{d}_t = t - \delta(q_t)$. These sequences will then be used in the process of lexical access.

7.2.2 Lexical Access

Lexical access is a procedure for locating words in the error-ridden state sequence. It is a search procedure, illustrated schematically in Fig. 7.7, that looks for optimal matches between the phonetic transcriptions of words given in the lexicon and the corrupted phonetic transcription obtained from (7.12).

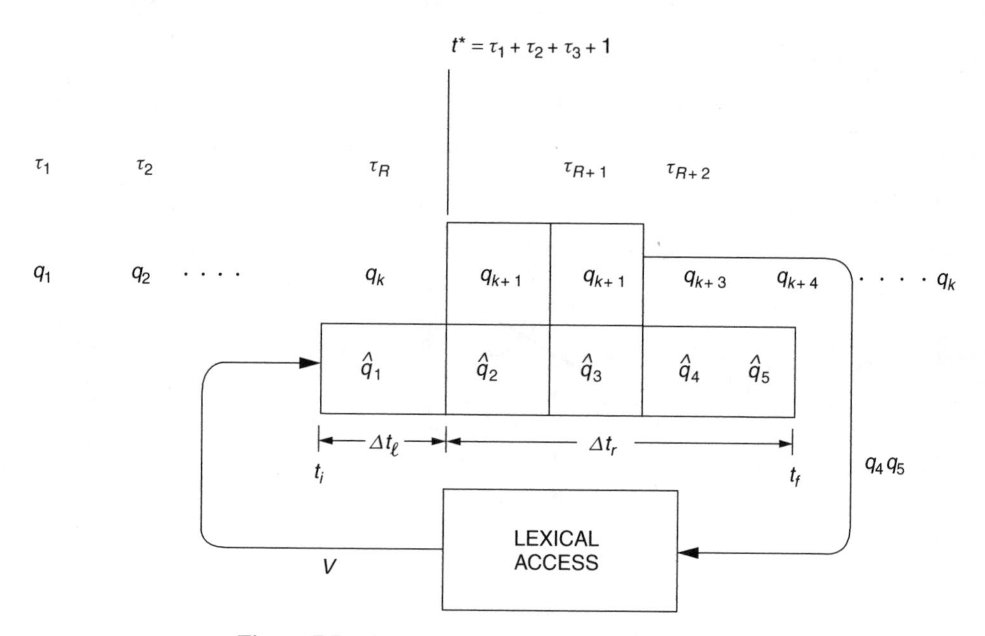

Figure 7.7 Operation of the lexical axis procedure

The matching process is a dynamic programming algorithm, illustrated in Fig. 7.8, in which the reference sequences from the lexicon are compared to the test sequence obtained from the phonetic transcription. Note that this is a comparison of symbol sequences in which insertions, deletions, and substitutions are allowed by the local constraints on the dynamic programming algorithm as shown in Fig. 7.9.

In order to implement the algorithm, we must first define the costs of substitution, deletion, and insertion of phones in the transcription. We begin with substitution since it is used to define the other two costs. Recall that each phonetic unit, w_j, is characterized by the Gaussian distribution $\mathcal{N}(\mathbf{x}, \boldsymbol{\mu}_j, \mathbf{U}_j)$. Thus it is natural to say that the cost of substituting one phone for another is just the dissimilarity between their respective distributions. In general, the dissimilarity between two arbitrary probability density functions $p_i(\mathbf{x})$ and $p_j(\mathbf{x})$ is measured by the Kullback–Leibler statistic [164]

$$D(p_i(\mathbf{x})||p_j(\mathbf{x})) = \int_{\mathbb{R}^d} p_i(\mathbf{x}) \log\left[\frac{p_i(\mathbf{x})}{p_j(\mathbf{x})}\right] d\mathbf{x}. \tag{7.15}$$

Note that although $D(p_i(\mathbf{x})||p_j(\mathbf{x}))$ is not a true metric, it is the case that $D(p_i(\mathbf{x})||p_j(\mathbf{x})) \geq 0$ with equality if and only if $p_i(\mathbf{x}) = p_j(\mathbf{x})$ for all $\mathbf{x} \in \Re^d$. This has the sensible interpretation that the cost of a correct transcription is zero.

In the Gaussian case, with

$$p_i(\mathbf{x}) = \frac{1}{\sqrt{2\pi^d}|\mathbf{U}_i^{-1}|^{1/2}} e^{-\frac{1}{2}(\mathbf{x}-\boldsymbol{\mu}_i)\mathbf{U}_i^{-1}(\mathbf{x}-\boldsymbol{\mu}_i)}, \tag{7.16}$$

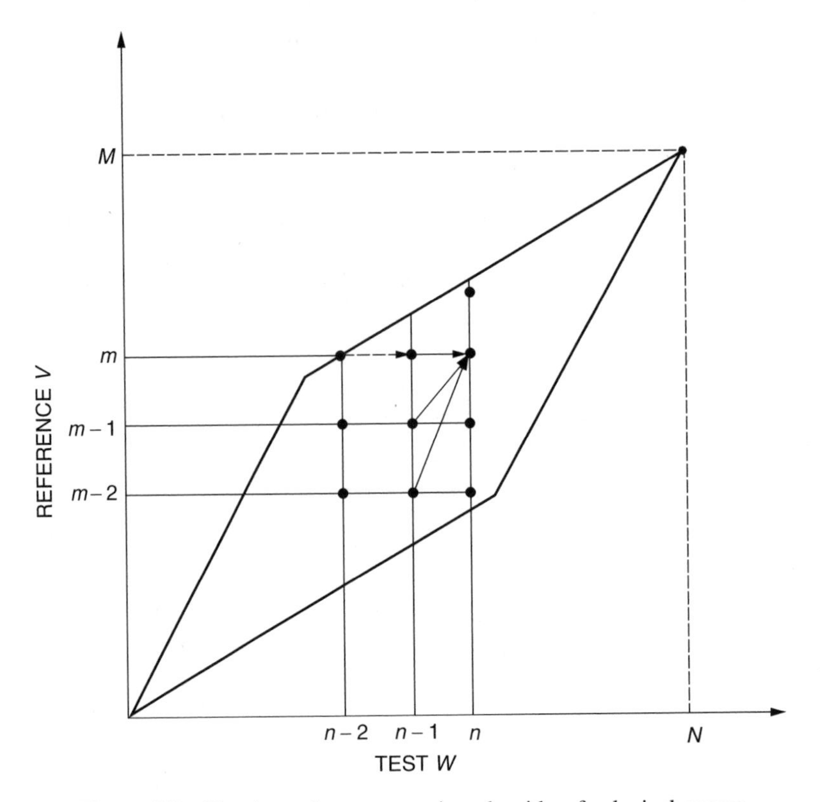

Figure 7.8 The dynamic programming algorithm for lexical access

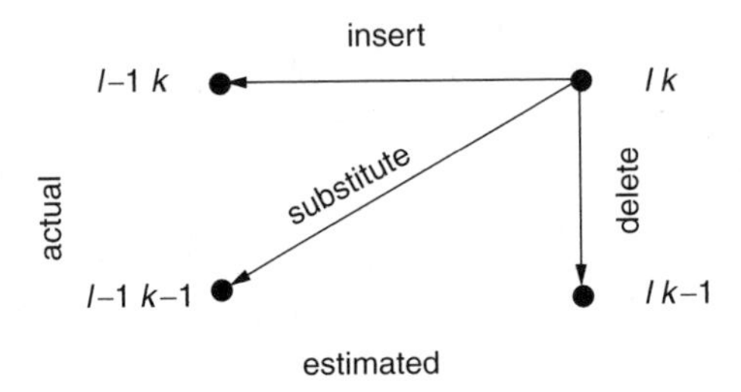

Figure 7.9 Dynamic programming algorithm local constraints

(7.15) becomes

$$D_{ij} = D(p_i(\mathbf{x})||p_j(\mathbf{x})) = \frac{1}{2}[\mathrm{tr}(\mathbf{U}_i\mathbf{U}_j^{-1}) + \log\left[\frac{|\mathbf{U}_j|}{|\mathbf{U}_i|}\right] - d + (\boldsymbol{\mu} - \boldsymbol{\mu}_j)'\mathbf{U}_j^{-1}(\boldsymbol{\mu}_i - \boldsymbol{\mu}_j)], \tag{7.17}$$

where D_{ij} is the dissimilarity between the ith phonetic unit and the jth phonetic unit.

Based on (7.17), we define the substitution cost of replacing the ith phonetic unit with the jth phonetic unit at the tth segment of the transcription as

$$S_{ij} = D_{ij}\hat{d}_t, \tag{7.18}$$

where $\hat{d}_t$ is the duration of the tth unit obtained from (7.12).

Now the following heuristic is used for the cost of insertions and deletions. A deletion is a substitution of silence (phonetic unit $i = 1$) for the decoded unit. Similarly, the cost of an insertion is that of a substitution of the inserted unit for silence. Thus the cost of deleting phonetic unit i, D_i, is

$$D_i = S_{1i}, \tag{7.19}$$

while the cost of inserting phonetic unit i is

$$I_i = S_{i1}. \tag{7.20}$$

Note that because (7.17) is not symmetric, $D_i \neq I_i$ in general.

Using the substitution, insertion, and deletion costs from (7.18), (7.20), and (7.19), respectively, the dynamic programming algorithm of Fig. 7.8 and 7.9 becomes

$$C_{lk} = \min\{C_{lk-1} + I_k, C_{l-1k} + D_k, C_{l-1k-1} + S_{lk}\}, \tag{7.21}$$

where C_{lk} is the cumulative cost of matching the corrupted phonetic transcription $\hat{q}_1\hat{q}_2\ldots\hat{q}_l$ with the phonetic spelling of word v taken from the lexicon, $q_1q_2\ldots q_k$. Then I_k is the cost of inserting $\hat{q}_k$ after q_l. Similarly, D_k is the cost of deleting $\hat{q}_k$ from the transcription and S_{lk} is the cost of substituting $\hat{q}_l$ for q_k.

The initial conditions are

$$\begin{aligned} C_{00} &= 0, \\ C_{l0} &= \sum_{j=1}^{l} D_1\hat{d}_j, \\ C_{0l} &= \sum_{j=1}^{l} I_1\hat{d}_j. \end{aligned} \tag{7.22}$$

7.2.3 Syntax Analysis

Syntax analysis is accomplished by adapting the asynchronous parser of (4.33) to the case where the log-likelihood function is replaced by the cost function C_{lk} described above. We thus have

$$\phi_k(B) = \min_{\{A\to vB\}}\{\min_l\{\phi_{k-l}(A) + C^{(v)}_{k-l,k}\}\}. \tag{7.23}$$

Note that there is a different cost for each hypothetical word v. All costs are computed from (7.21). Then, using the reconstruction procedure of (4.26)–(4.30), we get the optimal lexical transcription of the utterance and its syntactic structure.

7.3 Parameter Estimation from Fluent Speech

In order to implement the systems described in Sections 7.1 and 7.2, the parameters of their respective acoustic-phonetic models must be estimated. Because these models are intended to represent fluent speech, from any speaker, they cannot be trained from isolated utterances. Also, because of the large size of the models and the number of parameters they require, it is not practical to use speech data that has been segmented and labeled by expert phoneticians. Fortunately, there are nearly automatic procedures requiring a minimum of intervention by linguists. All that is needed is a phonetic transcription of a large corpus of fluent speech. In fact, as we shall see, even the transcription can be obtained automatically from ordinary text. Thus all that is ultimately required is speech data and its corresponding transcription.

7.3.1 Use of the Baum Algorithm

For the kinds of models used in the integrated architecture, the Baum algorithm is used to estimate the parameters of sequences of phones rather than individual ones. The training data is assumed to be a large set of spoken utterances, $\{x_i(t)\}_{i=1}^{N}$.

Each signal, $x_i(t)$, encodes a sentence W_i composed of the words $w_{i1}w_{i2}\ldots w_{in_i}$. Each w_{ij} is in a fixed lexicon V from which we can obtain the pronunciation of any entry in terms of a predetermined list of phones (e.g. the list in Table 2.2). Thus

$$w_{ij} = \phi_{k_1}\phi_{k_2}\ldots\phi_{k_{m_i}}. \tag{7.24}$$

Associated with each triphone, $\phi_{k_{j-1}}\phi_{k_j}\phi_{k_{j+1}}$ on the right-hand side of (7.24) there is an HMM of the form illustrated in Fig. 7.2. Each such model is characterized by a parameter vector, $\boldsymbol{\lambda}_{kj}$. Thus for any sentence, W_i, we can easily construct a corresponding HMM by concatenating all of the models in the correct sequence determined by the sequence of words (7.3) and the pronunciation of each word (7.24), taking into account the triphonic context for each phone including those that appear at word boundaries. Then the Baum algorithm (3.112)–(3.115) is used to estimate all the $\boldsymbol{\lambda}_{k_j}$ together from the observation sequence derived from $x_i(t)$.

This process is carried out for each W_i, $i = 1, 2, \ldots, N$. Whatever values are obtained for the $\boldsymbol{\lambda}_{k_j}$ from W_i become the initial values for the Baum iteration for W_{i+1}. Of course, in general, $W_i \neq W_{i+1}$ so different subsets of the parameters are reestimated for each W_i. Even with randomly assigned initial values, the process converges to a useful result provided the training sentences contain sufficiently many examples of each triphone. Hundreds of hours of speech data are needed, but the process is automatic and requires only computer time.

There are two ways to estimate the parameters of the model of Fig. 7.2. First, because the model has many fewer parameters than that of the integrated architecture, a far smaller training data set is required. In fact, a few hundred sentences will suffice. A data set of this size is small enough that it can be segmented and labeled by hand. Since each segment corresponds to one phone without regard for its context, all the data for a phone can be pooled and sample statistics can be calculated. These will give sufficiently good starting values for the means and covariances required by the model of (3.141) to use the reestimation procedure exactly as indicated in (3.151), (3.155), (3.158), and (3.163) on the entire training set without regard for the segments and labels.

Another approach is to use the triphonic models obtained by the procedure used to estimate the parameters of the integrated architecture in place of the observation densities. That is, use $\mathcal{L}(\mathbf{O}|\boldsymbol{\lambda}_j)$ instead of $b_j(\mathbf{O})$. The parameter $\boldsymbol{\lambda}_j$ is obtained by averaging the parameters for the jth phone over all phonetic contexts. According to this procedure, the model changes slightly. Instead of a single state for each phone, we now have a three-state, left-to-right HMM. The state transition matrix, $\mathbf{A}$, is unchanged but a_{ij} is the probability of a transition from the last state of the ith phone to the first state of the jth. It is this second method upon which the performance figures given in Section 7.4 are based. The first method has not been applied to vocabularies larger than 1000 words, so no claims for its performance can be made.

The transition parameters, a_{ij}, were obtained directly from the lexicon by counting the number of bigram sequences of phones in words and ignoring word junctures. Though a crude estimator, it is sufficient for good performance.

7.3.2 The Role of Text Analysis

The method used to obtain the parameters of the HMM used in the integrated architecture can be significantly refined by using the text analyzer of a text-to-speech synthesis system.

A text-to-speech synthesizer is a system for converting ordinary text into the equivalent acoustic signal. Such a system has two main parts: a text analyzer to generate a phonetic transliteration of the text, and a synthesizer to render the phone sequence acoustically. We will say more about the latter in Section 7.5.1.

A diagram of the text analyzer is shown in Fig. 7.10. The particular part of the system of interest here is the process by which ordinary text is transliterated into a phone sequence. This process can be used to select the triphone models to be concatenated and reestimated based on a given training sentence.

The part of the system that is relevant to this discussion is at the center of the figure and performs the function of determining the pronunciation of words from the text. There are four parts to the procedure, the pronunciation listing in a dictionary, morphological rules for deriving inflected forms from root words, rules for rhyming, and letter-to-sound rules.

The primary method of pronunciation is by direct search of the dictionary which contains a phonetic spelling for each entry. In order to keep the dictionary as small as possible, only root words are listed. Inflected forms derived from the addition of prefixes and suffixes indicating tense, number, and mood are omitted as are compound forms. In this way, the dictionary for English is reduced to approximately 40 000 entries.

The morphological rules give the pronunciations of the common inflected forms, mainly for nouns and verbs, to account for their declensions and conjugations, respectively. Irregular noun and verb forms for plurals and participles are listed in a table of exceptions.

The dictionary plus morphology will account for the pronunciation of most words. If, however, they fail to apply to a word encountered in a text, a set of rules about rhyming can be invoked. These rules will find a word which is likely to rhyme with a given word and from which the pronunciation can be determined.

Finally, if all else fails, the system resorts to letter-to-sound rules. Such rules are unreliable for English, the spelling of which is often not easily reconciled with pronunciation. However, as a last resort the rules are likely to yield an intelligible if not correct result.

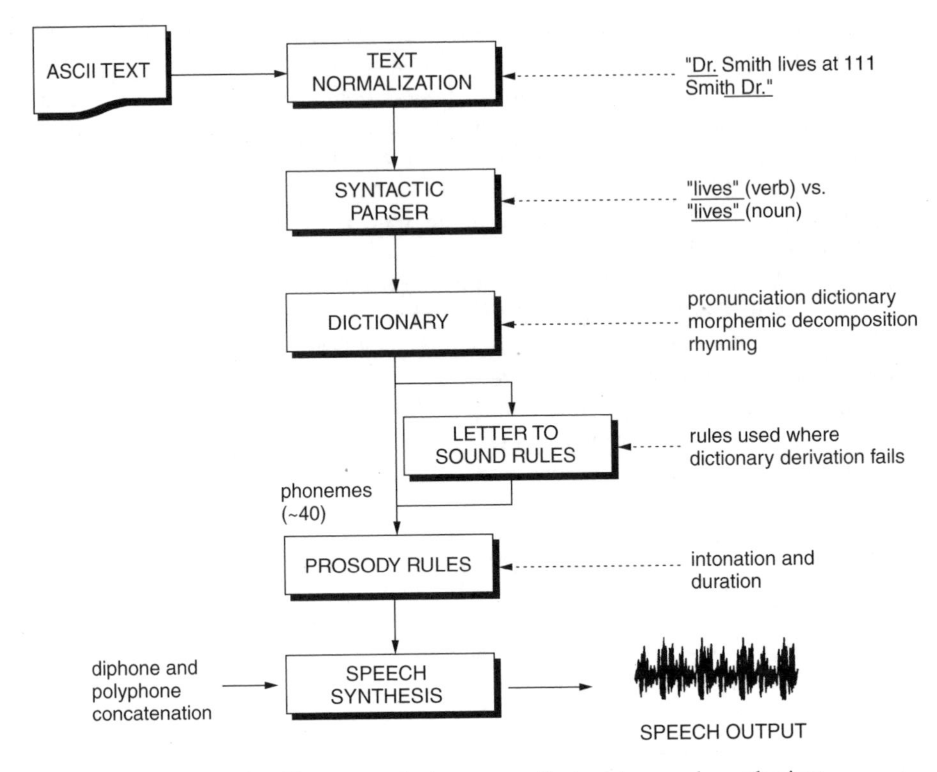

Figure 7.10 The text analysis system of a text-to-speech synthesizer

The combination of the four methods produces a surprisingly efficient and robust pronunciation. It is good enough so that a phonetic transcription laboriously generated by a linguist is not required and may safely be replaced by the automatically generated transcription. By doing so, the entire training procedure is automated. Thus large text corpora can simply be read aloud and the pronunciation obtained automatically from the text.

It is probably true that other components of the text analyzer could be used to make more refined HMMs of phonetic units. The use of prosodic features is very attractive in this regard. However, at the present time, the state-of-the-art technology does not make use of this information.

7.4 System Performance

Progress in automatic speech recognition has been significant over the past thirty years. Systems based on the integrated architecture described in Section 7.1 are commercially available as software for the now ubiquitous personal computer at prices suitable for the consumer electronics market. The same technology is embedded in large-scale telecommunications systems for use in the public switched telephone network. Most telephone subscribers have used such systems, some without even being aware they were talking to a machine, for a variety of information retrieval tasks. Laboratory prototypes of the system described in Section 7.2 have also been successfully tested.

Scientists and engineers who have witnessed the development of speech technologies over the decades are justified in noting the amazing improvement in the technology. The state-of-the-art devices are truly a marvel of engineering. Systems constrained to small vocabularies and highly circumscribed applications record accuracies above 90% averaged over unrestricted user populations. Makers of large-scale systems claim similar results for vocabularies of tens of thousands of words. However, all of the commercial systems and laboratory prototypes suffer from two deficiencies. First, performance is not robust. That is, it is subject to degradation and even catastrophic failure in just moderate signal-to-noise ratios (SNR <20 dB). Performance is also not robust with respect to accents, colloquial discourse, and variation of voice quality, especially children's voices. Second, the technology described in Sections 7.1 and 7.2 addresses only the transcription of speech into text. As such, it has no representation of meaning. We will consider the implications of these deficiencies in the final three chapters.

7.5 Other Speech Technologies

Thus far, the only technology we have considered is automatic speech recognition. There are, however, other aspects of speech communication that can be studied and used in practical applications based on the models used in automatic speech recognition. In fact these applications are easily understood in light of the methods used in automatic speech recognition. The following is a brief survey of such applications. The survey is oriented toward unsolved research problems in speech communication rather than the well-understood aspects of speech that have already found their way into applications.

7.5.1 Articulatory Speech Synthesis

There are two independent aspects of speech synthesis. First, there must be an orthographic representation (e.g. ordinary text) of the speech signal to be generated. Second, there must be a method for producing the desired acoustic signal from its orthography. The former has already been discussed in Section 7.3.2. The standard method for the latter is based on linear prediction as described in Section 2.3.2. This is really synthesis by analysis. That is, in order to perform the synthesis, a previous analysis must have been done on natural speech from which the parameters of the synthesizer are derived. While this method is suitable for many practical applications, it does not address the problem of synthesizing speech from first principles requiring no data extracted from speech signals. Speech synthesis from first principles is called articulatory synthesis and is based on the ideas developed in Section 2.1. The synthetic speech signal is the solution to the Webster equation for the boundary conditions governed by vocal tract geometry which is determined by the articulatory gestures required to make the desired phonetic elements. If a perfect model of the physical acoustics of the vocal apparatus and the articulatory dynamics were available, it should be possible to synthesize any desired voice and manner of speech. In particular, it should be possible to mimic a specific speaker based on only his individual vocal physiology. This capability would open the possibility of using an articulatory synthesizer to improve the speech of hearing-impaired children by getting them to imitate a visual display of the articulatory dynamics of normal speech by comparison with a simultaneous display of their own faulty articulation. An early attempt at this was made by Fallside [79].

Although articulatory synthesizers have been constructed by Coker [51], Hafer [115], and Huang [136], they have remained as curiosities of the laboratory because they are inferior to synthesis by analysis methods with respect to both voice quality and intelligibility. The relatively poor performance of articulatory synthesizers is a strong indication that our models of the speech signal are deficient and should be the subject of research.

7.5.2 Very Low-Bandwidth Speech Coding

High-fidelity speech can be stored and/or transmitted at approximately 50 kilobits per second (kb/s). Waveform coders can maintain the fidelity at 16 kb/s. There is some degradation in voice quality at 8 kb/s and speaker identity is lost though intelligibility is preserved at 4 kb/s. At data rates below that, speech quality is seriously compromised.

In contrast, text can be stored and/or transmitted at only 100 bits per second. This suggests the ultimate coding mechanism. Use a speech recognizer to convert speech to text, adding a few bits per second to encode individual qualities of the speaker's voice. Then decode the transmission using an articulatory synthesizer. Of course, no technology presently exists to implement this idea. It is, however, plausible and an interesting area for research.

7.5.3 Automatic Language Identification

An application that is little studied but more within the reach of existing technique is automatic language identification. The term is intended to mean naming a speaker's language from a brief sample of his natural fluent or even colloquial speech. One way to do this is to make an acoustic-phonetic model and lexicon for each language of interest. If the model is of the form described in Section 7.2, then a single decoder based on the modular architecture described there but omitting the parser can be implemented as shown in Fig. 7.11. Given an utterance in an unknown language, it could then be transcribed according to each of the models and identified as the language corresponding to the model

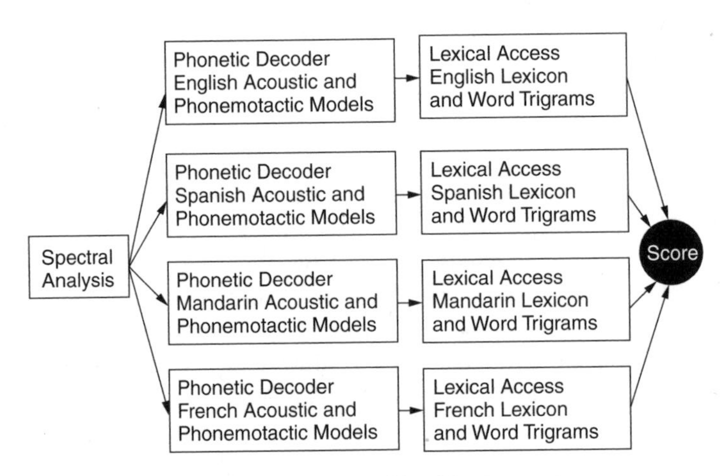

Figure 7.11 Language identification system

yielding the highest value of its likelihood function. An experiment by Hironymous [126] indicates the feasibility of this idea. More elaborate heuristics for language identification are discussed by Li [196].

7.5.4 Automatic Language Translation

Attempts to translate text from one language to another have been made from time to time since the 1950s. Such notions have inspired both writers of science fiction and comedians. Recently, text translation services have become commercially available, providing low but useful quality of translations of technical documents. There have also been some attempts to translate spoken messages from one language into another. Often referred to as translating telephony, the intended use is to allow monolingual speakers of different languages to converse over the telephone. One such experimental system built by Roe *et al.* [276] is shown in Fig. 7.12. The system comprises speech recognizers and synthesizers for both source and target languages, a bilingual dictionary, and a syntax transducer that transforms parse trees in the source language into equivalent ones in the target language. When restricted to a sufficiently limited domain such as banking and currency exchange, reliable, real-time translation of ordinary conversation is possible. A summary of the system performance is shown in Fig. 7.13.

In the distant future, one can imagine combining such a system with an advanced articulatory synthesizer, a language identification device and a low-bit-rate coder. One can then envision the following scenario. A telephone subscriber can call a long-lost relative in another country whose language he never learned. His speech and that of the

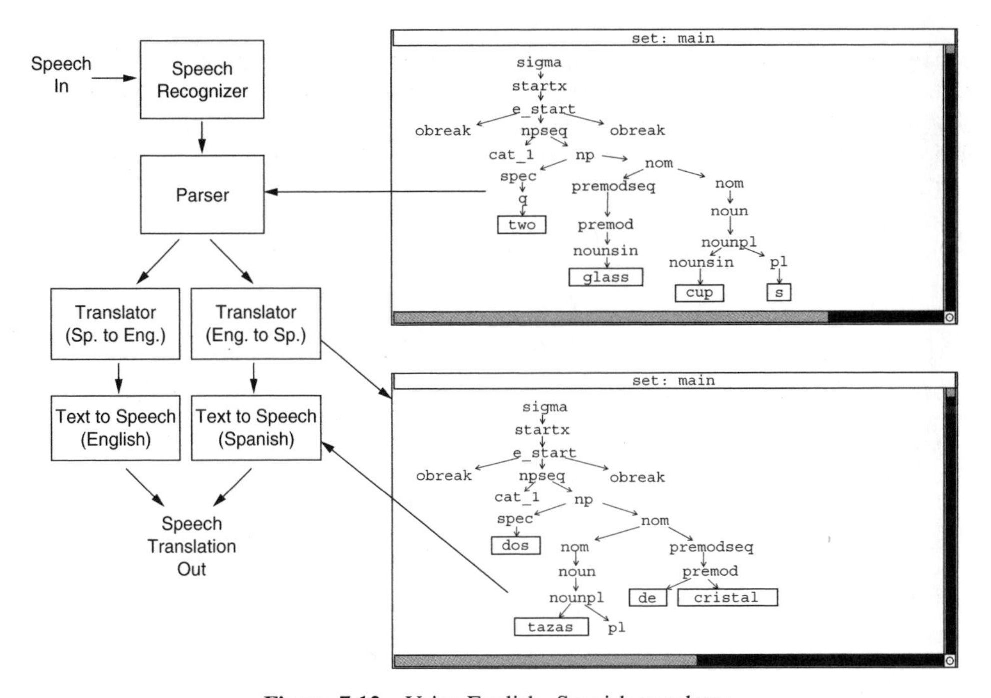

Figure 7.12 Voice English–Spanish translator

Translation		
Recognition	Translation	Percent
Correct	Good	86%
Semantically Exact	Good	4%
Semantically OK	Good	1%
Correct	OK	1%
Semantically OK	OK	3%
Bad	OK	1%
Bad	Bad	4%

Examples		
	Uttered	Translated
Sem. Ex./Good	*lire*	*lir***as**
Sem. OK/Good	*a two F com.*	*una com. de 2 F*
Correct/OK	*would i see*	*querría ver*
Sem. OK/OK	*Irish currency*	**el** *dinero irlandés*
Bad/OK	*for German marks*	**por** *marcos alem.*
Bad/Bad	*3 yen*	*a las 3 de la tarde*

96% at least OK

Figure 7.13 Performance of the spoken language translator

called party are automatically identified thereby invoking the appropriate translator. Both sides of the conversation are translated idiomatically, preserving emotional content and personality. The signals are then encoded to conserve bandwidth and decoded at the other end of the line so that each listener hears a perfect translation rendered in the actual voice of the speaker. More research is definitely needed.

8

Automatic Speech Understanding and Semantics

8.1 Transcription and Comprehension

The term "speech understanding", as it occurs in the electrical engineering literature, refers to the automatic understanding of natural spoken language by a machine such as a digital computer. It is important to realize at the outset that, as of this writing, no artificial device exists which is capable of "speech understanding" so defined. While there is no inherent reason why such a device could not be constructed, to date only machines of significantly reduced linguistic capability have been demonstrated.

The notion of "speech understanding" emerged in the early 1970s from the field of automatic speech recognition which preceded it by approximately 25 years. The two disciplines are, of course, closely related but, as Newell *et al.* [233] realized, are distinct in an important sense. Whereas recognition is the operation of transcribing speech into some conventional orthographic representation, understanding requires that the machine generate some formal symbolic representation of the meaning of the spoken message and perhaps even perform an appropriate physical activity based on the derived interpretation. Thus recognition is an abstract pattern recognition problem while understanding entails cognitive abilities.

One might well ask what the purpose of recognizing speech would be if not to understand it. It is quite possible that some of the pioneers of speech recognition envisioned speech understanding but felt that the latter was beyond the realm of any science they knew, whereas practical techniques for acoustic pattern recognition existed. They therefore restricted their efforts to the recognition of isolated words for the purpose of giving simple commands to a machine [63, 70]. For example, if it were possible to recognize a sequence of spoken numbers with distinct pauses between them, one could dial a telephone by voice. As useful as this might be, it does not qualify as speech understanding as defined here.

As soon as a spoken digit recognition machine was demonstrated, research embarked on a more ambitious recognition task, that of building a dictation machine or a voice-operated typewriter. This goal has yet to be accomplished. Arguably, the elusive nature of the voice-operated typewriter was one factor which prompted the coining of the term

Mathematical Models for Speech Technology. Stephen Levinson
 ISBN: 0-470-84407-8

"speech understanding". The differentiation between it and mere speech recognition was an implicit proposition that one must understand speech in order to transcribe it. Behind the proposition was a deep belief that understanding required an entirely different kind of computation, one that was the subject of research in artificial intelligence (AI).

This issue has yet to be resolved. As this chapter is being written there are programs commercially available that purport to take dictation. There are also experimental systems which display a rudimentary ability to understand some spoken dialogs. Neither of these is possessed of anything remotely like human linguistic and/or cognitive ability. In this brief section, we shall explore the fundamental theories and computational algorithms which are employed by existing speech understanding machines. We shall then consider some of the issues whose resolution might ultimately lead to the development of true speech understanding systems. The reader is reminded, however, that no definitive conclusions can be offered since the complexities of our human abilities to communicate in natural spoken language are still shrouded in mystery.

8.2 Limited Domain Semantics

If the vocabulary and subject matter are carefully circumscribed so as to apply only to a restricted domain of discourse and if the grammar is such that all well-formed sentences have a unique and unambiguous meaning, then automatic semantic analysis is computationally feasible. The general architecture of such a system is shown in Fig. 8.1 and its operation is explained below.

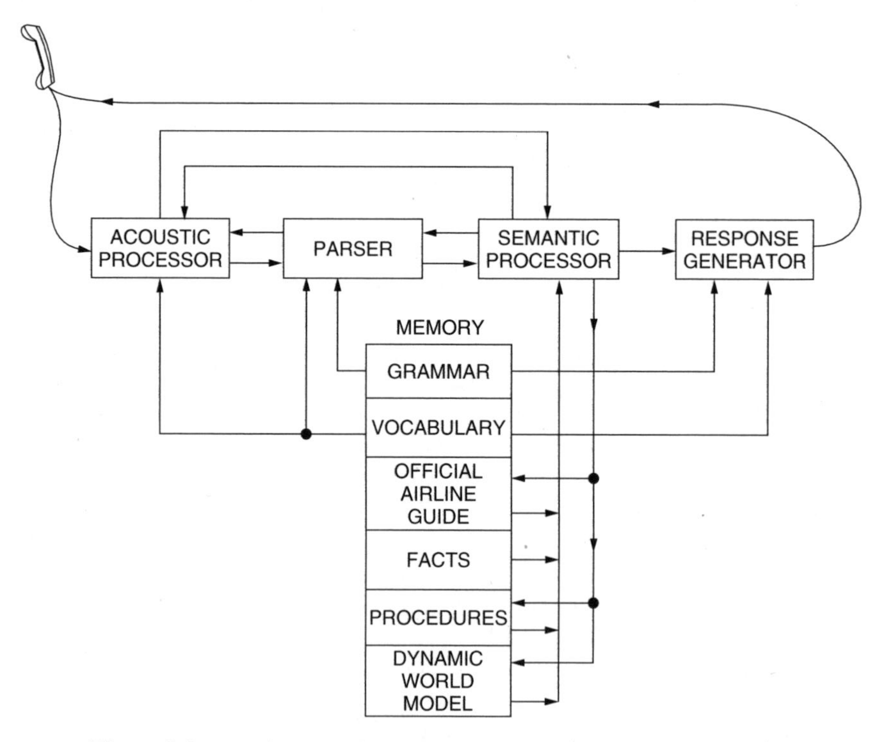

Figure 8.1 Architecture of an automatic speech understanding system

8.2.1 A Semantic Interpreter

The semantic processor of a conversational mode airline timetable and information system is essentially an interpreter for the formal language of Fig. 8.2. It may be thought of simply as a mapping

$$S : (Q \times V) \rightarrow A, \tag{8.1}$$

where Q is the set of states in the state diagram of the language, V is its terminal alphabet or vocabulary, and A is a set of actions which we define precisely below. The mapping in (8.1) is used in the following way. Let $\tilde{W}$ be the recognized input sentence and $\tilde{q}$ its state sequence, with

$$W = v_1 v_2 \cdots v_n, \quad v_i \in V, \quad \text{for } 1 \le i \le n, \tag{8.2}$$

and

$$\overline{q} = q_0 q_1 q_2 \cdots q_n, \quad q_i \in Q, \quad \text{for } 0 \le i \le n. \tag{8.3}$$

Then compute

$$S(q_i, v_i) = \alpha_i \in A, \quad \text{for } 1 \le i \le n. \tag{8.4}$$

Since S is not necessarily defined for all state–word pairs, some α_i may be Λ, the null action. The set of non-null actions determines the response, $R(\hat{W})$, to input $\hat{W}$, which we denote by

$$\{\alpha_i | \alpha_i \neq \Lambda => R(\hat{W})\}. \tag{8.5}$$

The semantic mapping, S, comprises 126 rules of the form of (8.4). To precisely define the actions, α_i, we must look at the communication aspect of semantics.

A well-known abstraction of the communication process is described by Fodor [91] and Minsky [222] as follows. For A to communicate with B, both must have a model or internal representation of the subject. A takes the state of his model and encodes it in a message which he transmits to B. B decodes the message in terms of his subject model and alters its state accordingly. Communication takes place to the extent that B's model is isomorphic to the state A's would be in had he received the same message. This is embodied in the task model, U, which is a finite universe of items which represent the categories in the database which the system understands. Actions, then, mediate between the input, the database, and the task model. An action α_i, is a 4-tuple (see Table 8.1),

$$\alpha_i = \alpha_i(X, U_j, K, U_k), \tag{8.6}$$

where $X \in V^*$ (usually $X \in V$), U_j is the present configuration of the task model, K is a response code, and U_k is the new configuration of the task model. Thus the α_i are instructions for a classical finite-state machine. The instructions correspond to the following actions: on input X with the present state of model U_j, respond with a sentence of form K and change the state of the model to U_k.

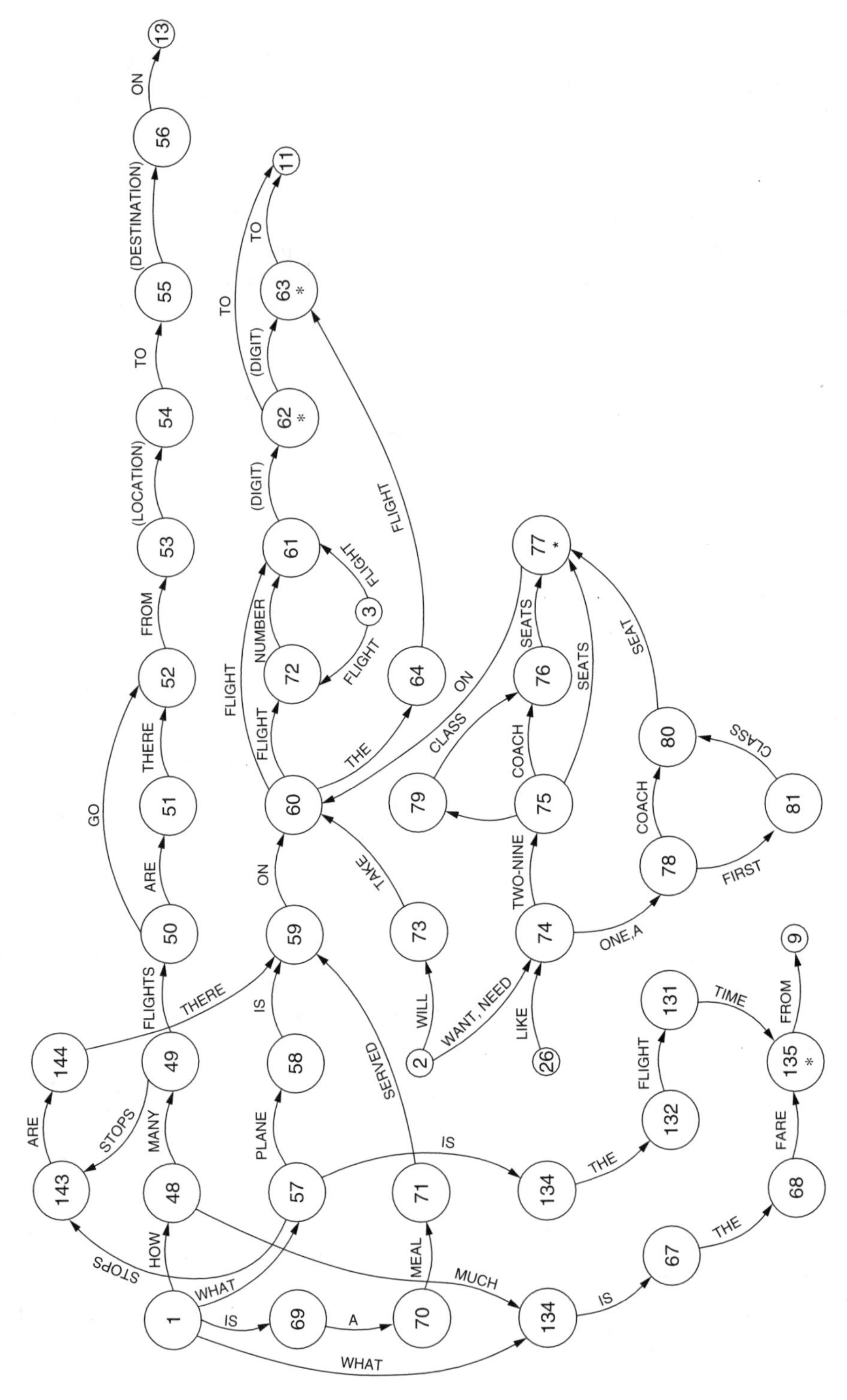

Figure 8.2 Formal language used by the semantic interpreter

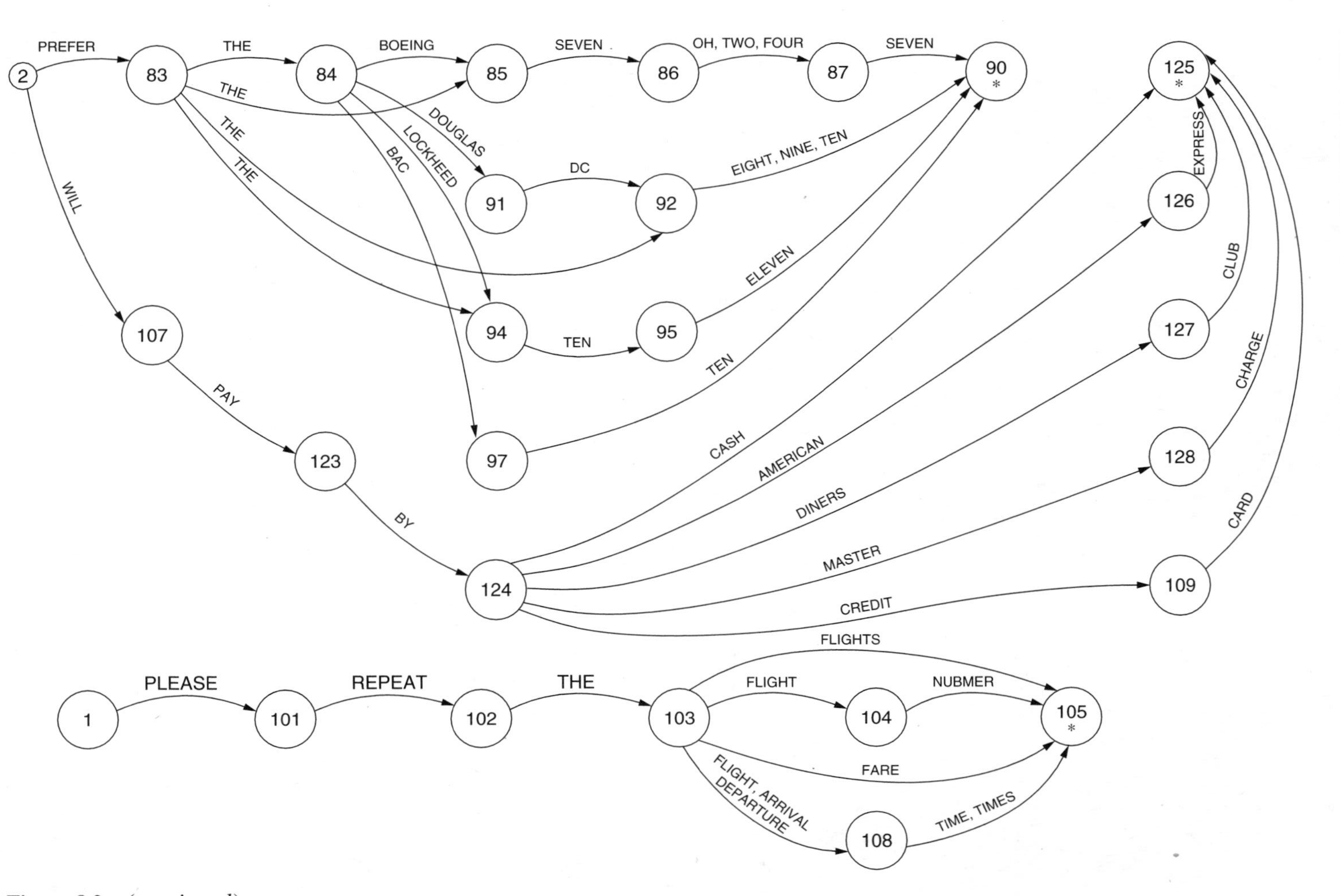

Figure 8.2 *(continued)*

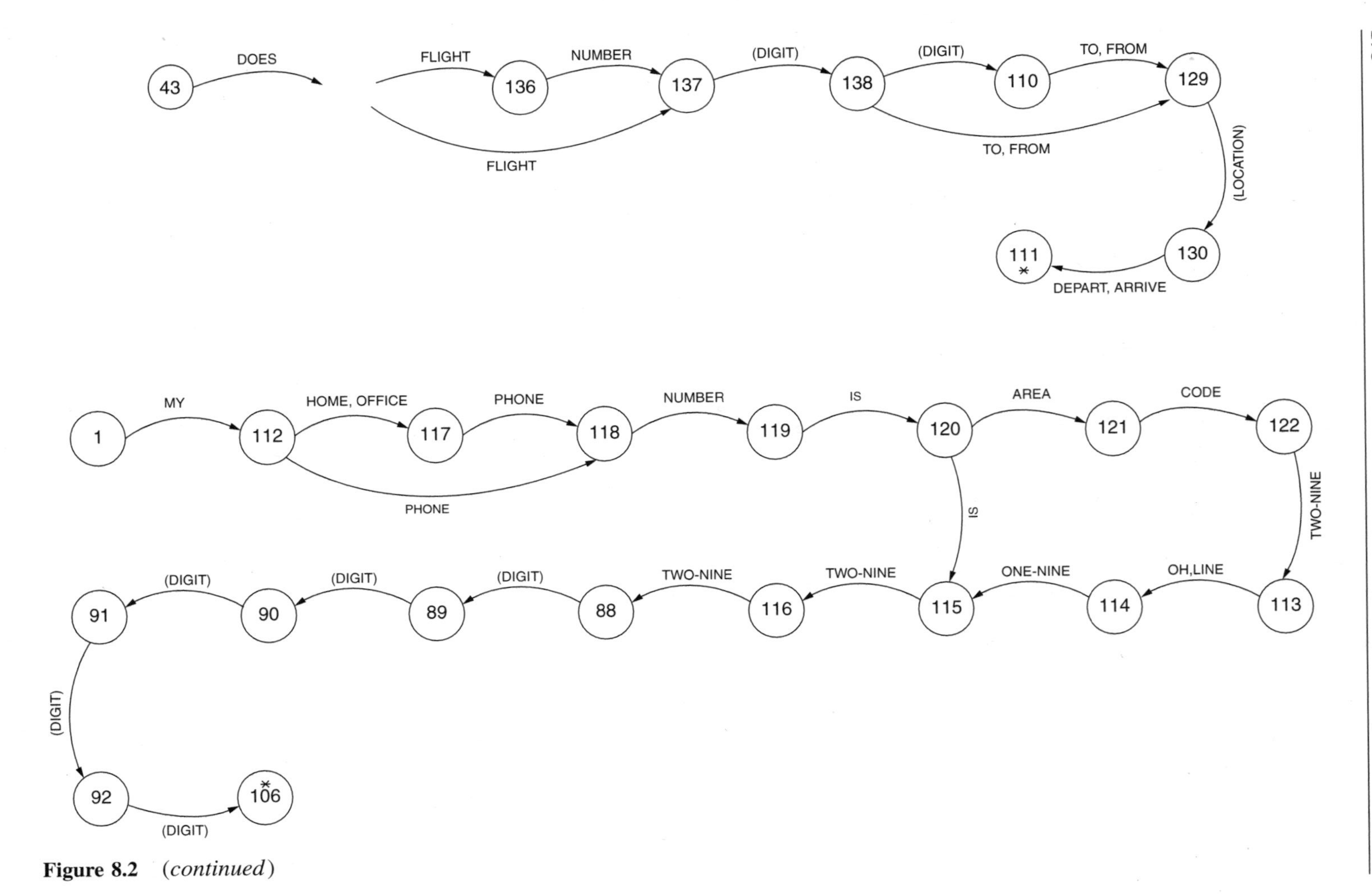

Figure 8.2 *(continued)*

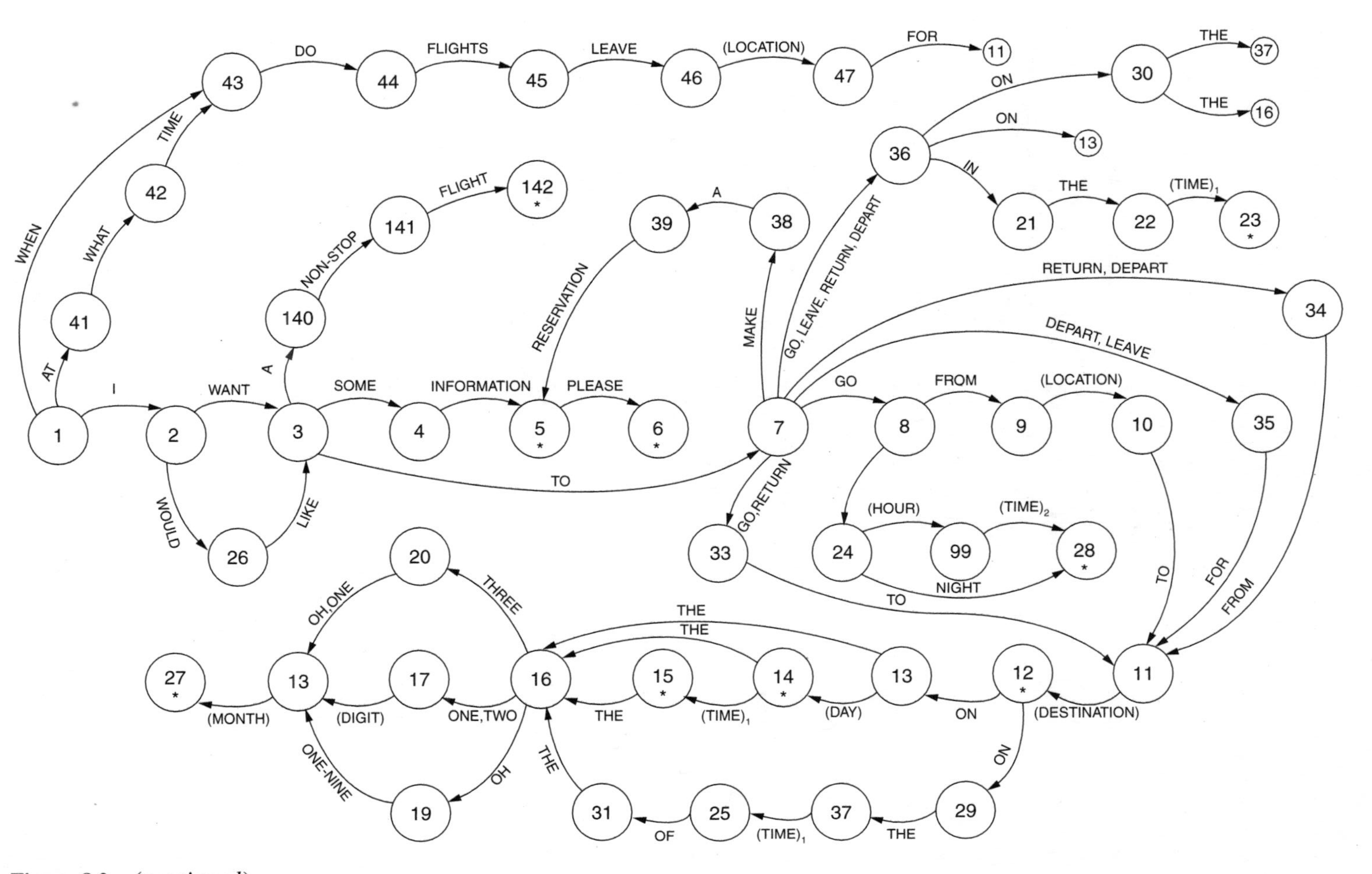

Figure 8.2 *(continued)*

Table 8.1 Semantic categories

Category	Sample
1 Information	I want some information, please.
2 Reservation	I would like to make a reservation.
3 Travel plans	I want to go to Boston on Monday evening.
4 General flight departure times	At what times do flights leave Chicago for Denver on Thursday afternoon?
5 Number of flights	How many flights go from Washington to Miami on the morning of the oh one May ?
6 Aircraft type	What plane is on flight number five?
7 Fare	How much is the fare from Detroit to Seattle on Sunday?
8 Meals	Is a meal served on the flight?
9 Flight choice	I will take flight six one to Philadelphia.
10 Seat selection	I need two first-class seats.
11 Aircraft choice	I prefer the Boeing seven oh seven.
12 Exact time specification	I want to leave at six a.m.
13 Repeat information	Please repeat the departure time.
14 Specific flight times	When does flight number one to Los Angeles arrive?
15 Method of payment	I will pay by American Express.
16 Phone number	My home phone number is five three six two one five two.
17 Non-stop flight request	I would like a non-stop flight.
18 Elapsed time	What is the flight time from New York to Denver on Wednesday night?
19 Stops	How many stops are there on the flight to Miami?

There are 15 elements of the task model; these are defined in Table 8.2. There are five ways to alter the state of the task model. Information can be directly given by the user; he can, for example, specify his destination, D. We can denote this by

$$u_1 \leftarrow D. \tag{8.7}$$

Next, we have default values which can be imposed. For example, unless otherwise specified, the number of tickets, N_t, is assumed to be one, and we have

$$u_{11} \leftarrow N_t = 1. \tag{8.8}$$

A database lookup can also alter the state of the u-model as follows. A flight number, N_f, a destination, D, and a class, C, provide sufficient information to look up the fare, F, in the Official Airline Guide. Thus

$$[u_1 = D \Lambda u_6 = N_f \Lambda u_7 = C] \stackrel{L}{=>} u_{10} = F. \tag{8.9}$$

An element of U can be computed from the values of other elements, for example, flight time, T_f, determined by point of origin, O, destination, D, arrival time, T_a, and departure

Table 8.2 Elements of the task model

Element	Symbol	Definition
u_1	D	Destination city
u_2	M	Meals served
u_3	D_w	Day of the week
u_4	T_d	Departure time
u_5	T_a	Arrival time
u_6	N_f	Flight number
u_7	C	Flight class
u_8	A	Aircraft type
u_9	N_s	Number of stops
u_{10}	F	Fare
u_{11}	N_t	Number of tickets
u_{12}	N_p	Telephone number
u_{13}	P	Method of payment
u_{14}	T_f	Elapsed (flight) time
u_{15}	O	Flight origin city

time, T_d. Origin and destination supply time zone information, while arrival and departure time give elapsed time. We say, then, that

$$[u_1 = D \Lambda u_4 = T_d \Lambda u_5 = T_a \Lambda u_{15} = 0] \stackrel{\oslash}{=>} u_{14} = T_f. \tag{8.10}$$

Finally, an element of U can be computed from user-supplied information which is not part of a flight description and is not stored as such. For instance, a departure date uniquely specifies a day of the week, D_w, by

$$[n_m \Lambda n_d \Lambda n_y] \stackrel{F}{=>} u_3 = D_w, \tag{8.11}$$

where n_m is the month, n_d is the date, n_y is the year, and F is a perpetual calendar function.

We can now give an example of a complete action. Suppose W was a request for the fare of a previously selected flight. Semantic decoding would enable action no. 14:

$$a_{14} = (\text{How much fare}, u_{10} = F \neq 0, K = 23, u_{10} = F). \tag{8.12}$$

That is, on a fare request, if u_{10} is some non-zero value, set the response code, K, to 23 and leave u_{10} unchanged. A value of $F = 0$ would indicate that a flight had not been selected as illustrated in (8.12), and a different response code would be issued, causing a message so indicating to be generated. The complete ensemble of actions which the system needs to perform its task is composed of 37 4-tuples similar to that of (8.6). This brings us to consideration of the response generation procedure. Responses in the form of English sentences are generated by the context-free grammar, G_s:

$$G_s = [V_N, V_T, \sigma_0, P], \tag{8.13}$$

where V_N is the set of non-terminal symbols, V_T is the set of terminal symbols (a vocabulary of 191 English words), σ_0 is the start symbol, and P the set of production rules. The production rules are of two forms:

$$\sigma_0 \to \gamma \in (V_N \cup V_T)^* \tag{8.14}$$

and

$$B \to b; \quad B \in V_N, \quad b \in V_T \text{ or } b = \lambda, \tag{8.15}$$

where λ is the null symbol. There are 30 productions of the form of (8.14) in P. Each one specifies the form of a specific reply and is designated by a response code, K. There are several hundred productions of the type of (8.15). Their purpose is to insert specific information into the skeleton of a message derived from a production of the other kind. As an example, consider an input requesting to know the number of stops on a specific flight. The appropriate response code is $K = 26$ and the production rule to which it corresponds is

$$\sigma_0 \to \text{THIS FLIGHT MAKES } B_1 B_2.$$

If $u_9 = N_s = 2$, then the following productions will be applied:

$$B_1 \to \text{TWO},$$
$$B_2 \to \text{STOPS},$$

resulting in the output string of symbols

$$\mathcal{S} = \text{THIS FLIGHT MAKES TWO STOPS.}$$

In the actual implementation, $\mathcal{S}$ is represented in the form of a string of ASCII characters. This is the form accepted by the text-to-speech synthesizer [236, 237] which produces an intelligible speech signal from $\mathcal{S}$ by rule.

8.2.2 *Error Recovery*

The components of the system described in Section 8.2.1 are integrated under a formal control structure shown in the flow chart of Fig. 8.3. It has two modes of operation, a normal mode and one for recovery from some error condition. The former is quite straightforward and is best illustrated by a complete example of the system operation. Consider the input sentence $\hat{W}$ = I WANT TO GO TO BOSTON ON TUESDAY MORNING. The state diagram of the sentence is Fig. 8.4, from which we immediately see that state sequence, $\overline{\mathbf{q}}$, is

$$\overline{\mathbf{q}} = (1, 2, 3, 7, 33, 11, 12, 13, 14, 15).$$

Four state–word pairs from S apply:

$$(33, \text{GO}) = \alpha_1,$$
$$(12, \text{BOSTON}) = \alpha_2,$$
$$(14, \text{TUESDAY}) = \alpha_3,$$
$$(15, \text{MORNING}) = \alpha_5.$$

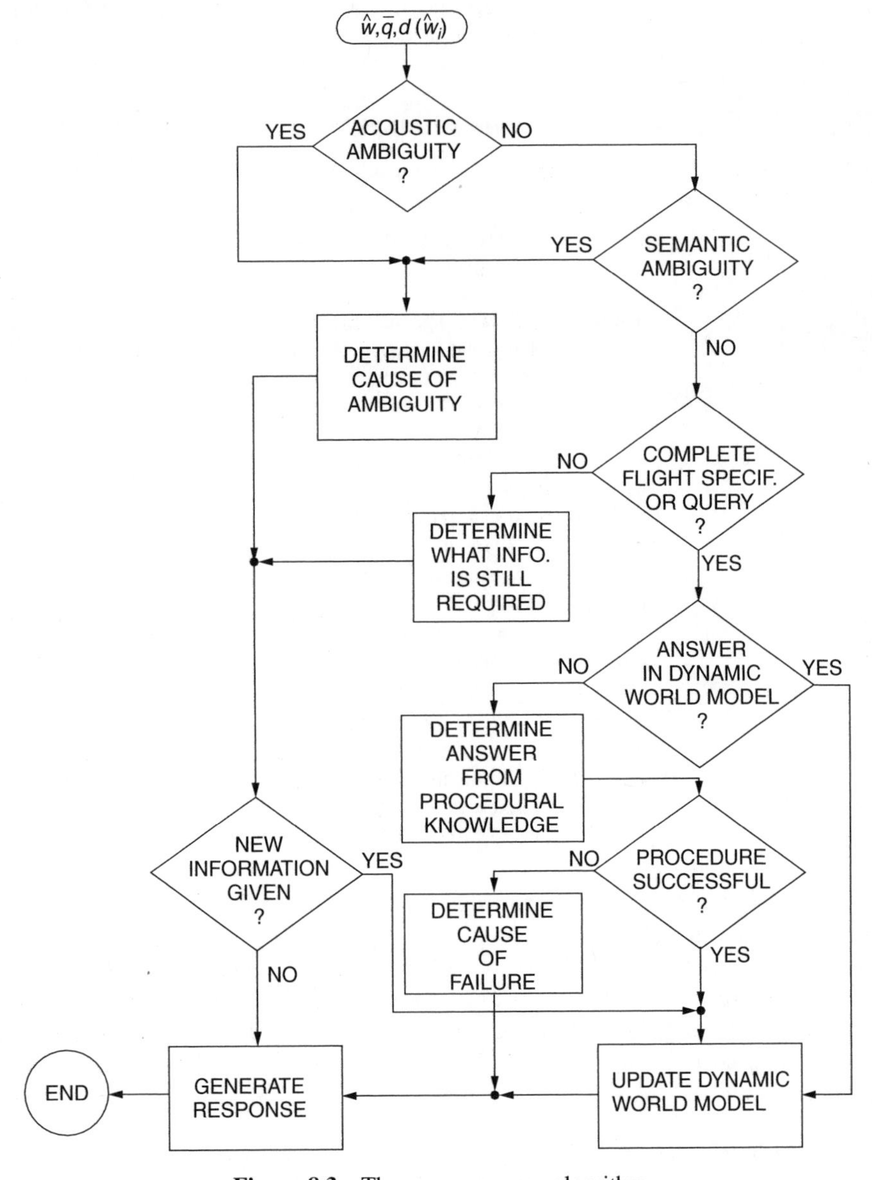

Figure 8.3 The error recovery algorithm

The actions invoked are the following:

$\alpha_1 = (\text{GO}, U, 0, U_0),$
$\alpha_2 = (\text{BOSTON}, U_0, 0, U_1 \leftarrow U_0 + u_1 \leftarrow D),$
$\alpha_3 = (\text{TUESDAY}, U_1, 0, U_2 \leftarrow U_1 + u_3 \leftarrow D_w),$
$\alpha_5 = (\text{MORNING}, U_2, 1, U_3 \leftarrow U_2 + U_4 \leftarrow T_d; U_3 \stackrel{L}{=}> C).$

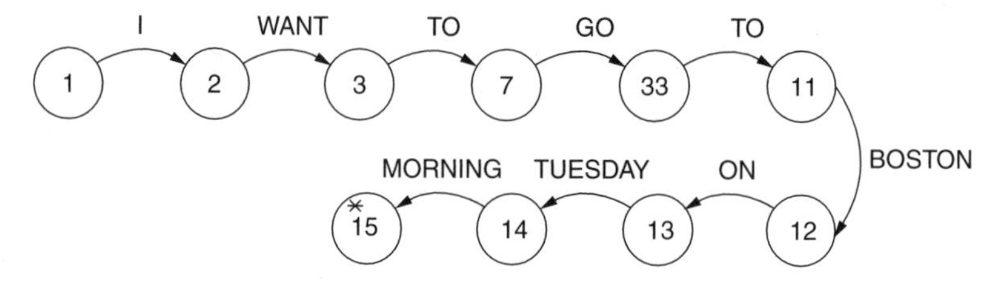

Figure 8.4 State sequence for semantic analysis

Table 8.3 State of the task model after processing the input sentence of Fig. 8.4

$u_1 = 1$	Boston
$u_2 = 0$	no meals
$u_3 = 2$	Tuesday
$u_4 = 1000$	ten a.m.
$u_5 = 1047$	ten forty seven a.m.
$u_6 = 3$	flight number three
$u_7 = 1$	coach (by default)
$u_8 = 208$	DC-9
$u_9 = 0$	no stops
$u_{10} = 56$	\$56.00
$u_{11} = 1$	1 seat (by default)
$u_{12} = 0$	phone number unknown
$u_{13} = 0$	method of payment unknown
$u_{14} = 0$	flight time not calculated
$u_{15} = 7$	New York (by default)

Action α_1 causes the task model in any state to be initialized to state U_0 and no response to be made. Next, α_2 changes the state from U_0 to U_1 by fixing the destination; no response is generated. Similarly, α_3 causes the day of the week to be defined. Finally, α_5 fixes an approximate hour of departure permitting a database lookup which gives a complete flight specification. The response code is set to 1. The state of the task model after the lookup is shown in Table 8.3. The response code, $K = 1$, causes application of the production rule

$$\sigma_0 \rightarrow \text{FLIGHT NUMBER } B_1 B_2 \text{ LEAVES } B_3 \text{ AT } B_4 B_5 B_6 B_7$$
$$\text{ARRIVES IN } B_8 \text{ AT } B_9 B_{10} B_{11} B_{12}.$$

From the meaning corresponding to u_6, we have

$$B_1 \rightarrow \lambda,$$
$$B_2 \rightarrow \text{THREE}.$$

From u_{15} we get

$$B_3 \rightarrow \text{NEW YORK}.$$

From u_4,

$$B_4 \rightarrow \lambda,$$
$$B_5 \rightarrow \lambda,$$
$$B_6 \rightarrow \text{TEN},$$
$$B_5 \rightarrow \text{A.M.}$$

From u_1,

$$B_8 \rightarrow \text{BOSTON}.$$

And finally, from u_5,

$$B_9 \rightarrow \text{TEN},$$
$$B_{10} \rightarrow \text{FORTY},$$
$$B_{11} \rightarrow \text{SEVEN},$$
$$B_{12} \rightarrow \text{A.M.}$$

Thus, $\mathcal{S}$ is FLIGHT NUMBER THREE LEAVES NEW YORK AT TEN A.M. ARRIVES IN BOSTON AT TEN FORTY SEVEN A.M. The voice response unit then utters the sentence.

An error condition occurs, putting the system in error recovery mode, for one of two reasons. Either the total distance for one or more content words exceeds a preset rejection threshold, or, due either to an error on the part of the user of the system or a catastrophic failure of the acoustic/syntactic processing, the $(\overline{q}, \hat{W})$ pair is inconsistent with the current state of the task model. The error recovery mode is essentially an elaborate heuristic, the purpose of which is to prevent communication from collapsing in the presence of ambiguity. The procedure is shown schematically in the flow chart of Fig. 8.3. The effect of this heuristic is to formulate a response to the input responsible for the error condition which will elicit from the user whatever information is required to resolve the ambiguity. The difficulty of this task is somewhat reduced by the fact that, by construction of the grammar, the appearance of a syntactic ambiguity is impossible. The decision blocks in the flow chart choose the sentential form of the response (i.e. production rules of the form (8.14)) while the processing blocks select the appropriate terminal symbols using rules of the form (8.15). Some examples of operation in this mode are given below.

By detaching the semantic processor, we can measure the accuracy with which the syntax-directed level-building algorithm can transcribe sentences. For this purpose, a set of 50 sentences using every vocabulary word and every state transition was constructed. These sentences were then spoken over dialed-up telephone lines by four speakers, two male and two female, at an average rate of 171 words/min. The test sentences ranged in length from 4 to 17 words. Two sentences containing telephone numbers were only weakly syntactically constrained, while others requesting or giving flight information were quite stylized. The utterances were band-pass-filtered from 200 to 3200 Hz, digitized at 6.67 kHz sampling rate and stored on disk files which were subsequently input to the syntax-directed level-building algorithm. The results of this test are summarized in Table 8.4.

In order to keep response time to a minimum, all on-line tests were performed in the speaker-dependent mode. Even under these conditions it takes about 1 minute to get a

Table 8.4 Effects of syntax on speech recognition

Speaker (sex)	Total sentences	Total words	Speaker-trained		Speaker-independent	
			Correct sentences	Correct words	Correct sentences	Correct words
S.E.L. (M)	59	478	56	470	55	470
A.E.R. (M)	50	424	46	417	42	413
K.L.S. (F)	50	424	42	393	43	413
C.E.S. (F)	50	424	35	349	41	408
Totals	209	1750	179	1629	181	1704
			86%	93%	87%	97%

response to a sentence 2 seconds length. To determine the correct response rate of the system, it was evaluated on a corpus of 351 sentences spoken by five male and two female speakers. After syntactic processing, 21 sentences contained word recognition errors. However, after semantic analysis 15 of these were correctly understood, despite the errors, and correct responses were made. In the remaining 6 cases, a semantic anomaly was detected, causing the machine to pose a question to the user who could then correct the problem. Thus we achieved a recognition accuracy of more than 98% and, when recognition errors occurred, communication did not break down.

These two tests give a statistical characterization of the system as one which is reasonably robust and fails gracefully. It is important, however, to look at some details of performance. For this purpose we constructed several 'problems' for the system to 'solve' each of which was supposed to reveal its ability to perform specific functions. Some sample transactions are listed below. The first is the simplest kind of interchange, a simple flight request.

Machine: THIS IS THE BELL LABORATORIES FLIGHT INFORMATION SYSTEM MAY I HELP YOU
User: I WOULD LIKE TO MAKE A RESERVATION PLEASE
M: PLEASE SPECIFY YOUR TRAVEL PLANS
U: I WANT TO GO TO BOSTON ON TUESDAY MORNING
M: FLIGHT NUMBER THREE LEAVES NEW YORK AT TEN A.M. ARRIVES IN BOSTON AT TEN FORTY SEVEN A.M.
U: I WANT TO LEAVE AT ELEVEN O'CLOCK
M: FLIGHT NUMBER FOUR LEAVES NEW YORK AT ELEVEN A.M. ARRIVES IN BOSTON AT ELEVEN FORTY FIVE A.M.
U: I WILL TAKE FLIGHT NUMBER FOUR
M: YOUR RESERVATION ON FLIGHT NUMBER FOUR TO BOSTON ON TUESDAY IS CONFIRMED THANK YOU

A slightly more interesting conversation is the following:

M: THIS IS THE BELL LABORATORIES FLIGHT INFORMATION SYSTEM MAY I HELP YOU

U: I WOULD LIKE TO GO TO CHICAGO
M: PLEASE SPECIFY THE DAY OF YOUR DEPARTURE
U: I WANT TO LEAVE AT FOUR P.M.
M: PLEASE SPECIFY THE DAY OF YOUR DEPARTURE
U: I WANT TO DEPART ON THE TWO FIVE MAY
M: FLIGHT NUMBER NINE TWO LEAVES NEW YORK AT FOUR P.M. ARRIVES IN CHICAGO AT FIVE TWENTY FIVE P.M.
U: WHAT PLANE IS ON THE FLIGHT
M: THE AIRCRAFT IS A BOEING SEVEN TWENTY SEVEN
U: HOW MUCH IS THE FARE
M: THE ROUND TRIP COACH FARE IS ONE HUNDRED FORTY TWO DOLLARS
U: I WANT A FIRST CLASS SEAT
M: ONE SEAT THANK YOU
U: WHAT IS THE FARE
M: THE ROUND TRIP FIRST CLASS FARE IS ONE HUNDRED EIGHTY FOUR DOLLARS
U: PLEASE REPEAT THE ARRIVAL TIME
M: THE ARRIVAL TIME IS FIVE TWENTY FIVE P.M.
U: I WILL TAKE THE FLIGHT
M: YOUR RESERVATION ON FLIGHT UMBER NINE TWO TO CHICAGO ON FRIDAY IS CONFIRMED THANK YOU

The following set of user inputs was carefully chosen to give the machine some difficulty:

M: THIS IS THE BELL LABORATORIES FLIGHT INFORMATION SYSTEM MAY I HELP YOU
U: I WANT ONE FIRST CLASS SEAT ON FLIGHT NUMBER THREE ONE TO DENVER ON SUNDAY
M: I AM SORRY THE FLIGHT IS NOT AVAILABLE ON SUNDAY
U: I WANT TO GO ON SATURDAY
M: FLIGHT NUMBER THREE ONE LEAVES NEW YORK AT NINE FORTY FIVE A.M. ARRIVES IN DENVER AT ONE TEN P.M.
U: WHAT IS THE FLIGHT TIME
M: THE FLIGHT TIME IS FIVE HOURS AND TWENTY FIVE MINUTES
U: HOW MANY STOPS ARE THERE ON THE FLIGHT
M: THIS FLIGHT MAKES ONE STOP
U: I WANT A NON-STOP FLIGHT
M: I AM SORRY THERE ARE NO MORNING NON-STOP FLIGHTS AVAILABLE
U: I WANT TO GO AT FOUR P.M.
M: FLIGHT NUMBER ONE SIX LEAVES NEW YORK AT FOUR FIFTY P.M. ARRIVES IN DENVER AT SEVEN P.M.
U: WHAT STOPS ARE THERE ON THE FLIGHT
M: THIS FLIGHT MAKES NO STOPS
U: IS A MEAL SERVED ON THE FLIGHT
M: DINNER IS SERVED

U: I WILL TAKE FLIGHT SIX FIVE
M: I AM SORRY THERE IS NO FLIGHT SIX FIVE TO DENVER
U: I WILL TAKE FLIGHT ONE SIX
M: YOUR RESERVATION ON FLIGHT NUMBER ONE SIX TO DENVER ON SATURDAY IS CONFIRMED THANK YOU

Questions need not occur in a logical sequence, as illustrated below:

M: THIS IS THE BELL LABORATORIES FLIGHT INFORMATION SYSTEM MAY I HELP YOU
U: AT WHAT TIME DOES FLIGHT FOUR FOUR TO LOS ANGELES DEPART
M: THE DEPARTURE TIME IS NINE P.M.
U: WHAT IS THE FARE FROM NEW YORK TO SEATTLE ON SUNDAY MORNING
M: THE ROUND TRIP COACH FARE IS THREE HUNDRED EIGHTY EIGHT DOLLARS

Finally, we observe that the machine can, to some extend, deal with non-sequiturs:

M: THIS IS THE BELL LABORATORIES FLIGHT INFORMATION SYSTEM MAY I HELP YOU
U: I WANT SOME INFORMATION
M: WHAT DO YOU WANT TO KNOW
U: I WILL TAKE THE FLIGHT
M: WHAT DID YOU SAY
U: IS A MEAL SERVED ON THE FLIGHT
M: FOR WHAT FLIGHT ARE YOU REQUESTING INFORMATION

From the above, the reader can easily observe that the dialogs which the system can sustain are not highly natural or sophisticated. The fact remains that speech communication, however stilted, has been achieved. Two results of this achievement are of significance. First, when one interacts with a system that communicates in a merely vaguely natural way, his perception of this machine is changed. Conventional notions of speech recognition accuracy and algorithms for data retrieval assume a secondary importance as attention is sharply focused on transmission of information. It is quite clear that the state of the art in speech recognition is advanced enough to support research in complete human–machine communication systems. Second, the synergistic effect of integrating several crude components into an interactive system is to produce a machine with greater capacities than might otherwise be expected. As the sophistication of the components increases and as their interaction becomes more complex, their behavior will at some point become a surprise even to their builders. In this last regard, one point should be emphasized. Unlike most systems which are reputed to be intelligent, the response of this one to a given input *cannot* be predicted, nor can a particular type of behavior be produced on demand. In this sense, the system can create surprises, even for its constructors.

8.3 The Semantics of Natural Language

The method of semantic analysis for limited domains discussed in Section 8.2 does not truly capture the semantics of natural language. The meanings of words are restricted to their use as information in the database but the general common-sense meanings are not present. Thus "Boston" is merely a page in the Official Airline Guide, not a city nor any of the things that we ordinarily associate with the notion of "city". The same is true of the semantics of "going" or "time".

Formalizing the semantics of natural language in all of its generality is a difficult problem to which there is presently no comprehensive solution. Most research on the subject rests on two principles. The first is that semantics depends on syntax. The second is that semantic analysis must generate a symbolic representation of the physical world that allows for predictions of reality by reasoning, is expressive enough to extend to all aspects of reality, and allows for different syntactic structures to generate the same meaning.

Syntax is connected to semantics in two principle ways. Structure-building rules of the form $A \longrightarrow BC$, $A, B, C \in V_N$, provide an abstraction of meaning. For example, the abstract meaning of $S \longrightarrow NP\ VP$ is that a sentence has an actor (*NP*), an action, and an object acted upon (*VP*). Then, the second syntactico-semantic relationship is captured by lexical assignment rules of the form $A \longrightarrow w$, $A \in V_N$, $w \in V_T$. Thus the meaning of the word, w, is the real concept to which it refers and its syntactic rule is determined by the part of speech represented by A. By applying lexcial semantics to the abstract structure, a specific meaning is obtained.

There are many variations on this idea, all of which fall into two categories: graphical methods and logical methods. In graphical methods, the nodes of a graph represent lexical semantics and the directed, labeled edges of the graph express relationships between the nodes they connect. Making the edges directed allows the notion of ordering the nodes in time, space or other scales.

Logical methods rest on the idea that sentences have meaning when they make true assertions about the world. The truth is verified by formal logical operations. If a sentence can be shown to be true then its meaning can be derived with the help of syntax.

8.3.1 *Shallow Semantics and Mutual Information*

Although it does not address the question of semantic structure or the means by which words are related to reality, information theory does give a means of capturing word sense. That is, words have different connotations when they are used in conjunction with other words. The word "bank" refers to different things when we speak of a "river bank" and a financial institution. The deep semantics of these usages is the existence of a common meaning, if there is one.

In the integrated architecture, semantics is only weakly represented by mutual information. That is the word pair $v_1 v_2$ has mutual information

$$I(v_1, v_2) = \log \frac{p(v_1, v_2)}{p(v_1)p(v_2)}. \tag{8.16}$$

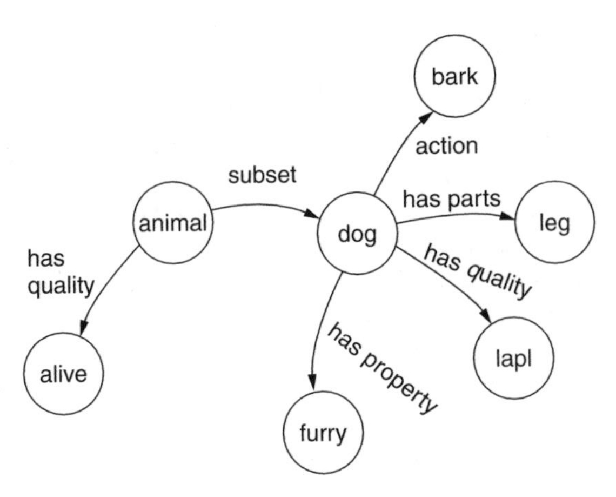

Figure 8.5 A typical semantic network

Values of $I(v_1, v_2)$ may be used to bias the n-gram probabilities. This is semantic information in the sense that two words that appear together frequently may have similar or complementary meanings.

8.3.2 *Graphical Methods*

Most graphical interpretations of semantics, of which there are numerous variants, can be traced back to the semantic net of Quillian [254]. In semantic nets, the relationships that label the edges are those of subset, quality, property, conjunction, disjunction, negation, instance, etc. The method is best explained by the diagram of Fig. 8.5 which represents the concept of "dog". The unstated assumption of this approach is that the required fidelity and expressiveness of this model can be achieved simply by exhaustively and laboriously making graphs of all the many objects in the world and connecting them appropriately. Unfortunately, to date, only toy examples have been implemented to demonstrate the principle.

8.3.3 *Formal Logical Models of Semantics*

Logical methods of semantic analysis rely on two types of formal logic, the propositional logic and the first-order predicate calculus, the latter being an extension of the former.

An important motivation for the development of mathematical logic was to address questions about the foundations of mathematics and particularly the nature of mathematical truth. Although mathematical truth is not the same as psychological truth, there is a strong intuitive sense that there is a close relationship between logical reasoning and how our minds know the "truth" about our quotidian existence and how that knowledge is expressed in natural language. It is this intuition that we will examine now. Then, in Chapter 9, we will return to a consideration of the consequences of formal logic in the development of mathematical models of cognition.

If there is to be a mathematical model of the general semantics of natural language – as opposed to the highly circumscribed domains of discourse exemplified by the method of

Section 8.2 – it must include a complete mental representation of physical reality. Formal logic is well suited to the task in that it has the following desirable properties.

First, it provides a verifiable model, the ability of which to represent reality can be empirically evaluated. It allows for a canonical meaning that can be expressed in different syntactic structures. It includes an inference procedure whereby conclusions can be drawn about specific or related ideas and events. Finally, it is expressive enough to encompass a complete model of the world.

A brief description of formal logic will suffice to demonstrate how a logical formulation of semantics displays these desirable properties. We begin with propositional logic which is concerned with the truth of statements called predicates. Thus in this formalism, there are two constants, T for "true" and F for "false". There are arbitrarily many predicates, $P, Q, R, \ldots$, each of which is either true or false. The predicates are formed according to the following syntax

$$\begin{aligned}
\langle atom \rangle &\longrightarrow T \\
\langle atom \rangle &\longrightarrow F \\
\langle atom \rangle &\longrightarrow P \\
\langle atom \rangle &\longrightarrow Q \\
&\vdots \\
\langle sentence \rangle &\longrightarrow \langle atom \rangle \\
\langle sentence \rangle &\longrightarrow \langle complexsentence \rangle \\
\langle complexsentence \rangle &\longrightarrow (\langle sentence \rangle) \\
\langle complexsentence \rangle &\longrightarrow \langle sentence \rangle \langle op \rangle \langle sentence \rangle \\
\langle complexsentence \rangle &\longrightarrow \neg\ \langle sentence \rangle \\
\langle op \rangle &\longrightarrow \wedge\ (logical\ \text{“}and\text{”}) \\
\langle op \rangle &\longrightarrow \vee\ (logical\ \text{“}or\text{”}) \\
\langle op \rangle &\longrightarrow \neg\ (negation) \\
\langle op \rangle &\longrightarrow \Longrightarrow (implication) \\
\langle op \rangle &\longrightarrow \Longleftrightarrow (if\ and\ only\ if)
\end{aligned} \tag{8.17}$$

Predicates joined by logical operators are evaluated according to the operator precedence ordering $\neg, \wedge, \vee, \Longrightarrow, \Longleftrightarrow$ unless enclosed in parentheses, in which case the parenthetical relations must be resolved first. Thus the predicate

$$\neg P \vee Q \vee R \Longrightarrow S \tag{8.18}$$

is equivalent to

$$((\neg P) V (Q \wedge R)) \Longrightarrow S. \tag{8.19}$$

Given truth values for P, Q, R, S, the predicate (8.19) can be evaluated as either T or F.

The propositional logic of (8.17) is not sufficiently rich for a general model of semantics. We can, however, obtain a sufficiently expressive model called the first-order predicate calculus by augmenting (8.17) with additional operators and constants, variables, quantifiers and functions. To accommodate these additions we use a similar syntax.

$$
\begin{aligned}
\langle formula \rangle &\longrightarrow \langle atom \rangle \\
\langle formula \rangle &\longrightarrow \langle formula \rangle \langle op \rangle \langle formula \rangle \\
\langle formula \rangle &\longrightarrow \langle quant \rangle \langle var \rangle \langle formula \rangle \\
\langle formula \rangle &\longrightarrow \neg \langle formula \rangle \\
\langle atom \rangle &\longrightarrow \langle pred\,(term) \rangle \\
\langle term \rangle &\longrightarrow \langle function\,(term) \rangle \\
\langle term \rangle &\longrightarrow \langle const \rangle \\
\langle term \rangle &\longrightarrow \langle var \rangle \\
\langle op \rangle &\longrightarrow = (equality) \\
\langle quant \rangle &\longrightarrow \forall (universal,\ i.e.\ \text{“}for\ all\text{”}) \\
\langle quant \rangle &\longrightarrow \exists\ (existential,\ i.e.\ \text{“}there\ exists\text{”}) \\
\langle quant \rangle &\longrightarrow \exists!\ (unique\ existential) \\
\langle const \rangle &\longrightarrow A \\
\langle const \rangle &\longrightarrow B \\
\langle const \rangle &\longrightarrow C \\
&\ \vdots \\
\langle var \rangle &\longrightarrow x \\
\langle var \rangle &\longrightarrow y \\
\langle var \rangle &\longrightarrow z \\
&\ \vdots \\
\langle function \rangle &\longrightarrow F \\
\langle function \rangle &\longrightarrow G \\
\langle function \rangle &\longrightarrow H \\
&\ \vdots
\end{aligned}
\tag{8.20}
$$

All other symbols are as defined in (8.17). Notice that in (8.20) both functions and predicates have arguments. A predicate can have a null argument, in which case it is as defined in the propositional logic. Predicates, functions and formulas all take values of either T or F.

As noted earlier, a desirable property of a semantic model is the capacity for inference or reasoning. Both propositional logic and first-order predicate calculus admit of formal procedures for inference. The following rules of inference allow the evaluation of complex formulae.

If P is true and $(P \Longrightarrow Q)$ is true then Q is true. This is the formal expression of "if–then" reasoning. If $\bigwedge_{i=1}^{n} P_i$ is true then P_i is true for $i = 1, 2, 3, \ldots, n$. If P_i is true for $i = 1, 2, \ldots, n$, then $\bigwedge_{i=1}^{n} P_i$ is true. If P_i is true for some i, then $\bigvee_{i=1}^{n} P_i$ is true for any subset $\{p_i | 1 \leq i \leq n\}$. Double negation means that $\neg(\neg P) \Longleftrightarrow P$ is always true for any P. Finally, one can use the notion of contradiction in the resolution rule; if $\neg P \Longrightarrow Q$ is true and $Q \Longrightarrow R$ is true, then $\neg P \Longrightarrow R$ is true. In general, these rules of inference are transitive, so that we can use chains of the form $P_0 \Longrightarrow P_1 \Longrightarrow \ldots \Longrightarrow P_n$.

In Section 8.3.4 we will see the way inference is used in semantic analysis. Of primary importance, however, is the expressiveness of the first-order predicate calculus. The foregoing discussion has been entirely abstract. We now must be specific and consider a particular ontology in order to apply the abstraction. That is, we need to choose a set of predicates and functions that will allow us to symbolically represent nearly all aspects of reality in natural language.

We start with lexical semantics which is just the meaning of isolated words or, alternatively, a mapping from a word to the object, action or idea to which it refers. For example, the predicate, **book**(x), is true if and only if the variable x is, in reality, a book. Similarly the verb "give" is represented by the function **give**(P, Q, R), where $P =$ **object**(x), $Q =$ **donor**(y) and $R =$ **recipient**(z). The function, **give**(P, Q, R), is defined to mean that the object x is transferred from the donor, y, to the recipient, z.

Also included amongst the necessary predicates are those that represent time by means of verb tense and aspect (i.e. event time relative to message time). In addition, there should be ways to indicate beliefs and imagination as putative but not necessarily real entities.

An example of the use of the first-order predicate calculus to express natural language is the following. Consider the sentence "Every dog has his day", rendered logically as follows:

$$\forall d \; \textbf{dog} \; (d) \Longrightarrow \exists \; a \; \textbf{day} \; (a) \wedge \; \textbf{owns}(d, a) \wedge \; \textbf{has} \; (d, a). \tag{8.21}$$

A direct translation of (8.21) is: For all d, where d is a dog, it is the case that there exists a day, a, and d owns a and the function **has**(d, a) is true, that is, d possesses a.

Two issues are immediately apparent. First, the symbolic representation implicit in the examples given above does not include any method for making the mapping between symbol and referent. The symbols are utterly abstract. We will consider this problem further in Chapter 10.

Second, as a purely practical matter, in order to make this method expressive, we need to exhaustively enumerate all elements of the ontology of common reality and devise a predicate or function for each in a way that allows for consistent verifiability. If we construe the ontology in a more circumscribed way, we are back to limited domain semantics and the logic might just as well be replaced by a finite-state machine. The CYC project of Lenat [177] is an attempt to codify the common-sense knowledge of lexical semantics. The project has met with dubious success for written language only. In Chapter 10, we shall also suggest a means to avoid the problem of exhaustive enumeration.

8.3.4 Relationship between Syntax and Semantics

Syntactic structure has a strong effect on meaning. Figure 8.6 shows two different parses for the sentence "John saw the man with the telescope". In the first case the adjectival prepositional phrase "with the telescope" is attached to the direct object "man", and the sentence is interpreted to mean that the man was carrying the telescope. In the second case the prepositional phrase is adverbial, modifying the verb "saw", yielding the interpretation that John used the telescope to see the man.

The syntactico-semantic connection is best illustrated by the famous example due to Chomsky [45]. He proposes that the sentence "Colorless green ideas sleep furiously" is syntactically well formed but semantically anomolous, that is, meaningless. It is easy to verify the syntactic validity of the sentence. It can be parsed with respect to the grammar of Section 4.3.3 yielding a single parse tree. However, application of the logical inference technique of Section 8.3.3 would discover the logical contradiction between the predicates **colorless**(x) and **green**(x) which cannot both be true for a given x. Furthermore, the

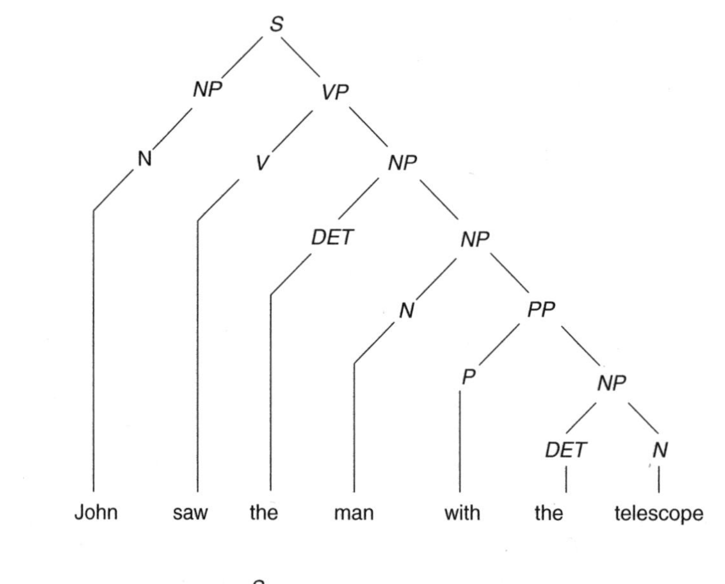

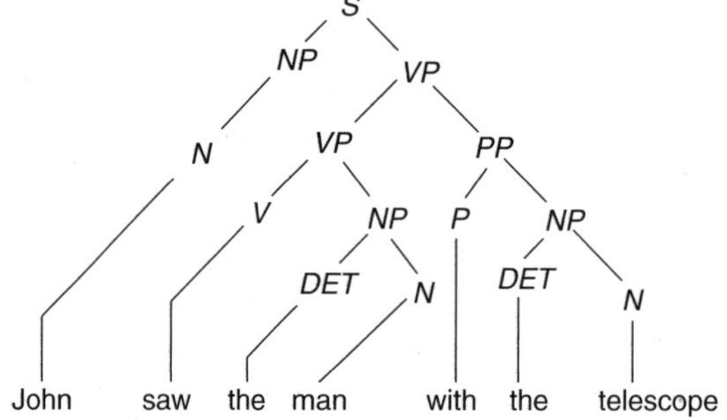

Figure 8.6 Syntactic structure affects meaning

function **sleep(animal** (x)) would be undefined since **animal(ideas)** would be false. Also, **furiously(action)** would be undefined since **action** = **sleep**(x) is an invalid argument of the function **furiously(action)**. Thus the sentence is logically false and/or undefined, hence meaningless.

However, there is a semantically valid interpretation of the sentence. Suppose we take "colorless" to be the opposite of "colorful", thus **colorless**(x) = **uninteresting**(x). Also let **green**(x) = **naive**(x). Under these definitions "colorless" and "green" are not logically inconsistent so **green(ideas)** would be true. Moreover, an idea can be "dormant" so **sleep(ideas)** = **dormant(ideas)** is well defined. Finally, **furiously(sleep)** could be interpreted to mean that the ideas were forced into dormancy and when they awake, they will do so resentfully. This is perhaps a poetic interpretation of the sentence, but it is far from absurd and it illustrates the difficulty of designing an exhaustive symbolic ontology.

Having considered lexical semantics, we must now examine the meaning of sentences. It is here that syntax becomes important. First, note that all lexical items are assigned a part-of-speech label. This is the rôle of the lexical assignment rules of Section 3.2.1. For example, we have production rules such as $\langle verb \rangle \longrightarrow run$ or $\langle noun \rangle \longrightarrow boy$.

The lexical assignment rules make a strong connection to semantics because nouns are specific objects or ideas that play the role of either agents or entities acted upon. Verbs are actions, functions of the agents. Adjectives are qualities or properties of nouns. Adverbs are qualities or properties of verbs, adjectives or other adverbs. Prepositions provide location, orientation or direction in space and time.

These syntactico-semantic constituents are combined according to a predicate argument structure, the most basic of which is the subject–verb–object (SVO) rule

$$S \longrightarrow NP\ \ VP\ \ = (\text{agent},\ \text{action},\ \text{object}). \tag{8.22}$$

The significance of (8.22) is that the basic syntactic rule $S \longrightarrow NP\ VP$ maps onto the semantic interpretation (SVO). Then the complete syntactico-semantic analysis proceeds as follows. First the syntactic structure for the sentence, "John gave the book to Mary" shown in Fig. 8.7 is obtained from a parser and grammar as explained in Sections 4.1.2 and 4.3.3. The parse tree is built up from the lexical entries to the root, S, of the tree. Then the λ-calculus is used to verify that the sentence is well formed. In this notation, the term λx simply means that the argument, x, of the function, $F(x, y)$, has not been bound to a particular value. When the value of x is determined, say A, then $\lambda x F(x, y)$ is replaced by $F(A, y)$. The result of this operation is shown in Fig. 8.8. A second pass through the parse tree from the lexical entries to the root resolves all of the λ meaning that the sentence is semantically valid. Had there been an unresolved λ then the sentence would be semantically anomolous. For the example as given, all arguments are bound and we get a logical expression of the meaning

In this simple case, the predicate is unambiguous so there is no need to use inference procedures to check for consistency. In general that is not the case, as was indicated in Chomsky's example.

8.4 System Architectures

At the time that the first speech understanding systems were built [180, 272], speech recognition was based on recognizing whole words as the fundamental acoustic patterns.

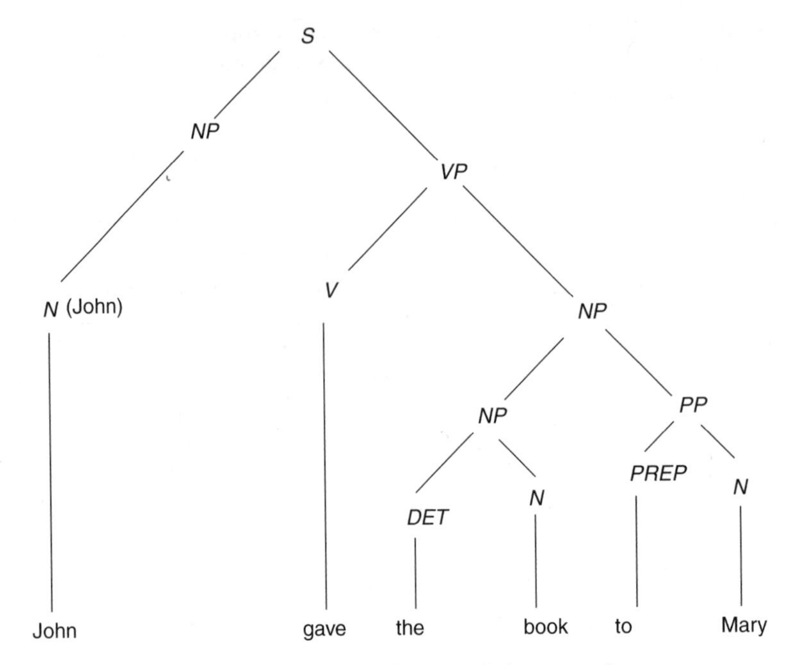

Figure 8.7 Syntactic framework for semantics

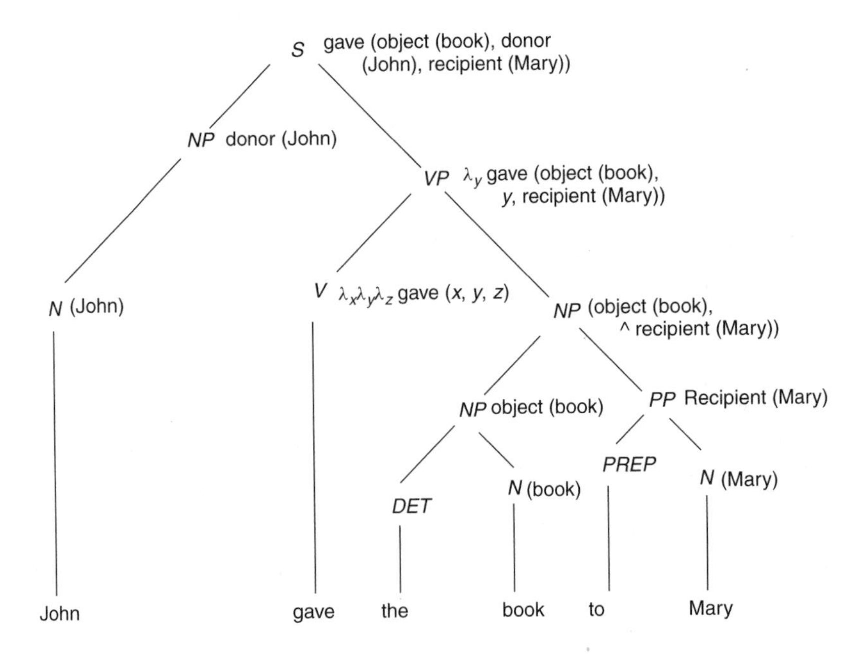

Figure 8.8 Semantic analysis using the λ-calculus

One consequence of this was that only small vocabularies – a few hundred words – could be reliably recognized, and consequently only highly stylized utterances pertaining to a highly circumscribed topic of conversation could be understood. From the early experiments, it soon became obvious that speech understanding systems needed larger vocabularies to demonstrate truly interesting behavior. The techniques of whole word recognition could not be extended to large vocabularies because they relied on a crude and implicit method of capturing the lower levels of linguistic structure in word- or phrase-length acoustic templates.

To attain the desired versatility, explicit representations of phonetics, phonology and phonotactics were required. That is, recognition had to be based on the inventory of fundamental sounds (phonetics) in speech as the primitive acoustic patterns to be recognized. Then rules about how these sounds change in different phonetic contexts (phonology) had to be applied and finally, rules specifying the order in which phonetic units can appear in sequences (phonotactics) had to be imposed. These aspects of linguistic structure were well understood and carefully documented by Chomsky [45], Chomsky and Halle [47] and others [238, 239]. However, the number and subtlety of these rules made their simple, direct incorporation in a computer program very cumbersome and fragile.

The solution to the conflicting requirements of large vocabularies and robust principles for representing known linguistic structure emerged from studies in the application of hidden Markov models to speech recognition (Chapters 3 and 7). Speech understanding systems are usually conceived as having two functionally separate parts: a "front end" which performs the signal processing and pattern recognition and a "back end" which takes the transcription produced by the "front end" and derives from it the intended meaning of the utterance. The "front end" may be thought of as the speech recognition part and the "back end" as the speech understanding part. The "back end" is based on the methods described in this chapter. The only connection between the two parts is a transcription of the speech input into text. As we shall see in Chapters 9 and 10, this is a weak model.

8.5 Human and Machine Performance

The foregoing discussion completes our consideration of the entire speech chain from acoustics through semantics. The section on general semantics is, of necessity, incomplete. While the methodology is straightforward, the details are absent. This is because they depend critically on an exhaustive implementation of all necessary predicates and functions. No doubt, different individuals would make different choices of ontologies and produce different representations of them. Even the best of such compilations would have, at best, a tenuous hold on reality but there are no known procedures to automatically generate the semantic constituents from data, as was done for all other aspects of linguistic structure. As a result, there are no general speech understanding systems that have a natural language ability even remotely comparable to human competence. Unfortunately, this remains an open problem. Chapters 9 and 10 offer some thoughts about how to solve it.

9

Theories of Mind and Language

9.1 The Challenge of Automatic Natural Language Understanding

The progression of mathematical analyses of the preceding pages may be used to construct machines that have a useful but limited ability to communicate by voice. It is now appropriate to ask what is required to advance the technology so that machines can engage in unrestricted conversation. The conventional answer is incremental improvement of existing methods. In these final two chapters, I offer an alternative. I suggest that machines will be able to use language just as humans do when they have the same cognitive abilities as humans possess, that is, they have a mind.

After centuries of study, the mind remains one of the most elusive objects of scientific inquiry. Among the many mysteries are how the mind develops and functions, how it engenders intelligent behavior, and how it is related to language. Such questions have been formulated in different ways and have been addressed from different perspectives. No general agreement amongst different schools of thought has yet emerged. Nonetheless, I wish to enter the fray. My answer is in two parts, a brief historiography and an experimental investigation derived from it.

The essential feature of my brief intellectual history is that it comprises both a diachronic and a synchronic analysis. That is, the subject is considered with respect to both its evolution across historical epochs and its development within a particular period. In the case of the sciences of mind, both perspectives are known to philosophers and historians but are often ignored by scientists themselves. They, and even those readers who are already steeped in the historical facts, may wish to endure yet another interpretation and the significance I ascribe to it.

9.2 Metaphors for Mind

The diachronic view of mind is best expressed by the founder of cybernetics, Norbert Wiener. In the introduction to his seminal 1948 treatise, *Cybernetics*, Wiener observed that over the entire history of Western thought, metaphors for mind have always been expressed in terms of the high technology of the day. He says:

Mathematical Models for Speech Technology. Stephen Levinson
 ISBN: 0-470-84407-8

> At every stage of technique since Daedalus or Hero of Alexandria, the ability of the artificer to produce a working simulacrum of a living organism has always intrigued people. This desire to produce and to study automata has always been expressed in terms of the living technique of the age. [330]

Wiener devised a particular metaphor for the mind which he regarded as spanning the entire historical trajectory. He explained:

> Cybernetics is a word invented to define a new field of science. It combines under one heading the study of what in a human context is sometimes loosely described as thinking and in engineering is known as control and communication. In other words, cybernetics attempts to find the common elements in the functioning of automatic machines and of the human nervous system, and to develop a theory which will cover the entire field of control and communication in machines and living organisms. The word cybernetics is taken from the Greek *kybernetes*, meaning steersman. If the 17th and early 18th centuries were the age of clocks, and the later 18th and 19th centuries the age of steam engines, the present time is the age of communication and control. [330]

Thus it is clear that Wiener construed mental function as the cooperation of many kinds of processes which he characterized as information flow and control and which he called cybernetics. For example, he generalized the very technical notion of the negative feedback principle to include the adaptation of complex systems and living organisms to changing environments. Similarly, he expanded the definition of stability to apply to any process, including cognition, used to maintain biological homeostasis. These ideas, he reasoned, would lead to an understanding of the incredible reliability of human cognitive functions such as perception, memory and motor control.

Although the synchronic history is best expressed in rigorous mathematical terms, it, too, can be faithfully summarized. The discipline with the unfortunate name of artificial intelligence – unfortunate because the word "artificial" cannot be cleared of its pejorative connotation of "fake" – was constructed out of the remains of the attempt to establish mathematics on an unassailable foundation. The failure of this effort led to a specific model of mental function. In 1937, Turing [319] resolved the decidability problem posed by Hilbert [273] by means of a model of computation which, despite its abstract nature, made a powerful appeal to a physical realization. Exactly when he came to appreciate the implications of his result is a subject of some debate [118, 128]. However, by 1950, his universal computer emerged as a "constructive" metaphor for the mind, allowing him to clearly set down what we today refer to as the strong theory of AI. This, in principle, could be experimentally verified by taking an agnostic position on the question of what the mind really is and requiring only that its behavior be indistinguishable from that of a computational mechanism by a human observer. In his seminal paper of that year which established the foundations for the field of AI, Turing asserted:

> The original question, "Can machines think?" I believe to be too meaningless to deserve discussion. Nevertheless I believe that at the end of the century the use of words and general educated opinion will have altered so much that one will be able to speak of machines thinking without expecting to be contradicted." [320]

The adventure starting from the crisis in the foundations of mathematics, around 1910, and ending with the emergence of the strong theory of AI is a fascinating journey which helps to explain why the Turing machine was so seductive an idea that it caused a revolution in the philosophy of mind.

9.2.1 Wiener's Cybernetics and the Diachronic History

I shall begin by elaborating upon Marshall's commentary [213] on Wiener's definition of cybernetics cited above. In so doing, I take the liberty of summarizing the entire history of thought about mind in the single chart of Figure 9.1.

Each row of Fig. 9.1 is a coarsely quantized time line on which the evolution of one of four mechanical metaphors for mind is traced by selecting specific examples. Each one represents a diachronic history that is simply an expansion of the notion expressed by Wiener about the significant inventions that became metaphors for mind in different centuries. Each individual entry in the chart represents an isolated history of the specific invention. There is also a migration of ideas, a synchronic history, as one proceeds along the columns. The right-hand column composes what I shall later describe in detail as the cybernetic paradigm.

The metaphor of "control" begins with the invention, perhaps too ancient to attribute, of the rudder. I know of no specific philosophical theory of mind that finds its roots in the nautical steering mechanism; however, allusions to rudders permeate our language in such expressions as "steering a course to avoid hazards", presumably by thinking. This suggests that the control of motion was recognized as a natural analogy to intellectual activity, a spirit controlling a body.

Equally plausible is the connection between mental function and the Archimedean science of hydraulics. The motivation might have been the need for sanitation in growing metropolitan areas. To early philosophers such as Herophilus, the control of flow of fluids through ducts provided a striking image of thoughts and emotions coursing through the channels of the human mind.

Pre-industrial period	Industrial period	Information age	Mathematical abstraction
Rudder	Governor Thermostat	Feedback amplifier	Control theory
Hydraulic systems	Telegraph	Internet	Communication Information theory
Wax tablets	Photographic plates	AUDREY	Pattern classification theory
Clocks	Analytical engine	ENIAC	Theory of computation

Figure 9.1 Metaphors for mind

In Section 2.5 we cited the Platonic theory of forms as a precursor to statistical pattern recognition. One could view early written symbols as designations for forms by stylized indentations made in wax or stone. Similarly, mental images were likened to impressions on wax tablets created by impinging sensory stimuli and manipulated by thought processes.

The ability to count and mechanical aids for that activity were also known in ancient times. The notion of generalized mechanical calculation seems not to emerge until later. European scholars fascinated by the wizardry of the French and German horologists envisioned the brain as a vast clockwork. In particular, Leibniz proposed his theory of monads as the fundamental calculators of the universe.

The industrial period witnessed a significant advance of the cybernetic paradigm toward a more recognizable and cogent form. Control mechanisms such as the governor and the thermostat, which tamed the temperamental steam engines, provided a vivid image of thought as the control of motion and power. Also in this period, thoughts ceased to be imagined as fluid flows controlled by pumps and valves but rather as electrical messages transmitted by telegraph wires. Wax tablets were discarded as an embodiment of memory in favor of the more refined photographic plate. And Babbage carried the identification of thought as clockwork to its mechanical extreme in his analytical engine.

In our own modern information age, the mastery of electromagnetic and electronic phenomena, enabled by the impressive power of classical mathematical analysis, resulted in the recasting of the cybernetic paradigm in electronics rather than mechanics. Control over steam engines was extended to intricate servomechanisms according to the principles of the feedback amplifier.

Telegraphy evolved into telephone and then into the modern global communication network in which information, whether it be text, image, audio, or video, is digitally encoded, packetized, multiplexed, routed, and reconstructed at its designated destination. The internet is so complex that it is often likened to the human central nervous system.

While photography is able to record visual images in exquisite detail, it is, by itself, incapable of analyzing the content of an image. It thus accounts for memory but not perception. Electrical circuits, however, offered the possibility of analyzing and identifying patterns that were stored in a memory. One of the earliest examples of this class of devices was AUDREY, which was capable of reliably recognizing spoken words [70]. The principles on which AUDREY was designed were easily generalized and extended to other problems of automatic perception.

While Babbage apparently did not recognize the full generality of his analytical engine, Turing did indeed understand the universality of his computer. Once again, modern electronic technologies provided an embodiment of an abstract theory. In the late 1940s and early 1950s, ENIAC [107], EDVAC [107], and several other computers were constructed, ushering in the era of the "electronic brain" and allowing, for the first time in history, serious entertainment of the notion of building a thinking machine.

Today, these histories are encapsulated in four mathematical disciplines: control theory, information theory, statistical decision theory, and automata theory. Collectively, these areas of applied mathematics formalize and rationalize problems of control, communication, classification, and computation. I call the unification of these ideas the cybernetic paradigm or, making an acronym of its components, C^4. I assert that the cybernetic paradigm is the ideal tool for the eventual construction of an artificial intelligence. It provides a quantitative, rational means for exploring the several intuitively appealing metaphors for mind that have persisted in Western culture. At the risk of overstating the argument, I remind the reader that C^4 is a union of different but related theories. Its utility derives from this collective property. One cannot hope to emulate human intelligence by considering only a part of the collection.

It is worth considering the components of the cybernetic paradigm in slightly more detail in their modern abstract form. This will enable me to describe how the constituent theories are related to each other, how they may be used to explain processes occurring in the human organism and why they are uniquely appropriate for the design of an intelligent machine.

The kinds of machines to which the four theories are applicable are represented by the following four canonical diagrams. Figure 9.2 shows the prototypical control system. The plant is required to maintain the response, $y(t)$, close to a dynamic command, $x(t)$. This is accomplished by computing a control signal, z, and comparing it with the input. The plant may be something as simple as a furnace, and its controller a thermostat. At the other extreme, the plant might be a national economy, and the controller the policies imposed upon it by a government.

Figure 9.3 is the famous model due to Shannon [295] of a communication system. A message represented by the signal, $x(t)$, is transmitted through some medium and is received as a signal, $y(t)$, which is decoded and interpreted as the intended message. To be useful, $y(t)$ should be similar, in a well-defined sense, to $x(t)$. The model applies to a spacecraft sending telemetry to earth or two people having a conversation.

Figure 9.4 depicts the canonical pattern recognition machine. The signal, $x(t)$, is measured and its relevant features extracted. These values are compared with prototypical values stored in memory and a decision is made by selecting the reference pattern, $y(t)$, to which the input signal is most similar. Once again, the range of processes fitting into

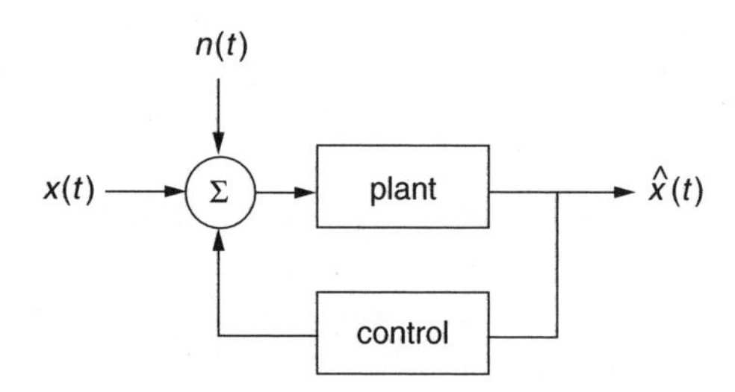

Figure 9.2 The feedback control system

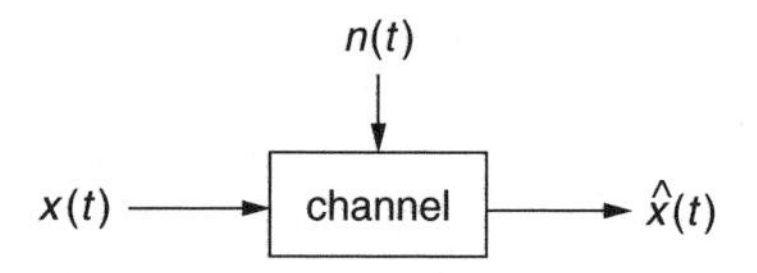

Figure 9.3 The classical model of a communication system (After Shannon)

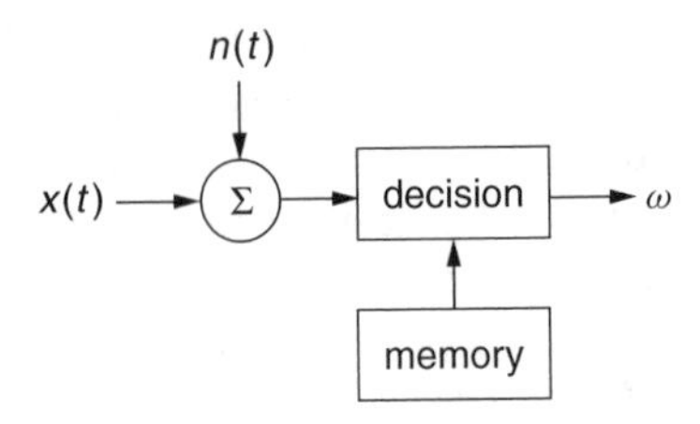

Figure 9.4 The classical pattern recognition system

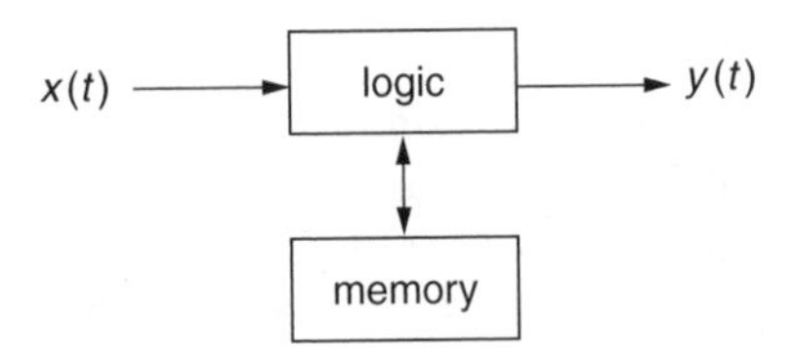

Figure 9.5 The canonical model of a computer (after Turing)

this schema is quite large. It encompasses simple problems such as automatic recognition of barcodes and subtle ones such as identifying an artist from his paintings.

Finally, Fig. 9.5 shows an abstract digital computer. Data and/or programs, $x(t)$, are read into a finite-state logic element that operates on $x(t)$ and produces an output, $y(t)$. Such devices may be highly specialized, such as a four-function arithmetic calculator, or very general in purpose, as in the case of a mainframe computer.

These four systems are interchangeable in some cases. For example, a control system can be used to perform pattern recognition if the output, $y(t)$, is quantized and is always the same for a well-defined class of command signals, $x(t)$. A pattern recognizer may be thought of as a communication channel whose fidelity is measured by the probability of a correct classification. And, as I shall remind the reader in the next section, a general-purpose computer can be programmed to simulate the other three systems.

It is also the case that implementations of any of these systems in electronic hardware have, as subsystems, one or more of the other systems. Thus, the control circuit in a servomechanism can be a microcomputer. Conversely, the auxiliary memory of a computer system such as a "hard drive" may contain several servomechanisms for fetch and store operations. The connection between the logic unit and the memory of a computer is a communication channel, while sophisticated communication channels usually contain computers to perform the coding functions. In pattern recognition systems, the features may be measured by instruments that are stabilized by feedback control systems and feature extraction may be accomplished by numerical computation. These are but a few examples of the interrelations among the four pillars of the cybernetic paradigm.

Figures 9.2 through 9.4 all show the symbol, **n**, signifying an unwanted signal called noise because any physical measurement has some uncertainty associated with it. All of the systems must be designed to function in the presence of corrupting noise. The computer depicted in Fig. 9.5 is not so afflicted, nor does it have any provision for dealing with ambiguity. I shall return to this important distinction when I discuss the role of the cybernetic paradigm in the mind.

Not only are the systems of Figs. 9.2 through 9.5 present in all of our modern machinery, they are also essential to the proper function of the human organism. Homeostasis and locomotion are accomplished by means of electrochemical feedback control systems. Neural transduction of the electrically encoded messages necessary for locomotion and distribution of biochemically encoded commands required for homeostasis through the circulatory system are well described as communication systems. Sensory perception is presumed to be achieved by exquisite feature extractors and decision mechanisms. And, of course, control of the entire organism in thought and action is assumed to be a computational process.

Wiener's historical perspective insists that intelligence is not simply a process of abstract thought but rather of the harmonious functioning of all aspects of our physical being and, as such, it requires all aspects of the cybernetic paradigm.

9.2.2 The Crisis in the Foundations of Mathematics

The isolated history of AI is the history associated with the single box in the lower right-hand corner of Fig. 9.1. It is the story of an insight which resulted from the glorious failure of an attempt to establish mathematics on an unimpeachable theoretical foundation once and for all.

Although cracks in the structure of mathematics had been visible since Hellenic times (e.g. Euclid's inability to eliminate the troublesome parallel postulate), the crisis did not become acute until the late nineteenth century as a result of thinking about large numbers encountered in the summation of infinite series. A brief outline of events is the following:

1. Cantor's theory of transfinite numbers [42] leads to controversy over the continuum hypothesis.
2. The ensuing debate causes a threefold schism of mathematics into the intuitionist, formalist, and logical schools.
3. Intuitionists, such as Brouer [39], deny the existence of infinite numbers and accept only arbitrarily large numbers.
4. Logicians, such as Russell and Whitehead [285], trace the problem to impredicative (i.e. self-referential) sets.
5. The formalists, such as Hilbert [273], assert that all mathematical questions can be resolved by proofs based on an appropriate set of axioms. There can be no "*ignorabimus*".
6. Gödel [104] proves the incompleteness theorem and thereby invalidates the formalist approach.
7. Turing [319] proves the undecidability theorem, thereby strengthening Gödel's result.
8. The mechanism of Turing's proof is a universal computer in the sense that it can emulate any possible computation.
9. Church [48] derives a similar result using lambda calculus.
10. The Church–Turing hypothesis is recognized as the theoretical foundation of AI.

A comprehensive treatment of the many aspects of this outline is well beyond the scope of this short essay. Such a discussion would require consideration of everything from the common thread of mathematical reasoning using the technique of diagonalization to the philosophical conflicts between realism and spiritualism and between free will and determinism. However, a bit more detail is required to support the position I am advocating.

Let me begin my rendition of the story with Cantor and his ideas about different degrees of infinity. Cantor was the son of a Lutheran minister who wanted his boy to pursue a life of service to God. Cantor, however, was not prepared to abandon his work in mathematics. He rationalized the conflict away by professing his hope that the beauty of his mathematics would attest to the glory of God and thus both appease his father and satisfy his intellectual aspirations.

Cantor's transfinite numbers are the cardinalities of different infinite sets. He began by comparing the set of "natural numbers" (i.e. positive integers) with the set of rational numbers. He observed that there are infinitely many naturals but, since the rationals are pairs of naturals, there should, in some sense, be more of them. But Cantor demonstrated by means of a diagonalization argument that this intuitively appealing notion is, after careful examination, flawed. The essence of his argument is illustrated in Fig. 9.6. By tracing along the indicated path, it is clear that there is a one-to-one mapping from naturals onto rationals. This means that the naturals and rationals have the same infinite cardinality which Cantor called the first transfinite number, aleph null ($\aleph_0$). Cantor then asked whether or not such a map exists from naturals onto reals. To answer the question, he constructed the matrix of Fig. 9.7 in which each row is a power series representation of a rational number, and the columns of the matrix correspond to the integral powers. Call the ijth entry in the matrix d_{ij}. Now consider the real number whose nth digit is $d_{nn} + 1$. In the example it would be 2.64117.... This number cannot be a row of the matrix because, by construction, it differs from the nth row in at least the nth column. Thus there cannot be a one-to-one mapping from naturals onto reals. This is interpreted to mean that there are more reals than naturals. This argument is particularly important because it invokes the summation of a power series such as a Fourier series which was the source of the argument about infinite numbers.

The cardinality of the reals is represented by symbol c, designating the continuum. In the sense of isomorphism illustrated in Figs. 9.6 and 9.7, $c > \aleph_0$. Then there must be some number, ω, that lies between the two. Cantor proposed that w is actually the largest integer and, using it, he constructed an entire, linearly ordered number system such that

$$\aleph_0 < \omega < \omega_1 < \omega_2 < \ldots < c, \tag{9.1}$$

	1	2	3	4	5 ......
1	1	1/2	1/3	1/4	1/5 .....
2	2	1	2/3	2/4	2/5 .....
3	3	3/2	1	3/4	3/5 .....
4	4	2	4/3	1	4/5 .....
5	5	5/2	5/3	5/3	1 .....
.	.	.	.	.	.
.	.	.	.	.	.
.	.	.	.	.	.

Figure 9.6 Mapping the rationals onto the integers

	1	2	3	4	5	6 ...
1	1.	0	0	0	0	0 ...
2	0.	5	0	0	0	0 ...
3	0.	3	3	3	3	3 ...
4	0.	2	5	0	0	0 ...
5	0.	2	0	0	0	0 ...
6	0.	1	6	6	6	6 ...
.	.	.	.	.	.	.
.	.	.	.	.	.	.
.	.	.	.	.	.	.

Figure 9.7 The reals cannot be mapped onto the integers

which he called the transfinite number system. Cantor was convinced that the beauty of his creation was a tribute to God's own handiwork, but his colleagues did not even accept its validity, let alone its sanctity. A bitter controversy ensued causing, or at least contributing to, Cantor's lapse into insanity.

The controversy centered around what came to be known as the "continuum hypothesis" which denies the existence of the system of (9.1). Intuitionists were opposed on the grounds that there is no infinite number let alone orders of infinite numbers. It was equally clear to the logicians that the whole enterprise made no mathematical sense because implicit in the diagonalization argument is the notion of impredicative sets which are based on self-reference. That is, an infinite set, the integers, is a subset of itself, the rationals. This kind of construction must lead to inconsistencies that they called antinomies. As we shall soon see, they were partially correct. The formalists, lead by Hilbert, took a more sympathetic approach. They said that the continuum hypothesis should admit of a definitive resolution; it should be possible to prove that either transfinite numbers exist or they do not exist. In fact, formalists generalized their position on the continuum hypothesis asserting that it must be possible to prove every true statement within a well-defined, suitably rich axiomatic (i.e. formal) system without recourse to some absolute physical interpretation, intuitive appeal, or lack thereof. This was a bold position since classical mathematics was inspired by physics. Moreover, the formalist's doctrine declared that there should be no constraint on mathematical thought as long as logic and an appropriate set of axioms are not violated. Under these conditions, there can be no *ignorabimus*. Nevertheless, no proof of the existence of Cantor's strange mathematical objects was forthcoming.

In 1931, Kurt Gödel destroyed Hilbert's hopes of establishing absolute certainty within any "interesting" axiomatic system by proving the incompleteness theorem.

Theorem 1 (Gödel). Any axiomatic system the structure of which is rich enough to express arithmetic on the natural numbers is either complete or consistent but not both.

An axiomatic system is complete if all true statements within it can be proven true and all false statements can be proven false. An axiomatic system is consistent if both a statement and its negation cannot be simultaneously true.

This remarkable theorem is proven by contradiction. In essence, the proof produces a version of the liar's paradox in which one is given a card the front of which asserts that the statement on the back of the card is false. The back of the card, however, states that the statement on the front of the card is true. Thus both statements are true if and only if they are false. Generating a formal contradiction of this type within an ordinary arithmetic system is arduous and a complete rendering of the proof cannot be given here. However, we can give a sketch of the proof in a manner that will provide a good intuition for the main ideas. The reader interested in a detailed rendering of the incompleteness theorem should consult Nagel and Newman [229].

To gain a better appreciation of Gödel's basic argument, consider the well-known Richard paradox. All arithmetic statements can be written, however awkwardly, in ordinary English. Then the statements can linearly ordered, in a manner analogous to the construction of Fig. 9.7, by simply arranging the statements in lexicographic order. Thus, every arithmetic statement has an integer associated with it. Next we define a special property of arithmetic statements. We will say that a statement is Richardian if and only if the statement is not true of its own ordinal number. For example, if the statement "N is prime" were to occur in the Nth position of the list of statements where N is not a prime number, then the statement would be Richardian. Alternatively, if N were, in fact, prime, then the statement would not have the Richardian property.

Notice that the Richardian property is, by definition, a statement about arithmetic and it obviously can be expressed in English. Eventually the following statement will appear on our lexicographically ordered list with ordinal number, N_r: The sentence of number N_r is Richardian. By definition of Richardian, the statement number N_r is Richardian if and only if it is not Richardian. In fact, this is not a paradox at all because the method used for generating the statements, namely English, has not been defined in the arithmetic system. However, the structure of the argument is useful if the Richardian property can be replaced by some other property that can be strictly defined within the arithmetic system.

Cantor's diagonalization technique, as it appears in the Richard paradox, can be used to prove the incompleteness theorem. To do so, the error of the Richard paradox must be avoided by making explicit that performing arithmetic operations and proving theorems about arithmetic are actually different encodings of the same fundamental symbolic process. Gödel devised an ingenious method for mapping statements about the integers onto the integers without going outside arithmetic itself. The method has come to be known as Gödel numbering or indexing.

Gödel's coding scheme begins with the primary operations of arithmetic shown in Fig. 9.8. This is essentially the logical system described in Section 8.3.3. The numbering system shown in Fig. 9.8 can be used to form the Gödel number for any arithmetic statement. Consider the example of

$$(\exists x)(x = Sy), \tag{9.2}$$

which asserts the existence of some number, x, that is the successor of some other number, y. First we assign an integer to each symbol according to the table of Fig. 9.8. Thus (9.2) is represented by the sequence of integers

$$8\ 4\ 11\ 9\ 8\ 11\ 5\ 7\ 13\ 9 \tag{9.3}$$

symbol	integer	example or meaning
~	1	negation
$\vee$	2	logical OR
=>	3	implication
$\exists$	4	existence
=	5	equality
0	6	zero
S	7	successor
(	8	left bracket
)	9	right bracket
,	10	delimiter
x	11	numerical variables
y	13	represented by primes
z	17	greater than 10
.		
.		
.		
p	11^2	arithmetic propositions
q	13^2	represented by squares
r	17^2	of primes greater than
.		10 such as $x = y$ or
.		$p = q$
.		
P	11^3	logical predicates
Q	13^3	represented by cubes
R	17^3	of primes greater than
.		10 such as prime or
.		composite
.		

Figure 9.8 Gödel numbering system

from which we construct the Gödel number, N, for (9.2) by using the integers in the sequence (9.3) as powers of successive primes. That is,

$$N = (2^8)(3^4)(5^{11})(7^9)(11^8)(13^{11})(17^5)(19^7)(23^{13})(29^9), \tag{9.4}$$

which is a large but finite number.

The next step is the definition of a sequence of predicates ending with **Dem**(x, y) which means that the sequence of arithmetic statements having Gödel number x is a proof of the statement with Gödel number y. The lengthy construction of this predicate is omitted here. The important point is that this predicate allows the formation, reminiscent of the Richard paradox, "The theorem, G, of Gödel number N is not provable". Once we have this statement, we can, in principle, form a matrix in which each statement is evaluated for every integer and shown to be either true or false. This enumeration procedure will, in principle, eventually lead to the diagonal element of the matrix corresponding to G being evaluated on its own Gödel number, N. This leads to the paradox that some G is provable if and only if it is not provable. Thus it must be the case that there is some G

that is not provable or else there is some G that is both provable and not provable. In other words, the system is either incomplete or inconsistent.

A corollary to the incompleteness theorem is that there is no escape from its effect. Suppose one were to find a true theorem that is not provable. Such a theorem could simply be added to the list of axioms but that would simply postpone the agony because Gödel's result ensures that another unprovable theorem would be created. In fact, to bring this long discussion back to Cantor, in 1963, Cohen [49] showed that the continuum hypothesis is independent of the axioms of set theory and can be appended or not, either choice leading to some other unprovable result. And so Hilbert's formalist goal was shown to be unattainable.

It is interesting to note that the failure to make mathematical reasoning absolute came quickly on the heels of a similar failure in physics to vindicate the Enlightenment philosophy by constructing an absolute interpretation of reality. Early twentieth-century physics was stricken by confusion resulting from thinking about small masses moving at high velocities. The confusion was resolved by the 1905 theory of special relativity and the 1927 theory of quantum mechanics with its intrinsic principle of uncertainty. Together these theories denied the possibility of absolute frames of reference and exact positions and momenta. This failure in physics has recently been seriously misinterpreted in a way that has an impact on theories of mind. I will return to this problem in the next chapter. First, however, I must finish this history with an account of its most important event.

9.2.3 Turing's Universal Machine

Although the formalists were devastated by Gödel's theorem, there was still hope. Hofstadter [130] explains the loophole, noting that in 1931 one could have imagined that there were only a few anomalous unprovable theorems or, better still, that there existed a formal procedure whereby one could decide whether or not any theorem was provable. Were that the case, then a weaker formalism could still be pursued in which all provable theorems could be proven.

Unfortunately for the formalists, Turing resoundingly quashed this hope in 1936 with his undecidability theorem. Although Church [48] proved the same result beautifully and elegantly using his recursive function theory, it was Turing's method with its decidedly mechanical flavor which led to the digital computer and AI.

Turing proposed the machine shown in Fig. 9.9. It comprises three main parts, a tape or memory arranged in cells in which one symbol may be written, a head or sensor capable of reading the symbols written on the tape, and a finite state controller. The operation of the Turing machine is completely specified by a set of instructions of the form

$$I = \{< q_i, a_j, q_k, d, a_l >\}. \tag{9.5}$$

A single instruction is the term of (9.5) enclosed in angle brackets and is understood to mean that when the machine is currently in state q_i and the head is reading symbol a_j, the state changes to q_k, the head moves as specified by d, and the symbol a_l is written on the tape. The states, q_i, are members of a finite set, Q. The symbols are selected from a finite alphabet, A, and the head movements are restricted allowing d to assume values of only $+1$, 0, or -1 corresponding to movement one cell to the right, no movement, or one cell to the left, respectively.

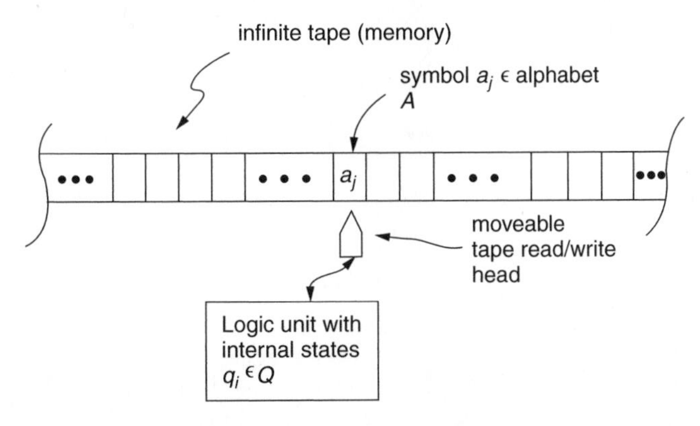

Figure 9.9 Turing machine

Ordinary Turing machines are those for which Q, A, and I are chosen so that the machine computes a specific function. We can then consider the entire ensemble of such machines indexed by their Gödel numbers defined as follows. Let $[-1] = 3$; $[0] = 5$; $[+1] = 7$; $[a_i] = 9 + 4i$; $[q_j] = 11 + 4j$. Then the jth instruction $< q, a, r, d, b >$ has the Gödel number

$$g_j = (2^{[q]})(3^{[a]})(5^{[r]})(7^{[d]})(11^{[b]}). \tag{9.6}$$

Finally, a Turing machine with instructions $\{I_1, I_2, \ldots, I_n\}$ has Gödel number G defined by

$$G = \prod_{j=1}^{n} p_j^{g_j}, \tag{9.7}$$

where p_j is the jth prime and g_j is the Gödel number of the jth instruction computed from (9.6).

Then Turing makes the stunning observation that one need not construct a special-purpose machine to evaluate a specific function because there is a universal machine that will emulate the behavior of any other Turing machine. One need only know the Gödel number of the desired machine. If one writes that Gödel number on the tape of the universal machine, it will exactly compute the corresponding function and write the answer on the tape just as the indexed machine would do. In modern parlance, we would call the Gödel number of the desired machine a program for the universal machine now known as a digital computer. We shall return to this crucial idea in a moment. Let us first examine how this universal machine was used to prove the undecidability theorem.

In the matrix of Fig. 9.10, the ijth element, a_{ij}, is the jth symbol of the result calculated by the machine with Gödel number G_i. Thus the nth row of the matrix, $a_{nm}, n = 1, 2, 3, \ldots$, is a computable number or a decidable theorem. Now form the number $\{a_{nn} + 1\}$ for $n = 1, 2, 3, \ldots, N$. By definition it is not in the matrix constructed thus far, hence we do not know whether or not it is computable. So we continue generating the matrix and we either find the number or not. If we do not find it, we still cannot

	1	2	3	4	5	...
1	a_{11}	a_{12}	a_{13}	a_{14}	a_{15}	...
2	a_{21}	a_{22}	a_{23}	a_{24}	a_{25}	...
3	a_{31}	a_{32}	a_{33}	a_{34}	a_{35}	...
.	.	.	.	.	.	...
.	.	.	.	.	.	...
.	.	.	.	.	.	...

Figure 9.10 Matrix of the digits of computable numbers

decide whether or not it is computable. If we do find it, it gives rise to a new number constructed as previously done which is not in the matrix. Thus we will never be able to test all the numbers and there must always be uncomputable ones. Hence we cannot separate the theorems into provable and unprovable classes.

In order to carry out this enumeration and diagonalization, we must have a universal machine to carry out all possible computations. The a_{ij} are computed by giving the universal machine the Gödel numbers G_i from (9.7). There is no machine more powerful than the universal machine in the sense that the addition of states, symbols, tapes, or initial data on the tape will not enable the augmented machine to enlarge the class of computable numbers it generates.

9.2.4 The Church–Turing Hypothesis

A technically correct if somewhat narrow interpretation of Turing's work is that it killed the formalist school of mathematics by depriving it of any sense of perfectibility. A slightly broader interpretation is that it dispatched the Enlightenment tradition, already left moribund by the successive shocks of relativity, uncertainty, and incompleteness. Indeed, undecidability might be seen as the fourth and final blow that rid the world, once and for all, of any illusion of perfectibility.

Nevertheless, an optimist will be quick to remind us that, in effect, Turing freed mathematics from the onerous burden of discovering God's one and only true mathematics and allowed mathematicians to invent new objects and theories having an aesthetic of their own as well as the possibility of future utility. The terror of being cut adrift from some of our most cherished moorings of intellectual security is also mitigated by the success of relativity and quantum mechanics.

I should like to propose, however, that the most far-reaching effect of undecidability was that it made possible the development of a constructive theory of mind. I choose my words carefully here because, though very tempting, it is, I shall soon argue, a serious error to make an immediate identification of the universal Turing machine and the human mind.

To understand the long and tortuous chain of reasoning stretching from the Turing machine to mind, we must first recall that there were other theories, contemporaneous with Turing's, from which the undecidability result might have emerged. Earlier we alluded to Church's lambda calculus, and there was also the Kleene [157] formalism and the McCulloch–Pitts networks [215]. What distinguishes Turing's work from that of his contemporaries is that Turing's theoretical vehicle was blatantly mechanical and fairly screamed for a physical embodiment, whereas the equivalent approaches of his colleagues were studiously abstract and not at all physically compelling. In fact, Hodges [128] suggests that it was Turing's practical and mechanistic world view that fostered his almost palpable approach to the seemingly obscure problems of meta-mathematics.

His biographer's interpretation notwithstanding, Turing's universal machine, however physically appealing, might have remained, in another era, a purely mental construct. In the early decades of the twentieth century, however, automatic reading and writing of symbols on tapes were well known as were devices that could rapidly switch among two or more stable states. Thus, one could easily imagine building a universal Turing machine capable of performing in an acceptably short interval the vast number of operations required for its utility.

Even so, there is yet another intuition required to bring forth a new metaphor for mind. Turing surely noticed that he had devised a single "machine" that could be instructed to emulate any member of an infinite class of "machines". It has been argued by Hamming [118] that Turing originally thought of his machine as merely a numerical calculator and did not appreciate its more general ability to manipulate abstract symbols in accordance with abstract rules. I regard this speculation as arrogant nonsense. Turing's letters [128] indicate that even as a schoolboy he had been captivated by mechanical analogies to human physiological functions.

It is this rational, tractable versatility that lends credence to the notion of a mechanism of thought. The leap from the Turing machine to mind is expressed in the Church–Turing hypothesis, which contends that any process admitting of a formal specification called an effective procedure can be effected by a universal Turing machine. The motivation for this hypothesis lies in the range of activities in which we humans engage and which are generally considered to be indicative of our intelligence. We can imagine that we could, if necessary, write out a sequence of directions describing how any given activity is performed. We also observe that human skills can be taught by formal instruction. We deduce from these observations that intelligent behavior, the consequence of mental function, can be formally specified and hence, in principle, realized by a suitably programmed universal Turing machine. The only missing piece of the puzzle is the Gödel number for the mind. This is the objective of the modern constructive theory of mind.

9.3 The Artificial Intelligence Program

9.3.1 Functional Equivalence and the Strong Theory of AI

The history that culminates in the Church–Turing hypothesis is the basis for the modern discipline of artificial intelligence. The hypothesis is restated in the so-called strong theory of AI which I paraphrase as "The mind is a program running on the computer called the brain". The clear implication is that thought processes in our minds are functionally equivalent to the manipulation of abstract symbols by a suitably constructed computer program.

9.3.2 The Broken Promise

When it first appeared, this new theory of mind was quite shocking. In fact, one could easily argue that it was more socially dislocating than the undecidability result that generated it. After all, it is one thing to imagine the mind as some immense clockwork, all the while secure in the knowledge that such a machine could never actually be built. It is quite a different matter to propose that an obviously constructible machine be capable of thought. Yet, it appeared quite likely that the new idea would succeed. Turing himself expended a substantial effort in convincing the scientific community and the public alike that an intelligent machine could now be built.

The strong theory of AI has some weaknesses. The term "effective procedure" as applied to human behavior is not rigorously defined, making it impossible to determine that there is an equivalent program for the universal machine. Then, even if "effective procedure" were well defined, the Church–Turing hypothesis still requires a proof. And then, even if a proof were available, it would still be required to constructively demonstrate that there are "effective procedures" for at least a large number of thought processes. Finally and most importantly, how shall we select those thoughts, skills, and actions that constitute intelligent behavior? Is intelligence just thinking or is there more?

On the other hand, the strong theory has a powerful intuitive appeal. Even if there are substantial gaps in the chain of reasoning, it is still the case that the digital computer is capable of performing an infinite number of different symbol manipulation processes and should, therefore, be sufficient for creating a mental model of the world and thereby displaying intelligent behavior. It is easy to understand how the first generation of scientists and engineers who developed the computer were easily persuaded that the intuitions were essentially correct and that the creation of a true thinking machine was not only inexorable but also close at hand.

Unfortunately, the theory was far more difficult to reduce to practice than anyone had imagined. The incipient field of AI found itself perpetually making ever more extravagant promises and irresponsible claims followed by rationalizations after regularly failing to realize them. It is not unfair to say that AI has shed a great deal of light on computation while doing little more than heat up debates on mind. One naturally wonders why the promise remains unfulfilled to this day.

9.3.3 Schorske's Causes of Cultural Decline

In his *magnum opus* on European intellectual history, Schorske [292] gives a highly instructive explanation for the unkept promises of AI. He convincingly argues that cultural endeavors stagnate and fail when they lose contact with their intellectual antecedents (i.e. their diachronic and synchronic histories) and become fixated in the technical details of contemporary thought.

> In the fields of greatest importance to my concern – literature, politics, art history, philosophy – scholarship in the 1950's was turning away from history as its basis for self understanding .... In one professional academic field after another, then, the diachronic line, the cord of consciousness that linked the present pursuits of each to its past concerns, was either cut or fraying. At the same time as they asserted their independence of the past, the academic disciplines became increasingly independent of each other as well. ... The historian seeks rather to locate and interpret the artifact temporally in a field where two lines

> intersect. One line is ... diachronic, by which he establishes the relation of a text or system of thought to previous expressions in the same branch of cultural activity. ... The other is ... synchronic; by it he assesses the relation of the content of the intellectual object to what is appearing in ... a culture at the same time. The diachronic thread is the warp, the synchronic one is the woof in the fabric of cultural history.

Although Schorske did not include science in his analysis, his thesis seems highly appropriate there, too, with AI as a striking instance. I submit that the chart of Fig. 9.1 is faithful to Schorske's notion, expressed in the excerpt above, of the intellectual cloth into which the history of mechanical metaphors for mind is woven. Applying Schorske's argument to AI, I conclude that the dismal outcome of the early experiments in AI can be attributed to the loss of the relevant synchronic and diachronic histories.

The digital computer is the technological *tour de force* of our age. In keeping with Wiener's observation, our high technology quickly generated a new metaphor for mind. In fact, soon after their commercial introduction, computers were publicly called "electronic brains" and became the heroes and villains of popular cinema.

The new metaphor was, indeed, more justifiable than any of its predecessors. The computer was versatile, reliable and fast, operating at electronic speeds capable of performing overwhelmingly lengthy calculations in heretofore unimaginably brief times. Surely, this was a thinking machine and we were utterly seduced by its power and beauty.

9.3.4 The Ahistorical Blind Alley

The difficulty arose only when, due to a subtle misinterpretation of a developing theory of computation, we fell into the trap of hubris from which we seemed unable to extricate ourselves. The result was that proper consideration was not accorded to history. Because we knew that the computer could carry out any computation, we concluded that we could concentrate exclusively on symbolic computation. This was an unfortunate but not unreasonable error. Since, according to the new metaphor, intelligence was understood to be the result of manipulating symbols and since the computer is the ultimate symbolic calculator, the simulation of intelligence must be reducible to entering the appropriate symbols into the computer and executing the appropriate programs which operate on those symbols. Everything that intelligence entails can be contained in the computer in an abstract, symbolic representation.

9.3.5 Observation, Introspection and Divine Inspiration

To many practitioners of the new art, the argument was incontrovertible. As we shall see, it is far from specious. The fallacy is discovered only when one asks where the symbols and programs originate. The obvious answer was that the programmer would chose the symbols and processes based on careful observation, introspection, argumentation, extreme cleverness, and divine inspiration. It was perfectly clear that the mind is representational, so why not simply discover the representations by any available means?

This methodology is inherently restricted to a particular moment in time. It has only two technical concerns, programs and data, which, in an automata-theoretic sense, are

Aspect	Computational Model	Cybernetic Model
representation	symbols	signals
coding	discrete	continuous
memory	local	distributed
stimuli	independent	integrated
focus	syntax	semantics
design	synthetic	adaptive
strategy	static	dynamic

Figure 9.11 Aspects of mind and language

equivalent. Thus, attention is focused on knowledge representation and organization. Programs are procedural knowledge, whereas data is declarative knowledge, and both are entirely symbolic.

AI systems that are built on this premise fall under the rubric of rule-based designs. As such, they have two distinct components. First, they utilize the rigorous techniques of computing for both algorithms and data structures. For example, algorithms may be based on fundamental principles of searching, sorting, parsing, and combinatorial optimization. Data structures often employ binary trees, directed graphs, linked lists, queues, and formal grammars. Sometimes, however, it is not clear how to use such techniques to simulate some aspects of intelligence, thus giving rise to the second component, heuristics. In this case, the programmer constructs rules on an *ad hoc* basis. Often expressed as logical predicates, the rules take the form: if the predicate P is true then execute function F. (Compare this with the explanation of semantics given in Chapter 7.) Both the predicates and the actions can be very complicated. Such rules often operate on lists having no discernible order. In instances when quantification is required, values on completely arbitrary scales are assigned. The only connection that rule-based systems have to physical reality is their author. Thus, intelligence is treated as a disembodied, abstract process. As such, it must be based on a priori choices of symbols representing whatever properties the programmer can detect in the real objects they signify and a loose collection of rules describing relations among the objects. The methods by means of which symbols are created, numerical values assigned, and rules inferred are largely subjective. This often results in inconsistent definitions, arithmetic absurdities and the competition – or even contradiction – of many rules to explain the same phenomenon. The extreme example of this approach is the CYC project of Lenat [177], and it should come as no surprise that after decades of exhaustingly detailed work, this and other such systems have met with little success.

9.3.6 Resurrecting the Program by Unifying the Synchronic and Diachronic

At this point it is instructive to compare the components of mind in the different perspectives espoused by Turing and Wiener. The simple chart of Fig. 9.11 suffices for the purpose.

The most common interpretation of the ENIAC box (in Fig. 9.1) and the one favored by Turing himself is an implementation in which the machine engages only in abstract thought and communicates with the real world via only a keyboard. However, in the penultimate paragraph of the 1950 paper Turing [320] suddenly offers an alternative approach from a more nearly Schorskian perspective that is much more aligned with the cybernetic paradigm:

> We may hope that machines will compete with men in all purely intellectual fields. But which are the best ones to start with? Even this is a difficult decision and people think that a very abstract activity like playing chess would be the best. It can also be maintained that it is best to provide the machine with the best sense organs that money can buy and then teach it to understand and speak English. This process could follow the normal teaching of a child. Things would be pointed out and named, etc. Again, I do not know what the right answer is but I think both approaches should be tried.

I argue that although the ahistorical approach could, in principle, succeed – we could guess the Gödel number for mind – it is highly unlikely. This approach ignores one of the most important purposes of intelligence, that of ensuring the survival of the relatively slow, relatively weak, relatively small human animal in a complex and often hostile environment. While it is true that some of the representations necessary for survival could be transmitted genetically or culturally, many critical behaviors are acquired by each individual through long periods of interaction with his environment. In order to acquire sufficient knowledge about and skills to function well within his surroundings, that is, to define symbols and build programs, sensorimotor function is required. This is the domain of the cybernetic paradigm, and thus I advocate an approach based on a synthesis of the synchronic and diachronic histories in the spirit of Turing's alternative.

10

A Speculation on the Prospects for a Science of Mind

10.1 The Parable of the Thermos Bottle: Measurements and Symbols

The story is told of a conversation that ensued when a small group of scientists representing different disciplines met over lunch. The discussion wandered politely, becoming intense upon reaching the matter of important unanswered questions. A physicist spoke passionately about the quest for a unified field theory. A biologist reminded his colleagues about the debate over the causes and timing of evolutionary development. A psychologist lamented the confusion over the nature of consciousness. And an anthropologist raised the issue of the environmental and genetic determinants of culture. An AI researcher who had been writhing uncontrollably in his desire to participate in the scientific braggadocio finally managed to insinuate himself into the conversation. "My colleagues and I", he ventured, "have been seeking to understand the thermos bottle." When this evinced only quizzical stares from his companions, he became impatient. "Well, look," he urged, "in the winter you put hot tea into a thermos and it stays hot." Puzzlement turned to incredulity. With frustration rising in his voice he persisted, "But in the summer you fill it with iced tea and it stays cold." Bemused silence descended. In total exasperation he cried, "Well... how does it know!!?"

What some, perhaps, will find deliciously and maliciously funny about this anecdote is its caricature of a far too prevalent proclivity to ascribe all phenomena to cognitive process rather than physical law. Buried beneath the perhaps mean-spirited nature of the parable, there lies a profound insight. Mental activity entails both physics and computation in complementary roles. To say, as I did in Chapter 9, that the mind is to be understood by a synthesis of its synchronic and diachronic histories is to recognize that while abstract, symbolic representation of the world is required, reality is presented to us only in the form of continuous, sensorimotor measurement. That is, the symbols, which are the basic elements of the contemporary perspective, derive from the measurements that are the province of the historical viewpoint. Cognition depends on symbolic representation, whereas measurement relies on physical sensors and actuators. Measurements may be transformed into symbols. Unless this transformation is well understood, there

Mathematical Models for Speech Technology. Stephen Levinson
 ISBN: 0-470-84407-8

can be no hope of formulating a sophisticated theory of mind without which, I argue, it is impossible to develop a useful technology for human–machine communication by voice.

Most engineers working on human–machine communication would argue that exactly the reverse of my argument is true. That is, if an advanced technology depends on understanding the mind, then there is little hope of achieving the goal. The mind is properly relegated to the realm of philosophy, and thus it can never be understood in a way that has any scientific basis or technological implications. I utterly reject this point of view. Just as the mathematical models presented in the first eight chapters of this volume are both a theory of language and the basis of speech technology, so can a rigorous theory of mind support a more advanced language processing technology. However, the new science we seek will not evolve from armchair research. Before I offer a concrete, experimental approach to the problem, I am obliged to make a brief digression to consider whether or not it is even sensible to search for a science of mind.

10.2 The Four Questions of Science

In order to speculate on how a science of mind might develop, it is helpful to consider other questions addressed by the sciences. As I suggested in the parable of the thermos bottle, science comprises four areas of inquiry that, taken together, completely cover its legitimate domain. Science asks about cosmos, life, mind, and society. As listed, the subjects of science are arranged in order of increasing complexity. That is, life results from the intricate composition of many small, inorganic, physical objects. As living organisms evolve to higher orders of complexity, minds emerge. Finally, societies may be considered to be large organized collections of minds.

10.2.1 Reductionism and Emergence

Each level of the scientific hierarchy emerges out of its predecessor in the sense that the science at one level must be distinct from but consistent with that of its predecessors. One might view this constraint as a generalization of reductionism in which successful theories at one level need not be directly expressed in terms of the essential constituents of the preceding levels but cannot result in contradictions of the theories that govern them. Thus, biological life develops from sufficiently complex combinations of inanimate matter, the behavior of which is well described by physics. Yet, it may not be useful to describe biological phenomena in terms of the behavior of the simplest physical objects involved in them. For example, it is possible, in principle, to explain the behavior of large biomolecules directly by solving the Schrödinger equation. Up to the present, this has only been accomplished for simple atoms, but there can be no doubt that the necessary solutions exist even if the techniques required to compute them are presently lacking [134, 131]. It appears, however, that even if such solutions were available to us, it is far more useful to consider the interactions of specific biomolecules such as proteins and amino acids to understand living organisms, remaining secure in the knowledge that these larger objects conform to the laws of physics the detailed expressions of which need not concern biologists. Not only do we obtain more parsimonious descriptions, but also one hopes that the relevant dynamics of biological systems are expressed in terms of a set of state variables that depend in a complicated way on physics but interact with each other in a simple and essential way for the characterization of biological laws. For example, this effect is evident within physics itself in the case of gases. The kinetic-molecular theory assures us that gases are composed of particles in continuous motion. In principle, the behavior of gases could be determined from the laws of mechanics governing the interactions of point masses. Due to the large numbers of particles involved, a trivial measure of complexity, it is, at best, impractical to study gases from this perspective. However, the state variables of temperature, pressure, and volume provide elegant descriptors of an ideal gas and we can understand, in principle, how they are related to the underlying kinetic-molecular theory without resort to knowledge of the dynamics of individual particles. Unfortunately, no such theories yet exist in biology.

The phenomenon of emergence as described above avoids, I believe, some of the difficulties that arise from theories of consilience [57], of which Wilson's sociobiology [331] is an instance. According to Wilson and his followers, human social behavior is genetically determined. While I am sympathetic to the idea that there are genetic influences in both personal and social interactions, the proposal of a direct genetic encoding of all behavior violates the spirit of emergence. That is, it ignores the effects of the specific, and as yet unknown, psychological and sociological dynamics. However, the principles of emergence do offer the possibility of a rigorous understanding of social behavior with which Wilson might well be satisfied.

Emergence also seems to provide a better way to study mind than does the blatantly reductionist theory proposed by Penrose [245], who views mind and consciousness as a result of quantum mechanics. Like consilience, the "quantum mind" skips two stages of emergence. Unlike consilience, there is absolutely no reason to suppose that mind is a quantum effect. I propose to elaborate on emergence as it relates to a science of mind in Section 10.3.1.

10.2.2 From Early Intuition to Quantitative Reasoning

I raised the issue of sociobiology only for the purpose of contrasting its premises with those of emergence. I do not intend to address sociological questions further. Rather, I would like to address the somewhat less ambitious question of what a sufficiently advanced psychology (i.e. science of mind) would have to be in order to support a technology of natural human–machine communication. To do so, I need to make a digression explaining how I construe science.

I subscribe to the doctrine of scientific realism which asserts, first and foremost, that science seeks to discover objective reality. This quest has a long tradition beginning with what Holton [132] calls the "Ionian Enchantment", Thales' idea that the world is comprehensible by means of a small number of natural laws. The subjects of these laws are assumed to be absolutely real. Thus, electrons, genes, and mental representations, listed in ascending order on my scale of scientific complexity, are assumed to exist even though they are not directly observable.

A contemporary rendering of this theory is given by Margenau [210] whose "constructual plane" shows how observation and reasoning establish that all matter and process is real and there is no place for the supernatural. This does not mean that everything is knowable or that all processes are possible. Prohibitions against perpetual motion or arbitrarily high velocities are not problematic, for example, because they violate established principles. The extent to which the universe is not knowable is knowable. The incompleteness of a theory is easily recognizable when there is widespread controversy about it.

The means by which science seeks explanations of objective reality are well known to every schoolchild. The so-called scientific method proceeds from observation to testable hypothesis leading to experimental evaluation. Although not considered part of the basic method, its motivation springs from curiosity and astute perception leading to early intuition, a critical form of reasoning in which relationships are postulated even though no causal argument can be formulated. The origins of early intuitions will be considered in Section 10.4.2.

What is often omitted from considerations of the scientific method is the requirement of quantitative reasoning expressed in a mathematical formalism. Many intellectual pursuits employ reasoning akin to the scientific method but they make no pretense of using mathematics. The relationship between science and mathematics is rarely explored by historians of science who leave the task to practitioners of science such as Hadamard [114], Poincaré [249], and Wigner who observed:

> The miracle of the appropriateness of the language of mathematics for the formulation of the laws of physics is a wonderful gift which we neither understand nor deserve. [328]

Dirac elucidated the role of mathematics in science, noting that:

> The physicist, in his study of natural phenomena, has two methods of making progress: (1) the method of experiment and observation, and (2) the method of mathematical reasoning. The former is just the collection of selected data; the latter enables one to infer results about experiments that have not been performed. There is no logical reason why the second method should be possible at all, but one has found in practice that it does work and meets with remarkable success. This must be ascribed to some mathematical quality in Nature, a quality which the casual observer of Nature would not suspect, but which nevertheless plays an important role in Nature's scheme.
>
> One might describe the mathematical quality in Nature by saying that the universe is so constituted that mathematics is a useful tool in its description. However, recent advances in physical science show that this statement of the case is too trivial ...
>
> The dominating idea in this application of mathematics to physics is that the equations representing the laws of motion should be of a simple form. The whole success of the scheme is due to the fact that equations of simple form do seem to work. The physicist is thus provided with a principle of simplicity which he can use as an instrument of research. [66]

Einstein made an even stronger claim for the role of mathematics in scientific discovery:

> [A]ny attempt to derive the fundamental laws of mechanics from elementary experience is destined to fail. [77]

After asserting that the axiomatic foundations of physics cannot be inferred from experience but that they can be correctly deduced by mathematical reasoning, he admits the need for experiment to guide the mathematics:

> [N]ature actualizes the simplest mathematically conceivable ideas. It is my conviction that through purely mathematical construction we can discover these concepts and the necessary

> connections between them that furnish the key to understanding the phenomena of nature. Experience can probably suggest the mathematical concepts, but they most certainly cannot be deduced from it. Experience, of course, remains the sole criterion of mathematical concepts' usefulness for physics. Nevertheless, the real creative principle lies in mathematics. Thus in a certain sense I regard it true that pure thought can grasp reality, as the ancients dreamed. [77]

Although science may have humble origins in curiosity regarding common experiences and intuitive explanations of them, much more is required. Science reaches its maturity only when it expresses laws of nature in mathematics, tests them against experiments based on quantitative measurements and finds them consistent with all other known laws. For example, one is tempted to say that the citation at the beginning of Section 2.5 shows that Plato understood pattern recognition. Upon further reflection, however, it becomes clear that this conclusion can only be reached by reading the ancient text with twenty-first century eyes. Looking back, we interpret the words with respect to our modern theories. But Plato had no way to quantify his ideas. He had no rigorous theory of probability. He did not even have the analytic geometry required to define a feature space. Plato had an early intuition about patterns but no mathematical expression of it and, hence, no science.

At this time, only physics is mature. As Hopfield notes, biology is maturing:

> [B]iology is becoming much more quantitative and integrated with other sciences. Quantification and a physical viewpoint are important in recent biology research – understanding how proteins fold, ... how contractile proteins generate forces, how patterns can be spontaneously generated by broken symmetry, how DNA sequences coding for different proteins can be arranged into evolutionary trees, how networks of chemical reactions result in "detection", "amplification", "decisions". This list could be a lot longer. [134]

Another important branch of mathematics relevant to biology is developing, the theory of dynamical systems operating far from equilibrium. Life may be seen as a collection of mechanisms preventing the descent into equilibrium or death by homogeneity.

Psychology and sociology are in their infancy but there is no reason to believe that they will not someday mature in the same sense that physics has done and biology is now doing. In fact, we have already had a glimpse of the mathematics needed to express the natural laws of biology, psychology, and sociology. The rudiments of a general theory of emergence are beginning to appear. One theme will certainly be that of properties of stochastic processes. Note that we have appealed to such mathematics in Chapters 3–6. General theories of complex systems will also borrow from thermodynamics, statistical mechanics, and information theory such notions as state variables, phase transitions, multivariate nonlinear dynamics – especially those operating far from equilibrium, and complexity measures. I shall return to this point in Section 10.5.

10.2.3 Objections to Mathematical Realism

My characterization of science will, no doubt, raise many objections. I cannot completely subdue them but I propose to comment as best I can on the more common ones.

The Objection from the Diversity of the Sciences

This argument rests on the proposition that biology, psychology, and sociology are intrinsically different from physics and need not follow the same path to success. The citation from Hopfield casts some doubt on the proposition as it applies to biology. The cases for psychology and sociology are not so easily made. In fact, none other than Wiener has expressed considerable pessimism in this regard.

> I mention this matter because of the considerable, and I think false, hopes which some of my friends have built for the social efficacy of whatever new ways of thinking this book may contain. They are certain that our control over our material environment has far outgrown our control over our social environment and our understanding thereof. Therefore, they consider that the main task of the future is to extend to the fields of anthropology, of sociology, of economics, the methods of the natural sciences, in the hope of achieving a like measure of success in the social fields. From believing this is necessary, they come to believe it possible. In this, I maintain, they show an excessive optimism, and a misunderstanding of the nature of all scientific achievement.
>
> All the great successes in precise science have been made in fields where there is a certain high degree of isolation of the phenomena from the observer. [330]

Wiener invokes the Maxwell demon problem, in which the observer has such a profound effect on the observed that no regularities can be reliably obtained, for the social sciences. While it is impossible to dismiss this argument entirely, it is essentially an argument from pessimism, the antidote to which is given by von Neumann and Morgenstern:

> It is not that there exists any fundamental reason why mathematics should not be used in economics. The arguments often heard that because of the human element, of the psychological factors etc., or because there is – allegedly – no measurement of important factors, mathematics will find no application, can all be dismissed as utterly mistaken. Almost all of the objections have been made, or might have been made, many centuries ago in fields where mathematics is now the chief instrument of analysis. [324]

It is worthwhile to consider a specific instance of the general argument. The geocentric theory of the solar system was gradually replaced with a heliocentric one by Copernicus, Galileo, and Kepler. It was not, however, until Newton and Leibniz introduced the calculus that the new theory could be completely vindicated. The problem with the geocentric theory was only partially due to its reliance on naive observation. The deeper difficulty was that, as a purely geometric theory, it concerned itself simply with the apparent locations of the sun and planets. There was no explanation of what caused the planets to move. The motions might just as well been the result of the planets being bolted to some giant clockwork. Until Newton, there was simply no mathematics to analyze quantities that change in time with respect to other quantities and thus no way to explain forces that might hold the planets in their orbits and determine their motions. The Ptolemaic solar system was adequate for timing religious rituals. With its epicycles upon epicycles it was quite accurate in its limited predictive abilities. But it could not support even the most elementary physics. That would have to wait until the completely new calculus vastly extended the existing static geometry. Centuries later, with the invention of several new branches of mathematics, we have come to understand that accurate measurements of time and position are impossible without a mature physics. Today, accurate navigation relies on relativistic corrections in the computation of satellite orbits. It takes a certain lack of imagination to believe that there will be no new mathematics that opens up biology, psychology, and sociology just as the calculus revealed physics.

The Objection from Cartesian Duality

The mind–body dichotomy is usually attributed to Descartes, who believed that humans alone are endowed with both a physical body and an incorporeal soul infused in it by God. The Greek word *psyche* is, in modern translations, alternatively rendered as either mind or soul, indicating some identification of our mental function with the supernatural. In essence, the objection is theological, placing the sacred soul, the seat of the mind, beyond the reach of science. This is, of course, in polar opposition to my definition of scientific realism and the conflict cannot be resolved. Even if one were to construct a machine, indistinguishable in its mental abilities from a human, a believer in dualism would reject it as a nothing but a superficial simulation of mind and perhaps a blasphemy.

The Objection from either Free will or Determinism

There are two other contradictory theological arguments that decry any exact science as a challenge to the essence of God. The conflict was framed as early as 1820 by Laplace, who wrote:

> An intelligent being who knew for a given instant all the forces by which nature is animated and possessed complete information on the state of matter of which nature consists – providing his mind were powerful enough to analyze these data – could express in the same equations the motion of the largest bodies of the universe and the motion of the smallest atoms. Nothing would be uncertain for him and he would see the future as well as the past at one glance. [172]

If such an idea is odious to the theologians when expressed only as a claim about physics, imagine how blasphemous it would be if extended to include mental activity. A science of mind must be viewed as an assault on the sanctity of the soul and thus as a sacrilegious violation of ethics.

But there is a contradictory aspect to determinism. Along with a soul, God has given man free will. If all of his actions are predetermined then volition is an illusion. We can perhaps escape from the paradox by asserting the existence, as a fundamental property of reality, of thermodynamic and quantum randomness. Thus the vision of Laplace must be amended to include the notion of stochastic process as a model for limitations on the precision of measurements and other phenomena that admit of only a probabilistic explanation.

There is some debate about the legitimacy of probabilistic theories in scientific realism. Perhaps the best-known rejection of probability as a valid scientific explanation comes from Einstein:

> Quantum mechanics is very impressive. But an inner voice tells me that it is not yet the real thing. The theory produces a good deal but hardly brings us closer to the theory of the Old One. I am at all events convinced that He does not play dice. [75]

Einstein was never able to satisfactorily resolve the issue. Quantum mechanics survives as a highly successful theory, and the mathematical theory of random processes permeates the cybernetic paradigm. Indeed, the several theories of linguistic structure discussed in Chapters 3–6 of this volume are stochastic models. No matter which side of the debate about free will and determinism one chooses, he runs afoul of religious doctrine. Perhaps Einstein has provided the solution with his enlightened faith:

> I believe in Spinoza's God who reveals himself in the orderly harmony that exists, and not in a God who concerns himself with the fates and actions of human beings. [76]

Dennett [64] has argued that thermodynamic and quantum randomness do not account for free will because at the level of psychological and social behavior these random fluctuations have zero mean and, hence, no effect. This does not diminish the value of stochastic models of behavior. It simply says that free will is not a consequence of a stochastic process.

We certainly feel as if we have free will. A better explanation of the phenomenon is that it is a consequence of our conscious minds (see Section 10.4). That is, we are able to decide amongst many possible courses of action and we hold ourselves responsible to make these choices.

Thus the world is, if not perfectly deterministic, at least predictable up to some statistically characterizable limits. We consciously use this fact to evaluate the consequences of our actions and make decisions accordingly. The conclusion is that neither free will nor determinism prohibit a science of mind.

The Postmodern Objection

Postmodernism traces it origins to French and German philosophy and is largely a reaction to the discontents of modern society. As a literary device, it is used to portray the confusion and wreckage of our era. In this role it can be quite effective. As a philosophy that denies the possibility of objective truth, insisting instead that truth flows from persuasion and coercive power, it is a strident opponent of scientific realism. Perhaps the postmodern school is an expression of a profound disappointment at our seeming inability to establish a sane society. Perhaps it resulted from a misinterpretation of either the intrinsic limitations on our knowledge of physics and mathematics as described in Section 9.2.4, or the empiricists' interpretation of science as merely a consistent explanation of reality. In any case, postmodernism mocks those claims it supposes science to be making with a counterclaim that all science is simply a social construction. Thus it should be no surprise that postmodernism reserves special contempt for definitive sciences of mind and society. The postmodern perspective is so alien to the scientist that it is difficult to imagine that anyone would actually advance such ideas. But these notions have, in fact, been seriously proposed. Latour is a leading advocate of the social construction of science. He begins with disingenuous praise for the goals of science, only to end by devaluing them as outmoded:

> We would like science to be free of war and politics. At least we would like to make decisions other than through compromise, drift and uncertainty. We would like to feel that

> somewhere, in addition to the chaotic confusion of power relations, there are rational relations ... surrounded by violence and disputation we would like to see clearings – whether isolated or connected – from which would emerge incontrovertible, effective actions ... The Enlightenment is about extending these clearings until they cover the world ... Few people still believe in the advent of the Enlightenment ... [173]

Next he dismisses the validity of scientific reasoning.

> We neither think nor reason. Rather we work on fragile materials – texts, inscriptions, traces, or paints – with other people. These materials are associated or dissociated by courage or effort; they have no meaning, value or coherence outside the narrow [social/political] network that holds them together for a time. [173]

He then goes on to deny the possibility of any universal scientific principle:

> Universality exists only "in potentia". In other words it does not exist unless we are prepared to pay a high price of building and maintaining costly and dangerous [social/political] liaisons. [173]

Aronowitz extends Latour's notion to mathematics noting that "[N]either logic nor mathematics escapes the contamination of the social" [7]. In particular, Campbell asserts that the contaminating social influences are capitalism, patriarchy and militarism, saying that "[M]athematics is portrayed as a woman whose nature desires to be the conquered Other" [218].

The postmodern attack on science is disingenuous. Its intellectual veneer is but a subterfuge for its virulent, if bankrupt, political agenda that abhors anything it regards as authoritarian and exalts all forms of cultural and intellectual relativism. It rejoices in the replacement of the canon of the Great Books with an anti-intellectual eclecticism. What could possibly be more authoritarian than a science that claims objective truth and requires of its practitioners the mastery of an imposing canon? Hence its unbridled derision of science.

Politics may be adversarial, but science is definitely not. Unlike a politician, nature does not change her design of the universe when science seems close to understanding aspects of it. Nor are the principles of scientific investigation subject to change even when nature consistently frustrates the efforts of science to discover her secrets. As Einstein once [74] remarked: "Subtle is the Lord but malicious He is not." Of course, his reference to the Lord is intended to mean the God of Spinoza, as cited above. From the scientific perspective, postmodernism has been completely discredited by the recent Sokal hoax in which the journal *Social Text* published, as a serious scholarly paper, a parody entitled "Transgressing the Boundaries: Towards a Transformative Hermeneutics of Quantum Gravity" [302]!

Beginning the New Science

I am certain that the answers to a few of the usual objections to my new positivism will not satisfy my critics. I fear that not even the realization of my proposal will accomplish that. However, I am not at all discouraged by the daunting task of producing the required new science. I am convinced that the ideas expressed above will eventually generate natural laws that we cannot yet even imagine. I must be content to be a participant in the early phases of development of psychology and sociology. I take comfort in the fact that physics needed more than three centuries to mature and it is not yet a finished product. While I cannot hope to invent the new mathematics that I posit exists, I can offer a working hypothesis, an experimental method, and a means of evaluating a novel constructive theory of mind.

10.3 A Constructive Theory of Mind

Proceeding on the assumption that a mature, quantitative science of mind can be discovered but painfully aware that neither the mathematics nor epistemology to support it yet exists, I can only offer a proposal to explore the terrain by means of a constructive theory. I use the term "constructive" not to mean helpful but rather to imply that in the process of building a machine with the desired behavior, an analytical theory of that behavior will become evident. I propose to build a stochastic model of mind with sufficiently many degrees of freedom that its detailed structure can be estimated from measurements and optimized with respect to a fidelity criterion. Then, a post-optimization interpretation of the model will, I predict, yield the desired analytical theory. The Poritz experiment recounted in Section 3.1.7 is, by this definition, a successful constructive theory of broad-category acoustic phonetics and phonotactics. If the original model is severely underdetermined, there is the risk that the model will capture artifacts present in the measurements. Fortunately, the real world is characterized by regularities too prominent and too consistent to be accidental. Wherever possible, the model should be constrained to reflect these regularities just as Poritz did (refer to Section 3.1.7).

10.3.1 Reinterpreting the Strong Theory of AI

As we noted in Section 9.3.2, the preferred demonstration of the strong theory of AI is a direct synthesis of a discrete symbolic model perfectly isomorphic to reality and unaffected by any uncertainty in the measurement of the physical correlates of the putative symbols. This method avers that the sensorimotor periphery may be safely ignored but it requires the full computational power of the universal Turing machine to be effective. In contrast, Turing's alternative as described in Section 9.3.6 uses the power of the universal machine only to simulate the physical processes from which the symbols, defined by probability distributions, derive. Thus cognitive function emerges from the solution to the presently unknown equations of motion underlying the mechanisms of mind. This will be the legitimate solution to the problem of the thermos bottle.

10.3.2 Generalizing the Turing Test

As noted in Section 9.2, Turing proposed that an artificial intelligence could, in principle, be evaluated experimentally. Taking an agnostic position on what mind really is, he suggested that the requirement should be only that the behavior of the machine be indistinguishable from that of a human by a human judge. This method of discrimination has come to be known as the "Turing test". Turing envisioned the "imitation game" would be conducted in the domain of abstract mental activity and communication would be only via a teletypewriter. If, however, we are to follow Turing's alternative discussed in Section 9.3.6, then the test must be appropriately modified.

Turing's alternative advocates connecting the machine to the real world via sense organs, and in Section 10.4 I shall propose augmenting the sensory function with a motor function. This implies a mechanical mind that is no longer restricted to engage in abstract thought alone. Rather, the machine is embodied, interactive, and adaptive. Embodiment allows for the symbols to be grounded in perception, locomotion, proprioception, and manipulation. Interaction suggests that the machine will continually both respond to and cause changes in a real physical environment. Adaptation refers to the ability of the machine to alter its perceptions and actions in response to observed changes in its environment in order to make its behavior as successful as possible.

Under Turing's alternative, the criterion for winning the imitation game must be modified accordingly. I propose that the criterion be one of interesting behavior in which cognitive process is evident. In particular, the acquisition and subsequent use of spoken language should be considered essential. By itself, language acquisition is not sufficient to conclude that there is a mind supporting it. However, if the experiment is conducted properly, the machine can be carefully instrumented, thereby allowing for the observation of internal changes corresponding to the development of significant and identifiable mental states. Thus, the generalized Turing test should evaluate both observable and internal behavior. I will return to this idea in Section 10.5.

10.4 The Problem of Consciousness

No serious treatment of a science of mind can long avoid the problem of consciousness, so before proceeding on to my proposed experiment based on Turing's alternative, I am obliged to comment on it. The literature on consciousness has a long history and is far too vast to even survey here. There are six recent books on the subject that give, at least, a modern perspective. Readers who are intrigued by this subject may wish to indulge in works by Damasio [56], Edelman and Tononi [73], McGinn [216], Tomasello [315], Fodor [92], and Penrose [245]. Collectively they comprise some 1300 pages and present quite different although related perspectives, including neurophysiology, connectionism, robotics, philosophy, psychology, and physics. For reasons that will become clear, I propose to summarize the problems in a few brief paragraphs.

Scholars differ on the definition of consciousness. There are several explanations based on information-theoretic models of perception, pattern recognition, sensory fusion, development and identification of mental states, attention, volition, and the awake–asleep distinction. Most philosophers agree that these concepts are closely related to consciousness but are not its essence. The crux of the issue is the experience, that is, the visceral feeling of our contacts with physical reality. There is, philosophers insist, a difference between function and experience. Thus, when a stimulus impinges on our sensory organs and we feel the sensation of color, sound, smell, or touch, something more than simple information processing is occurring. Moreover, it will be impossible to comprehend mind, let alone simulate it, without accounting for experience as a consequence of function and separate from it. In fact one of the best-known arguments against the symbolic computation theory of AI is due to Searle [293], who opines that a symbol processor, no matter how sophisticated it might appear, can never be conscious in the experiential sense.

Such a machine would be, to use the vernacular, a zombie, going through the motions of everyday activity without conscious awareness.

If one accepts this definition of consciousness and its implications for a theory of mind, one has few alternatives. One can become embroiled in questions of duality. One can either deny the problem or label it beyond our capacity to solve. Or one can postulate the existence of some additional natural process to explain it. The difficulty with these alternatives is that they all spring directly from the armchair. As such, they are the result of the same fallacy that plagued the symbolic computation approach to AI. The definition of consciousness is based on an informal description of a subjective process analogous to the introspective determination of the symbolic representations of reality. It is also an instance of the thermos bottle problem in which it is asked not "How does it know?" but rather "How does it experience?" while, of course, ignoring the underlying objective aspects of mind. Because of the intrinsic subjectivity of the conventional definition, there is no way to test its validity and, hence, no possible resolution of the issue.

There is a simple three-part answer to all of these questions. First, consciousness is nothing other than the mind's self-awareness. Regardless of the computations the mind is performing, it has a symbolic representation for itself that can enter into any or all of them. Second, experience is epiphenomenal. As such, it is an observable but irrelevant artifact of the particular machine in which the mind is implemented. According to my doctrine of functional equivalence, different machines will produce different, but still intelligent, behavior and will have different physical manifestations of experience. In the human and probably higher animals, the feeling of experience is a by-product of the electrochemical signals that constitute thought. Third, experience, like the mind that generates it, is emergent. When signals that mediate the complexity of an appropriately organized physical mechanism are great enough, the conscious mind emerges from them. Jaynes [145] argues that the emergence of mind was a behavior learned socially millennia ago that has been culturally transmitted since then. Jaynes argues further that the emergence was marked by the recognition that the internally audible thought process was actually one's own and did not originate outside one's body. This idea was the focus of acrimonious controversy. Because Jaynes made the tactical error of placing a date on the initial emergence of consciousness based on a literary analysis, he was criticized on historical grounds. Unfortunately, the issues of timing completely overshadowed the intriguing proposal that consciousness was acquired. I find this a fascinating conjecture because of its compatibility with the embodied, interactive, adaptive, emergent mind of the cybernetic paradigm. The problem then, as I argued in Section 9.3.6, is to find those quantities and the relationships among them that form the basis for mind.

10.5 The Role of Sensorimotor Function, Associative Memory and Reinforcement Learning in Automatic Acquisition of Spoken Language by an Autonomous Robot

It is tempting to say that I was inspired to undertake this research simply as a result of reading Turing's 1950 paper. Although I actually read it in an undergraduate psychology course, the original motivation is much more mundane. The methodology I am advocating here is based on a few early intuitions which arose from the difficulties I encountered in my research on speech recognition based on the material in Chapters 2–8. What I and many of my colleagues observed was that the signals we were analyzing seemed to have huge variabilities. Yet humans perceive these very signals as invariant. Humans rarely misunderstand a sentence or misclassify a visual scene, even a moving one. The obvious conclusion is that humans and machines are using quite different pattern recognition techniques based on quite different learning mechanisms. Machines rely on statistical models optimized with respect to preclassified data. The models are fixed upon completion of training, whereas humans optimize performance and are continuously adapting their strategies. Machines can achieve useful levels of performance only on artificially constrained tasks, whereas humans must achieve successful behavior in a world constrained only by the laws of nature. Therefore, the first intuition is that it may well be useful to try to simulate these aspects of human behavior. These early intuitions motivating my experiments were first outlined in [184]. The following six short sections explain my working hypothesis about mind and language. Section 10.5.7 then describes my experimental method for developing and testing my hypothesis.

10.5.1 Embodied Mind from Integrated Sensorimotor Function

The experiment described in Section 10.5.7 is predicated on the assertion that there is no such thing as a disembodied mind. While the mind is representational and thought processes well described as computational operations on the abstract symbolic representations, no such symbolic representations can arise without the sensorimotor function of the body. Theories of "the embodied mind" have existed since Turing himself proposed the idea in his seminal 1950 paper which described the first mathematically rigorous computational model of intelligence. More recently, Johnson [151], Lakoff and Johnson [167], and Jackendoff [141, 142] have given comprehensive treatments of the idea from the perspective of psycholinguistics. There have been other speculations on this subject, but they have not been in the mainstream of AI. More importantly, there has been very little experimental work on "embodied mind". One research project which definitely recognizes the importance of combined sensorimotor function in intelligent behavior is the COG project of Rodney Brooks at MIT [83]. There is also significant support for the importance of integrated sensorimotor abilities from psycholinguistics in Tanenhaus *et al.* [311, 312] showing the relationship of vision and language and new work in neurophysiology surveyed by Barinaga [23] demonstrating the existence of neural pathways from the motor areas of the brain to the cognitive areas.

The importance of the integration of all sensory and motor signals is revealed by thinking about perception and pattern recognition. If one examines the signals from individual sensory modalities independently, they appear to have large variances and hence overlapping distributions leading to large classification error probabilities. This is exactly because the signals have been projected onto a space of lower dimension from the space of higher dimension in which the integration of all sensorimotor modalities is properly represented. In that space of much larger volume, the signals are widely separated and robust classification can be achieved.

10.5.2 Associative Memory as the Basis for Thought

The mind is an associative pattern recognition engine that measures the proximity of one signal to another. For general pattern recognition proximity is a measure of similarity.

For vision, proximity implies continuity or connection. For reasoning, proximity leads to causality and prediction.

I submit that the primary mechanism of cognition is an associative memory that has the following properties. First, it must be able to relate input stimuli to desirable behavior. Second, the contents of the memory must be reliably retrieved when some stimuli are missing or corrupted. Third, the memory must be content addressable so that the presentation of any single stimulus will evoke all the related stimuli and their associated responses. The associative memory is capable of storing representations of stimuli of complex structure, thereby allowing for fusion of sensory modalities and motor functions and the recognition of intricate sequences thereof. The latter is particularly important for the grammar (i.e. phonology, morphology, and syntax) of language, although other complex memories may also be encoded in the same manner. However, rules governing the formation of such sequences are merely the code in which memories are represented. They serve two particular purposes, namely, to afford the memories some immunity to corruption and to make semantic processing robust and efficient.

There are many ways to implement an associative memory with the desired properties. The simplest is the nonparametric maximum likelihood pattern recognition algorithm, discussed in Section 2.5. According to this model, the extracted sensory features are simply time-stamped and concatenated in a long vector. This method has the advantage of computational simplicity but suffers from its inability to represent relationships among stimuli.

The preferred approach is to use a stochastic model that captures both probability distributions of stimuli and their underlying structure. Any of the mathematical models discussed in Chapter 3 are suitable for this purpose, but the most directly applicable is the hidden Markov model in which the observation densities are used to capture the statistics of the sensory data and the hidden state transitions become associated with structure. Details of the implementation are given in Section 10.5.7.

10.5.3 Reinforcement Learning via Interaction with Physical Reality

The content of the associative memory is acquired by reinforcement. I define "reinforcement" to mean that the probabilities of forming associations amongst stimuli and responses are increased when the responses are useful or successful and decreased otherwise. Initially, all associations except for a few instincts have probability zero. These probabilities can be changed based on perceived reinforcement signals. Reinforcement is provided in three ways. First, in the absence of any outside stimulus, the robot goes into autonomous exploration mode. In this mode, it moves randomly about the environment, scanning for events of interest. The pace of its motions is governed by a timer simulating attention span. While scanning, the robot compares its sensory inputs to memory. When good matches are found, the associated actions are carried out and the memory updated as required. If no relevant memories are retrieved, the robot looks for high correlations amongst its sensory inputs and stores the corresponding events in memory. Stability of the robot should allow for safe continuation of this mode for indefinite periods. This operation, which I call the cognitive cycle, is shown in Figure 10.1.

The most important mode of reinforcement is that of interactive instruction by a teacher. In this mode, the instructor will give explicit hard-wired reinforcement signals to the robot. Following the procedure established in the earlier pilot experiments, the instructor will

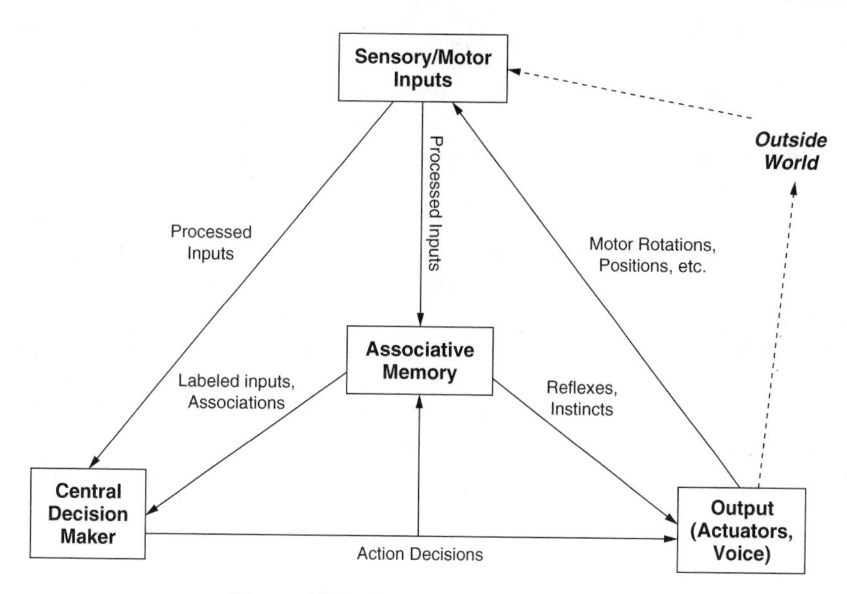

Figure 10.1 The basic cognitive cycle

give positive reinforcement by lightly flexing the robot's whiskers and negative feedback by preventing the motion in progress, thus stalling one or more of its actuators. Reinforcement of this type can be given while the robot is in autonomous exploration mode; however, its primary purpose is for language learning. Used this way, the instructor will initiate a sequence of verbal inputs, responses, and reinforcements. The robot's responses will be drawn from memory if an appropriate one exists. Failing that, a response will be generated at random. Correct setting of audio interest instincts and span of attention will facilitate this behavior.

The third form of reinforcement is direct demonstration. In this mode, the instructor will overhaul one or more of the robot's actuators, causing it to make some desired motion. The robot will record the sequence of positions and associate them with other sensory stimuli present during the operation, including speech. The intent is for the robot to learn verbs. For example, the instructor could turn the steering motors to the left while saying "turn left". Another example would be to turn the robot toward a red object while saying the word "red".

It is of utmost importance to distinguish this process from simple Skinnerian behaviorism. Although the robot is trained based on stimuli and responses to them, the mapping is not a simple one. The memory is symbolic, representational, and capable of learning complex relationships such as those needed to acquire and use language.

Once the mechanics of reinforcement are established, the main work of the project, the learning experiments, can begin. It should be obvious that significant ambiguity is present in the training as described above. Of course, we do not expect the robot to learn from single examples as humans often do. At least initially, instruction of the robot will require careful selection of stimuli and will result in slow progress. We expect that order of presentation will be important and we plan to experiment to see which sequences result in the most versatile, stable behavior.

10.5.4 Semantics as Sensorimotor Memory

Semantics is exactly the memorization of the correlation of sensorimotor stimuli of different modalities. A sufficiently large collection of such memories constitutes a mental model of the world. In particular, language is acquired by memorizing the associations between the acoustic stimulus we call speech and other sensorimotor stimuli. When language is acquired it enables the symbolic manipulation of most, but not all, of the mental model. The semantics of language is thus a symbolic representation of reality. However, the symbols and relations among them are not predetermined by a human creator but, rather, acquired and memorized in the course of interaction with the surroundings. The reinforcement regime will cause the contents of memory to converge to a configuration in which the acquired symbols reflect the observed regularities in the environment.

It is important to note that sensorimotor function forms the basis for much of semantics. In addition to concepts such as color which can only be understood visually, temperature and pain which are essentially tactile, force which is haptic, and time which can only be understood spatially, the meanings of many common words are derived from morphemes for direction or location concatenated with morphemes denoting specific physical actions. Such words can only be understood by direct appeal to a well-developed spatial sense and motor skills. As the associative memory grows, words formed in this manner can be associated with mental activities yielding meanings for abstract operations, objects, and qualities. For example, the sensations of force and balance, which are first defined in terms of their sensorimotor correlates, can later be used in a nonphysical sense to mean "persuade" or "compare", respectively. Obviously, the abstract words acquire their meanings only by analogy with their primary sensorimotor definitions.

Regardless of how the memory is implemented and trained, an important aspect of these experiments is to take frequent snapshots of it to analyze the development of memory associated with specific functions. Such data could serve as valuable diagnostics for the training procedure. I expect that as training progresses and the robot's behavior becomes more interesting, I might be able to identify the emergence of concept-like constructs.

10.5.5 The Primacy of Semantics in Linguistic Structure

Modern linguistics is dominated by the generative paradigm of Chomsky [45] that views grammar (i.e. acoustic phonetics, phonology, phonotactics, morphology, prosody, and syntax) as the core of language. Grammar is considered to be a complex system of deterministic rules unlike the probabilistic models studied in this book. In a now classic example (see Section 8.2), Chomsky considers the sentence "Colorless green ideas sleep furiously" which is grammatically well-formed but semantically anomalous. This sentence and the ungrammatical word sequence "Ideas colorless sleep furiously green" are said to have zero probability of occurrence and thus cannot be distinguished on that basis. What must, therefore, be important is the grammatical structure, or lack thereof, of the two sequences. It is this phenomenon to which traditional linguistics attends.

My working hypothesis is inherently non-Chomskian in its characterization of language. As defined in Section 10.5.3, successful behavior is the goal of reinforcement learning. Successful behavior requires intelligence which is just the procedure for extracting meaning from the environment and communicating meaningful messages. Thus language is a critical aspect of intelligence and semantics is the primary component of language. All other aspects of linguistic structure are simply the mechanisms for encoding meaning and are present in service to the primary function of language, to convey meaning and to make it robust in the presence of ambiguity.

10.5.6 Thought as Linguistic Manipulation of Mental Representations of Reality

It is important to emphasize that the primary mechanism of thought is mnemonic, not logical. This implies that the full computational power of the universal Turing machine is not required. The universal machine is used to implement the "fetch", "store", and "compare" operations of the associative memory. It is true that humans learn to reason, and this ability can contribute to successful behavior. However, logical reasoning is a very thin appliqué learned later in life, used only infrequently and then with great difficulty. It is not a native operation of the mind and, even when learned, it is based on memory. Humans simulate the formal logic of computers by analogic reasoning in which the unknown behavior of an object or organism is deduced from the memorized behavior of a similar object or system. Such reasoning is error-prone. But, even in competent adults, most cognitive function derives from associative memory.

Memory is built up from instincts by the reinforcement of successful behavior in the real world at large. As a cognitive model of reality is formed using appropriate computational mechanisms, a structure-preserving linguistic image of it is formed. When the language is fully acquired, most mental processes are mediated linguistically and we appear to think in our native language which we hear as our mind's voice.

10.5.7 Illy the Autonomous Robot

The experimental vehicle with which the ideas outlined above are explored is a quasi-anthropomorphic, autonomous robot, affectionately named Illy by her creators. However, before I undertook to build Illy, I did a pilot study with a simple device. In order to keep the engineering problems to a minimum, I used a child's toy called "Petster", a battery-operated platform in the form of a cat. Locomotion is provided by two wheels, each turned independently by a motor. The cat has photosensors, not cameras, for eyes, microphones in its ears, a piezoelectric sonar device on its collar, a loudspeaker in its head, and an RS232 communication port. Most important, however, are the microswitch in its tail, a touch sensor on the back of its neck and accelerometers on the motors. By connecting the port to a computer, one can send instructions to the on-board microcontroller to make the motors work and generate five different sounds through the speaker. The cat was connected to a computer equipped with a speech recognizer with a 2000-word vocabulary. That is, it had acoustic phonetic models for 2000 words but did not have a syntactic or semantic model. I built an associative memory using an information-theoretic multilayer perceptron described by Gorin [124] and used it to relate speech and other stimuli to actions. I trained the cat using the following reinforcement regime. I spoke to the machine in isolated words or short phrases. If I used words the recognizer knew, the computer would send a corresponding code to the cat which would then perform some action. If the action were appropriate I signaled by patting the cat on the neck and the perceptron weights would be changed accordingly. If the action were wrong, I signaled either by pulling the cat's tail or preventing it from moving, thus stalling out the motors. The activity would then stop but no change would be made in the memory. After about five minutes of patient training, the cat could learn a number of words and phrases which would cause it to move forward, backward, left, right, stop, and make a specific one of five available sounds (e.g. purr, meeow, hiss). I used any words or phrases I chose and various ways to express the same command. Despite this variation, the learning behavior was repeatable. Details of the memory architecture and training algorithms are given in Henis and Levinson [124].

Obviously, this was not a very sophisticated experiment. It did, however, show that lexical semantics and a simplistic word-order syntax could be acquired from virtually unconstrained speech in a real-time, on-line reinforcement training scheme when phonetics, phonology, and phonotactics are prespecified. The experiment also provided a better appreciation of the theoretical issues and practical difficulties that would be encountered in advancing this kind of investigation. Based on Petster, a more precise expression of the fundamental ideas was formulated [186, 187]. The Petster device, however, was inadequate because it did not have a visual sense nor arms and hands. Therefore, a new robot and control software was built in my laboratory.

Illy, The new machine, Illy (see Fig. 10.2), is based on the Trilobot platform manufactured by the Arrick Robotics company. My students and I have added audio [194, 195] and video [341, 342, 343] capabilities to the platform and have connected it to a network of small computers and workstations with large memories [186]. The robot communicates with the network via wireless ethernet [186]. I mounted two electret microphones and two small color-video cameras on the robot's movable head to provide for binaural hearing and binocular vision. The robot also has a single arm and hand with two degrees of freedom allowing for shoulder and thumb movement, enabling both lifting and gripping. A simple microprocessor and some auxiliary circuits govern all sensory and control functions [186]. A Pentium PC with video, audio and ethernet cards is mounted on-board. To meet the significantly larger power budget for the added hardware, the original power supply was replaced by a 12 volt motorcycle battery.

Illy is equipped with 14 sensors, including an array of touch-sensitive whiskers, compass, tilt sensor, thermometer, odometer, battery charge sensor, tachometer and steering

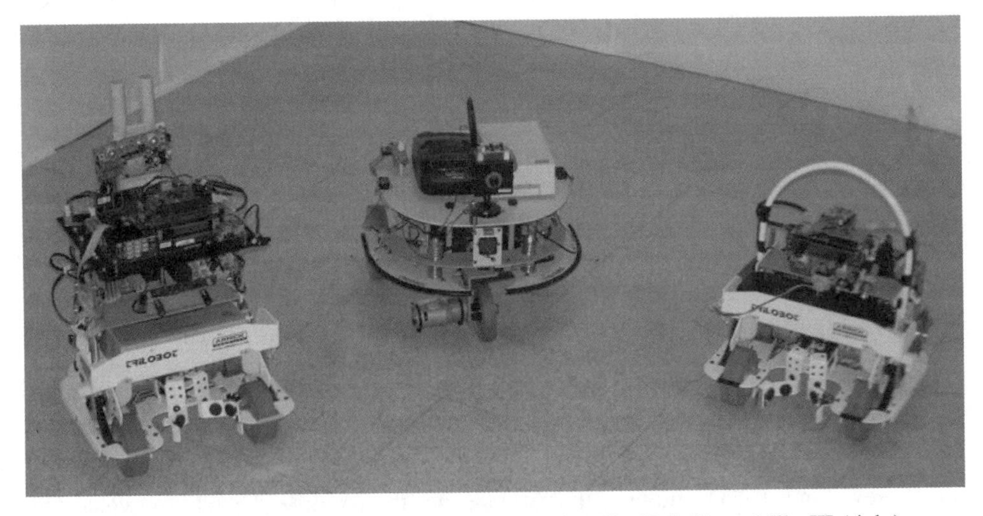

Figure 10.2 Three versions of Illy: Illy-I (center), Illy-II (left), and Illy-III (right)

angle indicator. There are two motors, one for each wheel, together providing both steering and locomotion. The status of all of these instruments is transmitted to the network via the radio link. There are provisions for several other user-provided control signals, including a standard RC servomotor and ample channel capacity to communicate them along with all the other data.

The on-board control system accepts commands to operate the motors and read the sensors in the form of a simple alphanumeric machine language. Instructions are transmitted to the controller via the radio link. In a similar fashion, codes indicating the results of the instructions and the status of the controller are returned to the network. Instructions may be combined to form programs.

Software

The control system for Illy is a distributed programming environment in which processes reside transparently on any of the networked computers or the on-board computer. The system allows for real-time, on-line operation at a rate of three complete executions of the cognitive cycle of Fig. 10.1 per second. The system is robust enough to allow for long periods of reliable operation in the autonomous exploration mode. This is essential for learning to take place.

The main program in this framework is called IServer. For a particular data stream, IServer sets up a ring buffer in shared memory. An example of a source process which reads audio data from the sound card and writes it to the ring buffer and a sink process for sound source localization, which accesses the audio data and uses it to determine the direction a sound is coming from is shown in the top left of Fig. 10.3.

Because of the demanding requirements of the input processing and the limited computing power available on Illy, much of the processing must take place on the networked computers. To support this, the IServer program includes a special sink process with the sole purpose of taking the data in the ring buffer and sending it across the network. On another machine, a corresponding source process receives this data and writes it to the ring buffer on its machine. A sink process on this other machine accesses this data in exactly the same manner as if it were on the original machine.

The lower right part of Figure 10.3 demonstrates this process, again using audio processing as an example. In this case, the sound source location program is running on Illy, and accesses audio from the ring buffer as before. A speech recognition program is running on another machine and needs access to the same audio data. For this to happen, an audio server running on Illy takes data from the ring buffer and sends it to the audio source process which writes it to its ring buffer just like any other source of audio data. The speech recognition program reads the data in the same manner as before. The ring buffer on a networked machine may also have an audio server which sends the audio data to other machines.

The ring buffer is divided into segments, the total number and size of which depends on the data type. Each segment is protected by a locking semaphore, so that a source process will not write to any block that is being read from, and a sink process will not read from a block that is being written to. Each segment of data includes a generic header specifying the byte count, a time stamp, and a sequence number.

This system, along with some general distributed shared memory and a separate server which manages connections among machines, forms the basis for our software framework,

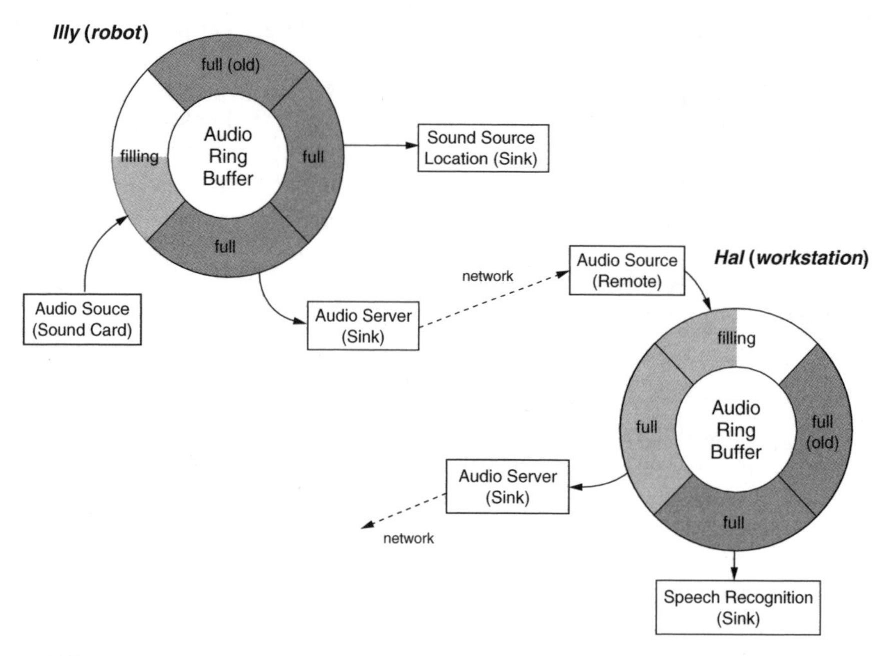

Figure 10.3 The IServer distributed computing architecture (on-board part, top left and remote part, bottom right)

upon which the aforementioned cognitive model is implemented including a central controller and a common centralized associative memory. The system has allowed us to take many disparate components and make them work together in a seamless fashion.

Associative Memory Architecture

The associative memory, the function of which is described in Section 10.5.2, is designed as follows. There is a separate group of HMMs for each modality – a set of HMMs for auditory inputs, a set of HMMs for visual inputs, etc. The states of these HMMs are used as the inputs to the next layer of HMMs. The structure is illustrated in Fig. 10.4. The benefits of this arrangement are that the initial layer of input processing can be reused by multiple models on the next level, and that each level of states has a particular meaning. In the example in Fig. 10.5, the states of the auditory HMMs are used by both the auditory-tactile HMM and the audio-visual HMM. The state of the audio-visual HMM might represent the simultaneous stimuli of the word "apple" and the image of an apple. Temporal sequences of the states of the HMMs are used as inputs to higher-level HMMs for more complex recognition. For example, we use sequences of phonemes or allophones as inputs to the multi-modality models at higher levels.

Performance

The autonomous exploration mode depends for its operation on the following programmed instinctual behaviors. First, Illy is irritable, that is, she will always make some response to

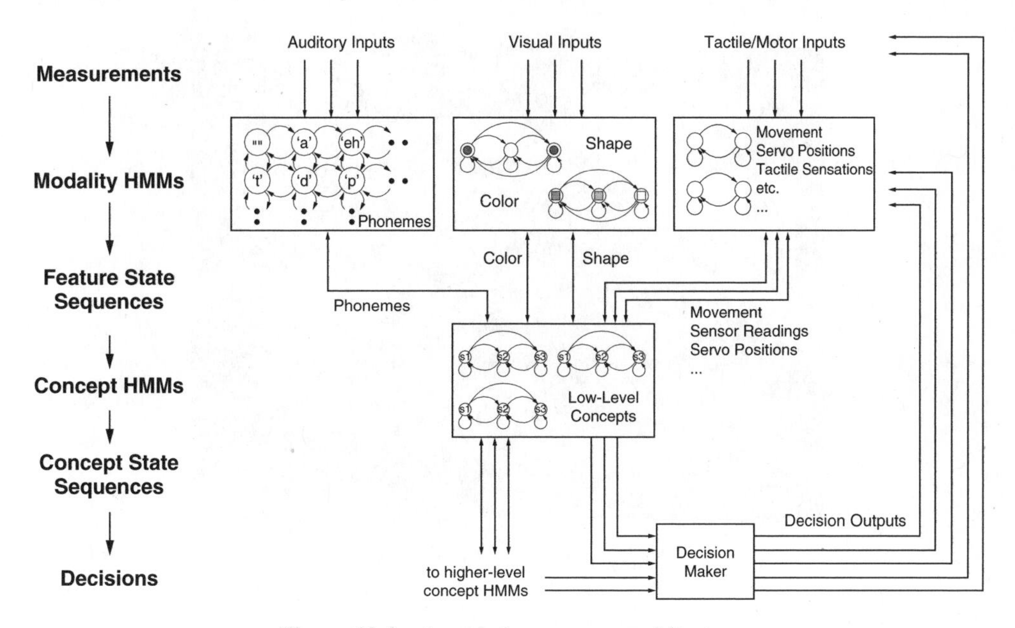

Figure 10.4 Associative memory architecture

every stimulus. Illy gets hungry when her battery is nearly drained. When that condition is detected she will sound an alarm and seek the nearest human to help her. Illy has some self-defense behaviors. She avoids collisions with large objects but will delay for a short time before initiating evasive action. She has a preference for brightly colored objects, medium intensity wideband noises, and localized rapid motions. She avoids obstacles in her path of motion, dark shadows, and high temperatures. Illy has a sleep instinct in which she is physically inactive but executes a clustering program to compress and combine data stored in memory. This idea is based on an interpretation of Robins and McCallum [275]. Finally, Illy has an instinct to imitate both speech and gestures, including arm, hand, and head motions.

Some of the more complex behaviors we have been able to demonstrate include sound source localization [194], object recognition and manipulation [342], navigation in response to voice command [202], visual object detection, identification, and location [343], visual navigation of a maze [197], and spoken word and phrase imitation [158], all of which, except for acoustic source localization, are learned behaviors. Illy is shown running a maze in Fig. 10.5.

While these functions hardly qualify Illy as a sentient being, I regard them as a firm foundation on which to support far richer cognitive activity. I anticipate the time when Illy will do something surprising and make a good showing on the generalized Turing test outlined in Section 10.3.2.

Obstacles to the Program

I am sure that critics can offer many possible reasons for what they foresee as an inevitable failure of the experiment described above. There are several aspects of my

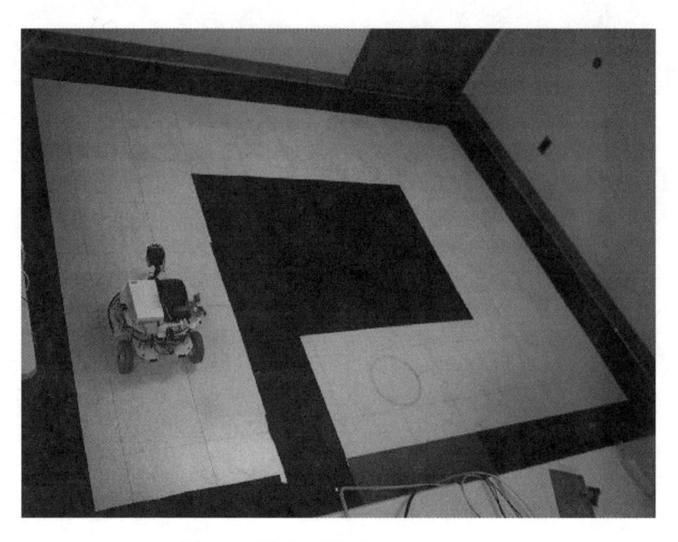

Figure 10.5 Illy-I runs a maze

working hypothesis that some consider suspect. My theory of functional equivalence is open to objection from those who believe that only biological brains, or at very least their mathematical isomorphs, can give rise to mind. My notions of associative memory, reinforcement, and semantics may also raise some eyebrows.

I regard all of these criticisms as the same in the sense that they all assert that some essential part of my hypotheses is inadequate or absolutely incorrect. I cannot provide any better support for my ideas than I have already done. I cannot refute any competing theories any more than I have already done. However, such armchair debates will never resolve the differences of opinion and I see no purpose in prolonging them.

There is a particular class of objections, however, that concern me deeply. These are problems that result from a usefully correct working hypothesis but an incomplete or inadequate experimental expression of it. I can imagine four such catastrophic errors. The first pitfall might result from an gross underestimate of the threshold of complexity required for the emergence of significant mental activity. The brain has some tens of trillions of neurons; perhaps that number of components is required to effect the kinds of behavior I hope to simulate.

The second problem might arise from a misunderstanding of the relative importance of adaptation on evolutionary time scales and learning on somatic time scales. Perhaps the billions of years of evolution were required to produce specific brain structures for specific cognitive purposes. Perhaps one cannot compensate for a lack of this optimization by clever adaptive processes over the relatively short period available for reinforcement learning.

A third issue might arise from a failure to appreciate the importance of acute perception and skilled motor control. Humans and animals are possessed of exquisite sensors and actuators. Perhaps these organs are required to bring perception to a level from which mental models of reality can be produced. Perhaps the cameras, microphones, servomotors, and other devices used in Illy are too crude to support intelligent behavior.

Finally, there is the ubiquitous problem known as the curse of dimensionality. Perhaps there are too many degrees of freedom and too little time to collect data for reinforcement learning to be effective.

I have no cogent replies to these objections. Nor is there any proof that they are fatal. Perhaps the experiments will inform the argument and indicate whether or not my working hypothesis and my enthusiasm for it are justified. Highly ambitious projects like the early attempts at flight are a venerable strategy for progress.

10.6 Final Thoughts: Predicting the Course of Discovery

The current euphoria about automatic speech recognition is based on the characterization of progress in the field as a "paradigm shift". This use of the term is inappropriate and misleading. The phrase was first used by Kuhn [163] to describe scientific revolution. As applied to automatic speech recognition, it casts incremental, technical advances as profound, conceptual scientific progress.

The difference is best understood by example. The change from a geocentric to a heliocentric model of the solar system alluded to in Section 10.2.3 is a Kuhnian "paradigm shift". Placing the sun rather than the earth at the center of the solar system may seem like a radical idea. Although it is counterintuitive to the naive observer, it does not, by itself, constitute a paradigm shift. The revolutionary concept arises from the consideration of another aspect of the solar system besides planetary position. The Ptolemaic epicycles do predict the positions of the planets as a function of time. In fact, they do so more accurately than the crude elliptical orbits postulated by Kepler. Indeed, the incremental improvement made by compounding epicycles on epicycles allows the incorrect theory to appear more accurate than the coarse but correct one. Clearly, heliocentricity alone is not the paradigm shift.

If, however, one asks what force moves the planets on their observed regular paths and how it accounts for their velocities and accelerations, the geocentric theory stands mute while the mechanical foundation of the heliocentric model turns eloquent. This, then, is the paradigm shift and its consequences are enormous. Epicycles may be acceptable for making ritual calendars but Newtonian mechanics not only opens new vistas but also, upon refinement, becomes highly accurate.

There is a very close analogy between astronomy and automatic speech recognition. At the present moment, we think of speech recognition as transcription from speech to some standard orthography. This decoding process corresponds to the computation of celestial location only. It ignores the essence of speech, its capacity to convey meaning, and is thus incomplete. The paradigm shift needed in our discipline is to make comprehension rather than transcription the organizing principle, just as force replaced location as the central construct in celestial mechanics. If one can calculate the forces acting on the planets, one can determine their orbits, from which the positions are a trivial consequence. Similarly, if one can extract meaning from an utterance, the lexical transcription will result as a by-product. Unfortunately, as we noted in Chapter 8, the process is often inverted by the attempt to use meaning to improve transcription accuracy rather than making meaning the primary aspect.

I view the experiment described in Section 10.5 as an attempt to instigate a paradigm shift in speech technology. This is an ambitious goal which runs counter to the prevailing ethos. We live in an age in which technique is prized over concept. This is not unreasonable because our daily lives depend on a technological cornucopia that has been slowly and steadily improving for several decades. It is not surprising that even the inventors are enchanted by the magic they themselves have wrought. Consequently, when technocrats predict the future of a promising new technology, they tend to be overly optimistic for the near term and overly pessimistic for the long haul. This happens because technical forecasting is always based on extrapolating what is presently known based on incremental improvement without regard for the possibilities of unproven, speculative approaches. Thus it is not surprising that the prediction for automatic speech recognition is that the existing performance deficits will be soon overcome by simply refining present techniques.

My prediction is that advances in our understanding of speech communication will come painfully slowly but eventually, perhaps many decades hence, automatic speech recognition at human performance levels will be ubiquitous. In the near term, incremental technical advances will result in a fragile technology of small commercial value in special markets, whereas major technological advances resulting from a true paradigm shift in the underlying science will enable machines to display human linguistic competence. This, in turn, will create a vast market of incalculable social and commercial value.

It is, of course, entirely possible that the technocrats are correct, that a diligent effort resulting in a long sequence of incremental improvements will yield the desired perfected automatic speech recognition technology. It is also possible that this strategy will come to grief because of the "first step fallacy" of Dreyfus [67] who warns of the impossibility of reaching the moon by climbing a tree. Such a strategy appears initially to head in the right direction but soon progress stops abruptly or tragically, far short of the goal, when the top of the tree is reached or the small upper limbs will no longer support the climber's weight. It seems obvious to me that the prudent plan is to openly acknowledge the risks of incrementalism and devote some effort to the plausible speculative approaches.

Perhaps more important, however, is recognition of the uniqueness of our particular technological goal. Unlike all other technologies that are integral parts of our daily lives because they provide us with capabilities otherwise unattainable, speech technology promises to increase the utility of a function at which we are already exquisitely proficient. Since using the present state of the art requires a serious diminution of our natural abilities and since we presently cannot leap the performance chasm between humans and machines, it seems only prudent that we should invest more in fundamental science in the expectation that it will eventually lead not only to a mature speech technology but also to many other things as yet unimagined. This strategy would, of course, alter the existing balance among science, technology, and the marketplace more in favor of precommercial experimentation while reducing the emphasis on immediate profit. There is good reason to believe, however, that ultimately this strategy will afford the greatest intellectual, social, and financial reward.

Bibliography

[1] *Proc. Int. Conf. on Development and Learning*. Michigan State University, 2000.

[2] M. Abromovitz and I. A. Stegun, editors. *Handbook of Mathematical Functions*. Dover Publications, New York, 1965.

[3] A. V. Aho, T. G. Szymanski, and M. Yannakakis. Enumerating the cartesian product of ordered sets. In *Proc. 14th Annual Conf. on Information Science and Systems*, Princeton, NJ, 1980.

[4] B. Aldefeld, S. E. Levinson, and T. G. Szymanski. A minimum distance search technique and its application to automatic directory assistance. *Bell Syst. Tech. J.*, 59:1343–1356, 1980.

[5] J. B. Allen. Cochlear micromechanics – a mechanism for transforming mechanical to neural tuning within the cochlea. *J. Acoust. Soc. Amer.*, 62:930–939, 1977.

[6] R. Alter. Utilization of contextual constraints in automatic speech recognition. *IEEE Trans. Audio Electroacoust.*, AU-16:6–11, 1968.

[7] S. Aronowitz. *Science as Power: Discourse and Ideology in Modern Society*. University of Minnesota Press, Minneapolis, 1988.

[8] B. S. Atal. Private communication, 1981.

[9] B. S. Atal, J. J. Chang, M. V. Mathews, and J. W. Tukey. Inversion of articulatory-to-acoustic transformation in the vocal tract by a computer-sorting technique. *J. Acoust. Soc. Amer.*, 63:1535–1555, 1978.

[10] B. S. Atal and M. R. Schroeder. Predictive coding of speech and subjective error criteria. *IEEE Trans. Acoust. Speech Signal Process.*, ASSP-27:247–254, 1979.

[11] L. R. Bahl, J. K. Baker, P. S. Cohen, A. G. Cole, F. Jelinek, B. L. Lewis, and R. L. Mercer. Automatic recognition of continuously spoken sentences from a finite state grammar. In *Proc. IEEE Int. Conf. on Acoustics, Speech, and Signal Processing*, pages 418–421, Washington, DC, 1979.

[12] L. R. Bahl, J. K. Baker, P. S. Cohen, N. R. Dixon, F. Jelinek, R. L. Mercer, and H. F. Silverman. Preliminary results on the performance of a system for the automatic recognition of continuous speech. In *Proc. IEEE Int. Conf. on Acoustics, Speech, and Signal Processing*, pages 425–429, April 1976.

[13] L. R. Bahl, J. K. Baker, P. S. Cohen, F. Jelinek, B. L. Lewis, and R. L. Mercer. Recognition of a continuously read natural corpus. In *Proc. IEEE Int. Conf. on Acoustics, Speech, and Signal Processing*, pages 442–446, Washington, DC, 1979.

[14] L. R. Bahl, R. Bakis, P. S. Cohen, A. G. Cole, F. Jelinek, B. L. Lewis, and R. L. Mercer. Further results on the recognition of a continuously read natural corpus. In *Proc. IEEE Int. Conf. on Acoustics, Speech, and Signal Processing*, pages 872–875, Denver, CO, 1980.

[15] L. R. Bahl, A. Cole, F. Jelinek, R. L. Mercer, A. Nadas, D. Nahamoo, and M. Picheny. Recognition of isolated word sentences from a 5000 word vocabulary office correspondence task. In *Proc. IEEE Int. Conf. on Acoustics, Speech, and Signal Processing*, pages 1065–1067, Boston, MA, 1983.

[16] L. R. Bahl and F. Jelinek. Decoding for channels with insertions deletions and substitutions with applications to speech recognition. *IEEE Trans. Inform. Theory*, IT-21:404–411, 1975.

Mathematical Models for Speech Technology. Stephen Levinson

ISBN: 0-470-84407-8

[17] L. R. Bahl, F. Jelinek, and R. L. Mercer. A maximum likelihood approach to continuous speech recognition. *IEEE Trans. Pattern Analysis and Machine Intelligence*, PAMI-5:179–190, 1983.

[18] L. R. Bahl, R. Bakis, P. S. Cohen, A. G. Cole, F. Jelinek, B. L. Lewis, and R. L. Mercer. Recognition results with several experimental acoustic processors. In *Proc. IEEE Int. Conf. on Acoustics, Speech, and Signal Processing*, pages 249–251, Washington, DC, 1979.

[19] J. K. Baker. The dragon system: an overview. *IEEE Trans. Acoust. Speech Signal Process.*, ASSP-23, 1975.

[20] J. K. Baker. Stochastic modeling for automatic speech understanding. In D. R. Reddy, editor, *Speech Recognition*, pages 521–542. Academic Press, New York, 1975.

[21] J. K. Baker. Trainable grammars for speech recognition. In J. J. Wolf and D. H. Klatt, editors, *Speech Comun. Papers of 97th Meeting of the Acoust. Soc. Amer.*, pages 547–550, 1979.

[22] G. H. Ball and D. J. Hall. Isodata: An interactive method of multivariate analysis and pattern classification. In *Proc. IFIPS Congr.*, 1965.

[23] M. Barinaga. The cerebellum: Movement coordinator or much more? *Science*, 272:482–483, 1996.

[24] S. L. Bates. A hardware realization of a PCM-ADPCM code converter. Unpublished thesis, 1976.

[25] L. E. Baum. An inequality and associated maximization technique in statistical estimation for probabilistic functions of a Markov process. *Inequalities*, III:1–8, 1972.

[26] L. E. Baum and J. A. Eagon. An inequality with applications to statistical estimation for probabilistic functions of a Markov process and to a model for ecology. *Bull. Amer. Math. Soc.*, 73:360–363, 1967.

[27] L. E. Baum and T. Petrie. Statistical inference for probabilistic functions of finite state Markov chains. *Ann. Math. Stat.*, 37:1559–1563, 1966.

[28] L. E. Baum, T. Petrie, G. Soules, and N. Weiss. A maximization technique in the statistical analysis of probabilistic functions of Markov chains. *Ann. Math. Statist.*, 41:164–171, 1970.

[29] L. E. Baum and G. R. Sell. Growth functions for transformations on manifolds. *Pacific J. Math.*, 27:211–227, 1968.

[30] G. Bekesy. *Experiments in Hearing*. McGraw-Hill, New York, 1960.

[31] R. E. Bellman. *Dynamic Programming*. Princeton University Press, Princeton, NJ, 1957.

[32] W. H. Beyer, editor. *Handbook of Tables for Probability and Statistics*. Chemical Rubber Co., Cleveland, OH, 1968.

[33] R. Billi. Vector quantization and Markov source models applied to speech recognition. In *Proc. Int. Conf. on Acoustics, Speech, and Signal Processing*, pages 574–577, Paris, France, 1982.

[34] T. L. Booth and R. A. Thompson. Applying probability measures to abstract languages. *IEEE Trans. Comput.*, C-22:442–450, 1973.

[35] H. Bourland, J. Wellekens, and H. Ney. Connected digit recognition using vector quantization. In *Proc. Int. Conf. on Acoustics, Speech, and Signal Processing*, pages 26.10.1–26.10–4, San Diego, CA, 1984.

[36] M. Braun. *Differential Equations and their Applications as an Introduction to Applied Mathematics*. Springer-Verlag, New York, 1975.

[37] J. S. Bridle and M. D. Brown. Connected word recognition using whole word templates. In *Proc. Inst. Acoust. Autumn Conf.*, pages 25–28, 1979.

[38] R. Brooks *et al.* The COG project: Building a humanoid robot. In C. L. Nehaniv, ed., *Computation for Metaphors, Analogy and Agents*. Springer-Verlag, Berlin, 1998.

[39] L. E. J. Brouer. Intuitionism and formalism. *Amer. Math. Soc. Bull.*, 20:81–96, 1913.

[40] J. S. Bruner, J. J. Goodnow, and G. A. Austin. *A Study of Thinking*. Wiley, New York, 1956.

[41] Mario Bunge. *Treatise on Basic Philosophy*. D. Reidel, Dordrecht and Boston.

[42] G. Cantor. *Transfinite Arithmetic*. Dover, New York, 1980.

[43] R. L. Cave and L. P. Neuwirth. Hidden Markov models for English. In J. D. Ferguson, editor, *Proc. Symp. on the Application of Hidden Markov Models to Text and Speech*, pages 16–56, Princeton, NJ, 1980.

[44] D. S. K. Chan and L. R. Rabiner. An algorithm for minimizing roundoff noise in cascade realizations of finite impulse response digital filters. *Bell Syst. Tech. J.*, 52:347–385, 1973.

[45] N. Chomsky. *Syntactic Structures*. Mouton, The Hague, 1957.

[46] N. Chomsky. On certain formal properties of grammars. *Inform. Contr.*, 2:137–167, 1959.

[47] N. Chomsky and M. Halle. *The Sound Patterns of English*. Harper and Row, New York, 1968.

[48] A. Church. *The Calculi of Lambda Conversion*, Ann. Math. Stud. 6. Princeton University Press, Princeton, NJ, 1951.

[49] P. J. Cohen. The independence of the continuum hypothesis. *Proc. Nat. Acad. Sci USA*, 50:1143–1148, 1963.

[50] P. S. Cohen and R. L. Mercer. The phonological component of an automatic speech recognition system. In D. R. Reddy, editor, *Speech Recognition*, pages 275–320. Academic Press, New York, 1975.

[51] C. H. Coker. A model of articulatory dynamics and control. *Proc. IEEE*, 64:452–460, 1976.

[52] C. H. Coker, N. Umeda, and C. P. Browman. Automatic synthesis from ordinary English text. *IEEE Trans. Audio Electroacoust.*, AU-21:293–298, 1973.

[53] T. M. Cover and P. Hart. The nearest neighbor decision rule. *IEEE Trans. Inform. Theory*, IT-13:21–27, 1967.

[54] R. E. Crochiere and A. V. Oppenheim. Analysis of linear digital networks. *Proc. IEEE*, 63:581–595, 1975.

[55] P. Cummiskey, N. S. Jayant, and J. L. Flanagan. Adaptive quantization in differential PCM coding of speech. *Bell Syst. Tech. J.*, 52:1105–1118, 1973.

[56] A. Damasio. *The Feeling of What Happens*. Harcourt Brace, New York, 1999.

[57] A. R. Damasio *et al.*, eds, *Unity of Knowledge: The Convergence of Natural and Human Science*. New York Academy of Sciences, New York, 2001.

[58] K. H. Davis, R. Biddulph, and S. Balashek. Automatic recognition of spoken digits. *J. Acoust. Soc. Amer.*, 24:637–642, 1952.

[59] S. B. Davis and P. Mermelstein. Comparison of parametric representation for monosyllabic word recognition in continuously spoken sentences. *IEEE Trans. Acoust., Speech, Signal Process.*, ASSP-28:357–366, 1980.

[60] R. DeMori. A descriptive technique for automatic speech recognition. *IEEE Trans. Audio Electroacoust.*, AU-21:89–100, 1973.

[61] R. DeMori, P. Laface, and Y. Mong. Parallel algorithms for syllable recognition in continuous speech. *IEEE Trans. Pattern Analysis and Machine Intelligence*, PAMI-7:56–69, 1985.

[62] A. P. Dempster, N. M. Laird, and D. B. Rubin. Maximum likelihood from incomplete data via the EM algorithm. *J. Roy. Statist. Soc. Ser. B*, 39:1–88, 1977.

[63] P. B. Denes and M. V. Mathews. Spoken digit recognition using time frequency pattern matching. *J. Acoust. Soc. Amer.*, 32:1450–1455, 1960.

[64] D. Dennett. *Consciousness Explained.* Little Brown, Boston, 1991.

[65] E. W. Dijkstra. A note on two problems in connection with graphs. *Num. Math.*, 1:269–271, 1969.

[66] A. M. Dirac. The relation between mathematics and physics. In *Proc. Roy. Soc. Edinburgh*, 59:122–129, 193.

[67] H. L. Dreyfus. *What Computers Can't Do: A Critique of Artificial Reason*. Harper and Row, New York, 1972.

[68] R. O. Duda and P. E. Hart. *Pattern Classification and Scene Analysis*. Wiley, New York, 1973.

[69] H. Dudley. The vocoder. *J. Acoust. Soc. Amer.*, 1930.

[70] H. Dudley and S. Balashek. Automatic recognition of phonetic patterns in speech. *J. Acoust. Soc. Amer.*, 30:721–739, 1958.

[71] H. K. Dunn, J. L. Flanagan, and P. J. Gestrin. Complex zeros of a triangular approximation to the glottal wave. *J. Acoust. Soc. Amer.*, 34, 1962.

[72] K. M. Eberhard, M. J. Spivey-Knowlton, J. C. Sedivy, and M. Tanenhaus. Eye movements as a window into real-time spoken language comprehension in natural contexts. *J. Psycholinguistic Res.*, 24:409–436, 1995.

[73] Gerald M. Edelman and Giulio Tononi. *A Universe of Consciousness*. Basic Books, New York, 2000.

[74] A. Einstein. *The Meaning of Relativity*. Princeton University Press, Princeton, NJ, 1921.

[75] A. Einstein. Unpublished letter to Max Borr. December 4, 1926.

[76] A. Einstein. Unpublished letter to Herbert Goldstein, New York, April 25, 1929.

[77] A. Einstein. *On the Method of Theoretical Physics*. Oxford University Press, New York, 1933.

[78] L. D. Erman, D. R. Fennell, R. B. Neely, and D. R. Reddy. The hearsay-1 speech understanding system: An example of the recognition process. *IEEE Trans. Comput.*, C-25:422–431, 1976.

[79] F. Fallside, R. V. Patel, and H. Seraji. Interactive graphics technique for the design of single-input feedback systems. *Proc. IEE*, 119(2): 247–254, 1972.

[80] K. Fan. *Les fonctions definies-postives et les fonctions completement monotones*. Gauthier-Villars, Paris, 1950.

[81] R. M. Fano. *Transmission of Information: A Statistical Theory of Communications*. Wiley, New York, 1961.

[82] G. Fant. *Acoustic Theory of Speech Production*. Mouton, The Hague, 1970.

[83] G. Fant. *Speech Sounds and Features*. MIT Press, Cambridge, MA, 1973.

[84] J. D. Ferguson. Variable duration models for speech. In J. D. Ferguson, editor, *Proceedings of the Symposium on the Application of Hidden Markov Models to Text and Speech*, pages 143–179, Princeton, NJ, 1980.

[85] H. L. Fitch. Reclaiming temporal information after dynamic time warping. *J. Acoust. Soc. Amer.*, 74, suppl. 1:816, 1983.

[86] J. L. Flanagan. *Speech Analysis Synthesis and Perception*. Springer-Verlag, New York, 2nd edition, 1972.

[87] J. L. Flanagan. Computers that talk and listen: Man–machine communication by voice. *Proc. IEEE*, 64:405–415, 1976.

[88] J. L. Flanagan, K. Ishizaka, and K. L. Shipley. Synthesis of speech from a dynamic model of the vocal cords and vocal tract. *Bell Sys. Tech. J.*, 544:485–506, 1975.

[89] H. Fletcher. *Speech and Hearing in Communication*. Van Nostrand, Princeton, NJ, 1953.

[90] R. Fletcher and M. J. D. Powell. A rapidly convergent descent method for minimization. *Computer*, 6:163–168, 1963.

[91] J. A. Fodor. *Language of Thought*, pp. 103ff. Crowell, New York, 1975.

[92] J. Fodor. *The Mind Doesn't Work That Way*. MIT Press, Cambridge, MA, 2000.

[93] K. S. Fu. *Sequential Methods in Pattern Recognition and Machine Learning*. Academic Press, New York, 1968.

[94] K. S. Fu. *Syntactic Methods in Pattern Recognition*. Academic Press, New York, 1974.

[95] K. S. Fu. *Syntactic Pattern Recognition*. Academic Press, NY, 1974.

[96] K. S. Fu. *Syntactic Pattern Recognition and Applications*. Prentice Hall, Englewood Cliffs, NJ, 1982.

[97] K. S. Fu, editor. *Digital Pattern Recognition*. Springer-Verlag, Berlin, 1976.

[98] K. S. Fu and T. L. Booth. Grammatical inference – introduction and survey. *IEEE Trans. Syst. Man Cybern.*, SMC-5:95–111 and 409–422, 1975.

[99] O. Fujimura, S. Kiritani, and H. Ishida. Computer controlled radiography for observation of articulatory and other human organs. *Comput. Biol. Med.*, 3:371–384, 1973.

[100] T. Fujisaki. A stochastic approach to sentence parsing. In *Proc. 10th Int. Conf. on Computer Linguistics*, pages 16–19, Stanford, CA, 1984.

[101] L. W. Fung and K. S. Fu. Syntactic decoding for computer communication and pattern recognition. *IEEE Trans. Comput.*, C-24:662–667, 1975.

[102] M. R. Garey and D. S. Johnson. *Computers and Intractability: A Guide to the Theory of NP-Completeness*. W. H. Freeman, San Francisco, 1979.

[103] R. A. Gillman. A fast frequency domain pitch algorithm. *J. Acoust. Soc. Amer.*, 58:562, 1975.

[104] K. Gödel. Über formal unentscheidbare Sätze der Principia Mathematica und verwandter Systeme. *Monatsh. Math. Phys.*, 38:173–198, 1931.

[105] E. M. Gold. Language identification in the Limit. *Inf. Control*, 10:447 ff., 1967.

[106] A. Goldberg. *Constructions: A Construction Grammar Approach to Argument Structure*. University of Chicago Press, Chicago, 1995.

[107] H. Goldstine. *The Computer from Pascal to von Neumann*. Princeton University Press, Princeton, NJ, 1972.

[108] R. G. Goodman. Analysis of languages for man–machine voice communication. Technical report, Department of Computer Science, Carnegie Mellon University, May 1976.

[109] I. S. Gradshteyn and I. M. Ryzhik. *Tables of Integrals Series and Products*. Academic Press, New York, 1980.

[110] A. H. Gray and J. D. Markel. Distance measures for signal processing. *IEEE Trans. Acoust., Speech, Signal Processing*, ASSP-24:380–391, 1975.

[111] A. H. Gray and J. D. Markel. Quantization and bit allocation in speech processing. *IEEE Trans. Acoust. Speech Signal Process.*, ASSP-24:459–473, December 1976.

[112] S. Greibach. Formal languages: Origins and directions. In *Proc. 20th Annual Symp. on Foundations of Computer Science*, pages 66–90, 1979.

[113] S. Grossberg, editor. *The Adaptive Brain*. North-Holland, Amsterdam, 1987.

[114] J. Hadamard. *Scientific Creativity*. Dover, New York, 1960.

[115] E. H. Hafer. Speech analysis by articulatory synthesis. Master's thesis, Northwestern University, 1974.

[116] J. L. Hall. Two tone suppression in a non-linear model of the basilar membrane. *J. Acoust. Soc. Amer.*, 61:802–810, 1977.

[117] M. Halle and K. M. Stevens. Speech recognition: A model and a program for research. *IRE Trans. Inform. Theory*, IT-8:155–159, 1962.

[118] R. W. Hamming. We would know what they thought when they did it. In N. Metropolis, J. Howlett, and G. Rota, editors, *A History of Computing in the Twentieth Century*. Academic Press, New York, 1980.

[119] Z. Harris. *A Grammar of English on Mathematical Principles*. Wiley, New York, 1982.

[120] M. A. Harrison. *Introduction to Formal Language Theory*. Addison-Wesley, Reading, MA, 1979.

[121] J. A. Hartigan. *Clustering Algorithms*. Wiley, New York, 1975.

[122] J. P. Haton and J. M. Pierrel. Syntactic-semantic interpretation of sentences in the Myrtille-II speech understanding system. In *Proc. Int. Conf. on Acoustics, Speech, and Signal Processing*, pages 892–895, Denver, CO, 1980.

[123] D. O. Hebb. *The Organization of Behavior*. Wiley, NY, 1949.

[124] E. A. Henis and S. E. Levinson. Language as part of sensorimotor behavior. In *Proc. AAAI Symposium*, Cambridge, MA, November 1995.

[125] M. R. Hestenes. *Optimization Theory*. Wiley, New York, 1975.

[126] J. Hironymous. Automatic language identification from acoustic phonetic models. Bell Laboratories technical memorandum 1995.

[127] Y. C. Ho and A. K. Agrawal. On pattern classification algorithms: introduction and survey. *Proc. IEEE*, 56, 1968.

[128] A. Hodges. *Alan Turing, the Enigma*. Simon and Schuster, New York, 1983.

[129] A. L. Hodgkin and A. F. Huxley. A quantitative description of membrane current and its application to conduction and excitation in nerves. *J. Physiol.*, 117:500–544, 1952.

[130] D. R. Hofstadter. *Gödel, Escher, Bach: an Eternal Golden Braid*. Basic Books, New York, 1979.

[131] P. Hohenberg and W. Kohn. *Phys. Rev. B*, 136:864, 1964.

[132] G. Holton. *Einstein, History and Other Passions*. AIP Press, Woodbury, NY, 1995.

[133] J. E. Hopcroft and J. D. Ullman. *Formal Languages and Their Relation to Automata*. Addison-Wesley, Reading, MA, 1969.

[134] J. J. Hopfield. Form follows function. *Physics Today*, pages 10–11, November 2002.

[135] M. J. Hopper, editor. *Harwell Subroutine Library: A Catalog of Subroutines*, volume 55. AERE Harwell, Oxfordshire, England, 1979.

[136] J. Huang. Computational fluid dynamics for articulatory speech synthesis. Unpublished Ph.D. dissertation, University of Illinois at Urbana-Champaign, 2001.

[137] E. B. Hunt. *Artificial Intelligence*. Academic Press, New York, 1975.

[138] M. J. Hunt, M. Lennig, and P. Mermelstein. Experiments in syllable based recognition of continuous speech. In *Proc. Int. Conf. on Acoustics, Speech, and Signal Processing*, pages 880–883, Denver, CO, 1980.

[139] K. Ishizaka, J. C. French, and J. L. Flanagan. Direct determination of vocal tract wall impedance. *IEEE Trans. Acoust. Speech Signal Process.*, ASSP-23:370–373, 1975.

[140] F. Itakura. Minimum prediction residual principle applied to speech recognition. *IEEE Trans. Acoust. Speech Signal Process.*, ASSP-23:67–72, 1975.

[141] R. Jackendoff. Parts and boundaries. *Cognition*, 41:9–45, 1991.

[142] R. Jackendoff. The architecture of the linguistic–spatial interface. In P. Bloom, M. A. Peterson, L. Nadel, and M. F. Garrett, editors, *Language and Space: Language, Speech and Communication*, pages 1–30. MIT Press, Cambridge, MA, 1996.

[143] R. Jakobson. Observations on the phonological, classification of consonants. In *Proc. 3rd Int'l Congress of Phonetic Sciences*, pages 34–41, 1939.

[144] W. James. *Talks to Teachers on Psychology and to Students on Some of Life's Ideals*. Holt, New York, 1899.

[145] J. Jaynes. *The Origins of Consciousness in the Breakdown of the Bicameral Mind*. Princeton University Press, Princeton, NJ, 1975.

[146] F. Jelinek. A fast sequential decoding algorithm using a stack. *IBM J. Res. Devel.*, 13:675–685, 1969.

[147] F. Jelinek. Continuous speech recognition by statistical methods. *Proc. IEEE*, 64:532–556, 1976.

[148] F. Jelinek, L. R. Bahl, and R. L. Mercer. Design of a linguistic statistical decoder for the recognition of continuous speech. *IEEE Trans. Inform. Theory*, IT-21:250–256, 1975.

[149] F. Jelinek, L. R. Bahl, and R. L. Mercer. Continuous speech recognition: Statistical methods. In P. R. Krishnaiah and L. N. Kanal, editors, *Classification, Pattern Recognition, and Reduction in Dimensionality*, Handbook of Statistics 2. North-Holland, Amsterdam, 1982.

[150] F. Jelinek and R. L. Mercer. Interpolated estimation of Markov source parameters from sparse data. In E. Gelsema and L. Kanal, eds, *Pattern Recognition in Practice*, pages 381–397. North-Holland, Amsterdam, 1980.

[151] M. Johnson. *The Body in the Mind: The Bodily Basis of Meaning, Imagination and Reason*. University of Chicago Press, Chicago, 1987.

[152] B. H. Juang, S. E. Levinson, and M. M. Sondhi. Maximum likelihood estimation for multivariate mixture observations of Markov chains. *IEEE Trans. Inform. Theory*, IT-32:307–310, March 1986.

[153] W. V. Kempelen. *The Mechanism of Speech Follows from the Description of a Speaking Machine*. J. V. Degen, 1791.

[154] J. Kittler, K. S. Fu, and L. F. Pau, editors. *Pattern Recognition Theory and Application*. D. Reidel, Dordrecht, 1982.

[155] D. H. Klatt. Review of the ARPA speech understanding project. *J. Acoust. Soc. Amer.*, 62:1345–1366, 1977.

[156] D. H. Klatt and K. N. Stevens. On the automatic recognition of continuous speech: Implications from a spectrogram reading experiment. *IEEE Trans. Audio Electroacoust.*, AU-21:210–217, 1973.

[157] S. C. Kleene. *Introduction to Mathematics*. Van Nostrand, Princeton, NJ, 1952.

[158] M. Kleffner. A method of automatic speech imitation via warped linear prediction. Master's thesis, University of Illinois at Urbana-Champaign.

[159] D. E. Knuth. *The Art of Computer Programming. Volume 1: Fundamental Algorithms*. Addiston-Wesley, Reading, MA, 1968.

[160] D. E. Knuth. *The Art of Computer Programming. Volume 3: Sorting and Searching*. Addison-Wesley, Reading, MA, 1973.

[161] T. Kohonen. *Self Organization and Associative Memory*. Springer-Verlag, Berlin, 1988.

[162] T. Kohonen, H. Rittinen, M. Jalanko, E. Reuhkala, and S. Haltsonen. A thousand word recognition system based on learning subspace method and redundant hash addressing. In *Proc. 5th Int. Conf. on Pattern Recognition*, pages 158–165, Miami Beach, FL, 1980.

[163] T. S. Kuhn. *The Structure of Scientific Revolutions*, 2nd ed. University of Chicago Press, Chicago, 1970.

[164] S. Kullback and R. A. Leibler. On information and sufficiency. *Ann. Math. Statist.*, 22:79–86, 1951.

[165] G. Lakoff. *Women, Fire and Dangerous Things*. University of Chicago Press, Chicago, 1987.

[166] G. Lakoff. *Moral Politics*. University of Chicago Press, Chicago, 1996.

[167] G. Lakoff and M. Johnson. *Metaphors We Live By*. University of Chicago Press, Chicago, 1980.

[168] G. Lakoff and M. Johnson. *Philosophy in the Flesh: The Embodied Mind and its Challenges to Western Thought*. Basic Books, New York, 1999.

[169] G. Lakoff and R. Nunez. *Where Mathematics Comes From: How the Embodied Mind Brings Mathematics into Being*. Basic Books, New York, 2000.

[170] B. Landau and R. Jackendoff. "What" and "where" in spatial language and spatial cognition. *Behavioral and Brain Sciences*, 16:217–265, 1993.

[171] R. W. Langacker. *Foundations of Cognitive Grammar*. Stanford University Press, Stanford, CA, 1986.

[172] P. Laplace. *A Philosophical Essay on Probabilities*. Dover, New York, 1951.

[173] B. Latour. *The Pasteurization of France*. Harvard University Press, Cambridge, MA, 1988.

[174] E. Lawler. *Combinatorial Optimization.* Holt, Rinehart and Winston, New York, 1976.

[175] W. A. Lea, M. F. Medress, and T. E. Skinner. A prosodically guided speech understanding strategy. *IEEE Trans. Acoust. Speech Signal Process.*, ASSP-23:30–38, 1975.

[176] W. A. Lea and J. E. Shoup. Gaps in the technology of speech understanding. In *Proc. IEEE Int. Conf. on Acoustics, Speech, and Signal Processing*, pages 405–408, Tulsa, OK, 1978.

[177] D. B. Lenat. CYC: A large-scale investment in knowledge infrastructure. *Comm. ACM*, 38(11): 33–38, 1995.

[178] V. R. Lesser, R. D. Fennell, L. D. Erman, and D. R. Reddy. Organization of the Hearsay-II speech understanding system. *IEEE Trans. Acoust., Speech Signal Process.*, ASSP-23:11–24, 1975.

[179] S. E. Levinson. An artificial intelligence approach to automatic speech recognition. In *Proc. IEEE Conf. on Systems, Man and Cybernetics*, pages 344–345, Boston, MA, November 1973.

[180] S. E. Levinson. The vocal speech understanding system. In *Proc. 4th Int. Conf. on Artificial Intelligence*, pages 499–505, Tbilisi, USSR, 1975.

[181] S. E. Levinson. Cybernetics and automatic speech understanding. In *Proc. IEEE ICISS*, pages 501–506, Patras, Greece, August 1976.

[182] S. E. Levinson. The effects of syntactic analysis on word recognition accuracy. *Bell Syst. Tech. J.*, 57:1627–1644, 1977.

[183] S. E. Levinson. Improving word recognition accuracy by means of syntax. In *Proc. IEEE Int. Conf. on Acoustics, Speech, and Signal Processing*, Hartford, CT, May 1977.

[184] S. E. Levinson. Implications of an early experiment in speech understanding. In *Proc. AAAI Symposium*, pages 36–37, Stanford, CA, March 1987.

[185] S. E. Levinson. Speech recognition technology: A critique. In D. B. Roe and J. G. Wilpon, editors, *Voice Communication between Humans and Machines*, pages 159–164. National Academy Press, Washington, DC, 1994.

[186] S. E. Levinson. The role of sensorimotor function, associative memory and reinforcement learning in automatic speech recognition. In *Proc. Conf. on Machines that Learn*, Snowbird, UT, April 1996.

[187] S. E. Levinson. Mind and language. In *Proc. Int'l Conf. On Development and Learning*, Michigan State University, April 2000.

[188] S. E. Levinson, A. Ljolje, and L. G. Miller. Large vocabulary speech recognition using a hidden Markov model for acoustic/phonetic classification. *Speech Technology*, pages 26–32, 1989.

[189] S. E. Levinson, L. R. Rabiner, A. E. Rosenberg, and J. G. Wilpon. Interactive clustering techniques for selecting speaker independent reference templates for isolated word recognition. *IEEE Trans. Acoust. Speech Signal Process.*, ASSP-27:134–140, 1979.

[190] S. E. Levinson, L. R. Rabiner, and M. M. Sondhi. An introduction to the application of the theory of probabilistic functions of a Markov process to automatic speech recognition. *Bell Syst. Tech. J.*, 62:1035–1074, 1983.

[191] S. E. Levinson and A. E. Rosenberg. Some experiments with a syntax direct speech recognition system. In *Proc. IEEE Int. Conf. on Acoustics, Speech and Signal Processing*, pages 700–703, 1978.

[192] S. E. Levinson and A. E. Rosenberg. A new system for continuous speech recognition – preliminary results. In *Proc. IEEE Int. Conf. on Acoustics, Speech and Signal Processing*, pages 239–243, 1979.

[193] S. E. Levinson and K. L. Shipley. A conversational mode airline information and reservation system using speech input and output. *Bell Syst. Tech. J.*, 59:119–137, 1980.

[194] D. Li. Computational models for binaural sound source localization and sound understanding. PhD thesis, University of Illinois at Urbana-Champaign, 2003.

[195] D. Li and S. E. Levinson. A Bayes rule-based hierarchical system for binaural sound source localization. In *Proc. IEEE Int. Conf. on Acoustics Speech and Signal Processing*, Hong Kong, April 2003.

[196] K. P. Li and T. J. Edwards. Statistical models for automatic language identification. *IEEE Int. Conf. on Acoustics, Speech and Signal Processing*, Denver, CO, pp. 884–887, 1980.

[197] R.-S. Lin. Learning vision-based robot navigation. Master's thesis, University of Illinois at Urbana-Champaign, 2002.

[198] Y. Linde, A. Buzo, and R. M. Gray. An algorithm for vector quantizer design. *IEEE Trans. Commun.*, COM-28:84–95, 1980.

[199] L. A. Liporace. Linear estimation of non-stationary signals. *J. Acoust. Soc. Am.*, 58:1288–1295, December 1975.

[200] L. R. Liporace. Maximum likelihood estimation for multivariate observations of Markov sources. *IEEE Trans. Inform. Theory*, IT-28:729–734, September 1982.

[201] R. J. Lipton and L. Snyder. On the optimal parsing of speech. Research Report 37, Dept. of Computer Science, Yale University, New Haven, CT, 1974.

[202] Q. Liu. Interactive and incremental learning via a multisensory mobile robot. PhD thesis, University of Illinois at Urbana-Champign, 2001.

[203] S. P. Lloyd. Least squares quantization in PCM. *IEEE Trans. Inform. Theory*, IT-28:129–136, 1982.

[204] D. O. Loftsgaarden and C. P. Quesenberry. A nonparametric estimate of a multivariate density function. *Ann. Math. Statist*, 36:1049–1051, 1965.

[205] G. G. Lorentz. The 13th problem of Hilbert. In F. E. Browder, editor, *Mathematical Developments Arising from Hilbert Problems*. American Mathematical Society, Providence, RI, 1976.

[206] B. T. Lowerre. The HARPY speech understanding system. PhD thesis, Carnegie Mellon University, Pittsburgh, PA, 1976.

[207] J. MacQueen. Some methods for classification and analysis of multivariate observations. In *L. LeCam and J. Neyman, eds, Proc. 5th Berkeley Symposium on Mathematical Statistics and Probability*, volume 1, pages 281–298. University of California Press, Berkeley, 1967.

[208] J. Makhoul. Linear prediction: A tutorial review. *Proc. IEEE*, 63:561–580, 1975.

[209] M. P. Marcus. *Theory of Syntactic Recognition for Natural Language*. MIT Press, Cambridge, MA, 1980.

[210] H. Margenau. *The Nature of Physical Reality*. Yale University Press, New Haven, CT, 1958.

[211] J. D. Markel and A. H. Gray. *Linear Prediction of Speech*. Springer-Verlag, New York, 1976.

[212] A. A. Markov. An example of statistical investigation in the text of "Eugene Onyegin" illustrating coupling of "tests" in chains. In *Proc. Acad. Sci*, volume 7, pages 153–162, St. Petersburg, 1913.

[213] J. C. Marshall. Minds, machines and metaphors. *Soc. Stud. Sci.*, 7: 475–488, 1977.

[214] T. R. McCalla. *Introduction to Numerical Methods and FORTRAN Programming*. Wiley, New York, 1967.

[215] W. S. McCulloch and W. Pitts. A logical calculus of the ideas imminent in nervous activity. *Bull. Math. Biophys.*, 5:115–133, 1943.

[216] C. McGinn. *The Mysterious Flame*. Basic Books, New York, 1999.

[217] W. S. Meisel. *Computer Oriented Approaches to Pattern Recognition*. Academic Press, New York, 1972.

[218] C. Merchant. *The Death of Nature: Women, Ecology and the Scientific Revolution*. Harper and Row, New York, 1980.

[219] G. Mercier, A. Nouhen, P. Quinton, and J. Siroux. The keal speech understanding system. In J. C. Simon, editor, *Spoken Language Generation and Understanding*, pages 525–544. D. Reidel, Dordrecht.

[220] G. A. Miller, G. A. Heise, and Lichten. The intelligibility of speech as a function of the context of the test materials. *J. Exp. Psych.*, 41:329–335, 1951.

[221] M. Minsky and S. Papert. *Perceptrons: An Introduction to Computational Geometry*. MIT Press, Cambridge, MA, 1969.

[222] M. L. Minsky. Matter, mind and models. In *Semantic Information Processing*, pages 425–432. MIT Press, Cambridge, MA, 1968.

[223] P. M. Morse and K. U. Ingard. *Theoretical Acoustics*. McGraw-Hill, New York, 1968.

[224] C. S. Myers and S. E. Levinson. Speaker independent connected word recognition using a syntax directed dynamic programming procedure. *IEEE Trans. Acoust. Speech Signal Process.*, ASSP-30:561–565, 1982.

[225] C. S. Myers and L. R. Rabiner. Connected digit recognition using a level building DTW algorithm. *IEEE Trans. Acoust. Speech Signal Process.*, ASSP-29:351–363, 1981.

[226] C. S. Myers and L. R. Rabiner. A level building dynamic time warping algorithm for connected word recognition. *IEEE Trans. Acoust. Speech Signal Process.*, ASSP-29:284–297, 1981.

[227] A. Nadas. Hidden Markov chains, the forward-backward algorithm and initial statistics. *IEEE Trans. Acoust. Speech Signal Process.*, ASSP-31:504–506, April 1983.

[228] A. Nadas. Estimation of probabilities in the language model of the IBM speech recognition system. *IEEE Trans. Acoust. Speech Signal Process.*, ASSP-32:859–861, 1984.

[229] E. Nagel and J. R. Newman. *Gödel's Proof.* New York University Press, New York, 1958.

[230] G. Nagy. State of the art in pattern recognition. *Proc. IEEE*, 56:836–60, 1967.

[231] R. Nakatsu and M. Kohda. An acoustic processor in a conversational speech recognition system. *Rev. ECL*, 26:1505–1520, 1978.

[232] W. L. Nelson. Physical principles for economies of skilled movements. *Biol. Cybern.*, 46:135–147, 1983.

[233] A. Newell, J. Barnett, J. Forgie, C. Green, D. Klatt, J. C. R. Licklieder, J. Munson, R. Reddy, and W. Woods. *Speech Understanding Systems – Final Report of a Study Group.* North Holland, Amsterdam, 1973.

[234] H. Ney. The use of a one stage dynamic programming algorithm for connected word recognition. *IEEE Trans. Acoust. Speech Signal Process.*, ASSP-32:263–271, 1984.

[235] N. Nilsson. *Problem Solving Methods in Artificial Intelligence.* McGraw-Hill, New York, 1971.

[236] J. P. Olive. Rule synthesis of speech from dyadic units. In *Proc. IEEE Int. Conf. on Acoustics, Speech, and Signal Processing*, pages 568–570, Hartford, CT, 1977.

[237] J. P. Olive. A real time phonetic synthesizer. *J. Acoust. Soc. Amer.*, 66:663–673, 1981.

[238] B. T. Oshika. Phonological rule testing of conversational speech. In *Proc. IEEE Int. Conf. on Acoustics, Speech, and Signal Processing*, page 577, Philadelphia, PA, 1976.

[239] B. T. Oshika, V. W. Zue, R. V. Weeks, H. Nue, and J. Auerbach. The role of phonological rules in speech understanding research. *IEEE Trans. Acoust. Speech, Signal Process.*, ASSP-23:104–112, 1975.

[240] D. S. Passman. The Jacobian of a growth transformation. *Pacific J. Math.*, 44:281–290, 1973.

[241] E. A. Patrick. *Fundamentals of Pattern Recognition.* Prentice Hall, Englewood Cliffs, NJ, 1972.

[242] E. A. Patrick and F. P. Fischer. A generalized k-nearest neighbor rule. *Inform. Control*, 16:128–152, 1970.

[243] A. Paz. *Introduction to Probabilistic Automata.* Academic Press, New York, 1971.

[244] C. S. Peirce. *Collected Papers of Charles Sanders Peirce*, C. Hartstone and P. Weiss, eds. Harvard University Press, Cambridge, MA, 1935.

[245] R. Penrose. *The Emporor's New Mind.* Oxford University Press, 1990.

[246] G. Perennou. The Arial II speech recognition system. In J. P. Haton, editor, *Automatic Speech Analysis and Recognition*, pages 269–275. Reidel, Dordrecht, 1982.

[247] G. E. Peterson and H. L. Barney. Control methods used in a study of the vowels. *J. Acoust. Soc. Amer.*, 24:175–185, 1952.

[248] Plato. *The Republic.* Penguin, Harmondsworth, 1955.

[249] H. Poincaré. *Discovery in Mathematical Physics.* Dover, New York, 1960.

[250] A. B. Poritz. Linear predictive hidden markov models. In J. D. Ferguson, editor, *Proc. Symp. on the Application of Hidden Markov Models to Text and Speech*, pages 88–142, Princeton, NJ, 1980.

[251] A. B. Poritz. Linear predictive hidden Markov models and the speech signal. In *Proc. Int. Conf. on Acoustics, Speech, and Signal Processing*, pages 1291–1294, Paris, France, 1982.

[252] M. R. Portnoff. A quasi-one-dimensional digital simulation for the time varying vocal tract. Master's thesis, MIT, 1973.

[253] R. K. Potter, G. A. Koop, and H. G. Kopp. *Visible Speech.* Dover, New York, 1968.

[254] R. Quillian. Semantic nets. In M. Minsky, editor, *Semantic Information Processing.* MIT Press, Cambridge, MA, 1968.

[255] L. R. Rabiner. On creating reference templates for speaker independent recognition of isolated words. *IEEE Trans. Acoust. Speech Signal Process.*, ASSP-26:34–42, 1978.

[256] L. R. Rabiner, A. Bergh, and J. G. Wilpon. An improved training procedure for connected digit recognition. *Bell Syst. Tech. J.*, 61:981–1001, 1982.

[257] L. R. Rabiner and B.-H. Juang. *Fundamentals of Speech Recognition.* Prentice Hall, Englewood Cliffs, NJ, 1993.

[258] L. R. Rabiner, B. H. Juang, S. E. Levinson, and M. M. Sondhi. Recognition of isolated digits using hidden Markov models with continuous mixture densities. *AT & T Tech. J.*, 64:1211–1234, 1985.

[259] L. R. Rabiner and S. E. Levinson. Isolated and connected word recognition – theory and selected applications. *IEEE Trans. Commun.*, COM-29:621–659, 1981.

[260] L. R. Rabiner and S. E. Levinson. A speaker independent syntax directed connected word recognition system based on hidden Markov models and level building. *IEEE Trans. Acoust. Speech Signal Process.*, ASSP-33(3):561–573, 1985.

[261] L. R. Rabiner, S. E. Levinson, A. E. Rosenberg, and J. G. Wilpon. Speaker-independent recognition of isolated words using clustering techniques. *IEEE Trans. Acoust. Speech Signal Process.*, ASSP-27:336–349, 1979.

[262] L. R. Rabiner, S. E. Levinson, and M. M. Sondhi. On the application of vector quantization and hidden Markov models to speaker independent isolated word recognition. *Bell Syst. Tech. J.*, 62:1075–1105, 1983.

[263] L. R. Rabiner, S. E. Levinson, and M. M. Sondhi. On the use of hidden Markov models for speaker independent recognition of isolated words from a medium-size vocabulary. *AT & T Bell Lab. Tech. J.*, 63:627–642, 1984.

[264] L. R. Rabiner, A. E. Rosenberg, and S. E. Levinson. Considerations in dynamic time warping for discrete word recognition. *IEEE Trans. Acoust. Speech Signal Process.*, ASSP-26:575–582, 1978.

[265] L. R. Rabiner and R. W. Schafer. *Digital Processing of Speech Signals*. Prentice Hall, Englewood Cliffs, NJ, 1978.

[266] L. R. Rabiner and C. E. Schmidt. Application of dynamic time warping to connected digit recognition. *IEEE Trans. Acoust. Speech Signal Process.*, ASSP-28:337–388, 1980.

[267] L. R. Rabiner, M. M. Sondhi, and S. E. Levinson. A vector quantizer combining energy and LPC parameters and its application to isolated word recognition. *AT & T Bell Lab. Tech. J.*, 63:721–735, 1984.

[268] L. R. Rabiner and J. G. Wilpon. Application of clustering techniques to speaker-trained word recognition. *Bell Syst. Tech. J.*, 5:2217–2233, 1979.

[269] L. R. Rabiner and J. G. Wilpon. Considerations in applying clustering techniques to speaker independent word recognition. *J. Acoust. Soc. Amer.*, 66:663–673, 1979.

[270] L. R. Rabiner and J. G. Wilpon. A simplified robust training procedure for speaker trained isolated word recognition systems. *J. Acoust. Soc. Amer.*, 68:1271–1276, 1980.

[271] L. R. Rabiner, J. G. Wilpon, and J. G. Ackenhusen. On the effects of varying analysis parameters on an LPC-based isolated word recognizer. *Bell Syst. Tech. J.*, 60:893–911, 1981.

[272] D. R. Reddy. Computer recognition of connected speech. *J. Acoust. Soc. Amer.*, 42:329–347, 1967.

[273] C. Reid. *Hilbert*. Springer-Verlag, Berlin, 1970.

[274] M. D. Riley. *Speech Time Frequency Representations*. Kluwer Academic, Boston, 1989.

[275] A. Robins and S. McCallum. The consolidation of learning during sleep: Comparing pseudorehearsal and unlearning accounts. *Neural Networks*, 12:1191–1206, 1999.

[276] D. B. Roe and R. Sproat. The VEST spoken language translation system. In *Proc. ICASSP-93*, May 1993.

[277] J. B. Rosen. The gradient projection method for nonlinear programming – Part I: Linear constraints. *J. Soc. Indust. Appl. Math.*, 8:181–217, 1960.

[278] A. E. Rosenberg and F. Itakura. Evaluation of an automatic word recognition over dialed-up telephone lines. *J. Acoust. Soc. Amer.*, 60, Suppl. 1:512, 1976.

[279] A. E. Rosenberg, L. R. Rabiner, J. G. Wilpon, and D. Kahn. Demisyllable based isolated word recognition system. *IEEE Trans. Acoust. Speech Signal Process.*, ASSP-31:713–726, 1983.

[280] F. Rosenblatt. *Principles of Neurodynamics: Perceptrons and the Theory of Brain Mechanisms*. Spartan, Washington, DC, 1962.

[281] L. H. Rosenthal, L. R. Rabiner, R. W. Schafer, P. Cummiskey, and J. L. Flanagan. A multiline computer voice response system using ADPCM coded speech. *IEEE Trans. Acoust. Speech Signal Process.*, ASSP-22:339–352, 1974.

[282] D. E. Rumelhart, G. E. Hinton, and R. J. Williams. Learning internal representations by error propagation. In D. E. Rumelhart and J. L. McClelland, *Parallel Distributed Processing: Explorations in the Microstructure of Cognition*. MIT Press, 1986.

[283] D. E. Rumelhart and J. L. McClelland. *Parallel Distributed Processing*. MIT, Cambridge, MA, 1986.

[284] G. Ruske and T. Schotola. The efficiency of demisyllable segmentation in the recognition of spoken words. In J. P. Haton, editor, *Automatic Speech Analysis and Recognition*, pages 153–163. Reidel, Dordrecht, 1982.

[285] B. Russell and A. N. Whitehead. *Principia Mathematica*. Cambridge University Press, Cambridge, 1962.

[286] M. J. Russell and R. K. Moore. Explicit modeling of state occupancy in hidden Markov models for automatic speech recognition. In *Proc. IEEE Int. Conf. on Acoustics, Speech, and Signal Processing*, pages 5–8, Tampa, FL, March 1985.

[287] M. J. Russell, R. K. Moore, and M. J. Tomlinson. Some techniques for incorporating local timescale information into a dynamic time warping algorithm for automatic speech recognition. In *Proc. IEEE Int. Conf. on Acoustics, Speech, and Signal Processing*, pages 1037–1040, Boston, MA, 1983.

[288] H. Sakoe. Two level DP-matching – a dynamic programming based pattern matching algorithm for connected word recognition. *IEEE Trans. Acoust. Speech Signal Process.*, ASSP-27:588–595, 1979.

[289] H. Sakoe and S. Chiba. A dynamic programming approach to continuous speech recognition. In *Proc. 7th Int. Congr. on Acoustics*, volume 3, pages 65–68, Budapest, 1971.

[290] H. Sakoe and S. Chiba. Dynamic programming algorithm optimization for spoken word recognition. *IEEE Trans. Acoust. Speech Signal Process.*, ASSP-26:43–49, 1978.

[291] C. Scagliola. Continuous speech recognition without segmentation: Two ways of using diphones as basic speech units. *Speech Commun.*, 2:199–201, 1983.

[292] C. E. Schorske. *Fin-de-Siècle Vienna: Politics and Culture*. Knopf, New York, 1979.

[293] J. R. Searle. Minds, brains and programs. *Behavioral and Brain Sciences*, 3:417–457, 1980.

[294] G. Sebestyen. *Decision Making Processes in Pattern Recognition*. McMillan, New York, 1962.

[295] C. E. Shannon. A mathematical theory of communication. *Bell Syst. Tech. J.*, 27:379–423, 1948.

[296] J. F. Shapiro. *Mathematical Programming Structures and Algorithms*. Wiley, New York, 1979.

[297] K. Shikano and M. Kohda. A linguistic processor in a conversational speech recognition system. *Trans. Inst. Elec. Commun. Eng. Japan*, E61:342–343, 1978.

[298] D. W. Shipman and V. W. Zue. Properties of large lexicons: Implications for advanced isolated word recognition systems. In *Proc. Int. Conf. on Acoustics, Speech, and Signal Processing*, pages 546–549, Paris, France, 1982.

[299] J. E. Shore and R. W. Johnson. Axiomatic derivation of the principle of maximum entropy and the principle of minimum cross entropy. *IEEE Trans. Inform. Theory*, IT-26:26–36, 1980.

[300] H. F. Silverman and N. R. Dixon. A parametrically controlled spectral analysis system for speech. *IEEE Trans. Acoust. Speech Signal Process.*, ASSP-23:369–381, 1974.

[301] J. C. Simon. *Patterns and Operators*. McGraw Hill, New York, 1986.

[302] A. D. Sokal. Transgressing the boundaries: Towards a transformative hermeneutics of quantum gravity. *Social Text*, 46/47:217–252, 1996.

[303] M. M. Sondhi. Model for wave propagation in a lossy vocal tract. *J. Acoust. Soc. Amer.*, 55:1070–1075, 1974.

[304] M. M. Sondhi. Estimation of vocal-tract areas: The need for acoustical measurements. *IEEE Trans. Acoust. Speech Signal Process.*, ASSP-27:268–273, 1979.

[305] M. M. Sondhi and S. E. Levinson. Computing relative redundancy to measure grammatical constraint in speech recognition tasks. In *Proc. Int. Conf. on Acoustics, Speech, and Signal Processing*, pages 409–412, Tulsa, OK, 1978.

[306] H. W. Sorenson and D. L. Alspach. Recursive Bayesian estimation using Gaussian sums. *Automatica*, 7:465–479, 1971.

[307] P. F. Stebe. Invariant functions of an iterative process for maximization of a polynomial. *Pacific. J. Math.*, 43:765–783, 1972.

[308] K. N. Stevens. Acoustic correlates of some phonetic categories. *J. Acoust. Soc. Amer.*, 68:836–842, 1980.

[309] I. Tan. Unpublished MSEE thesis, University of Illinois at Urbana-Champaign, June 2002.

[310] E. Tanaka and K. S. Fu. Error correcting parsers for formal languages. *IEEE Trans. Comput.*, C-27:605–615, 1978.

[311] M. Tanenhaus, M. J. Spivey-Knowlton, K. Eberhard, and J. C. Sedivy. Integration of visual and linguistic information in spoken language comprehension. *Science*, 268:1632–1634, 1995.

[312] M. Tanenhaus, M. J. Spivey-Knowlton, K. Eberhard, and J. C. Sedivy. Using eye movements to study spoken language comprehension: Evidence for visually mediated incremental interpretation. In T. Inui and J. L. McClelland, editors, *Attention and Performance 16: Information Integration in Perception and Communication*, pages 457–478. MIT Press, 1996.

[313] C. C. Tappert. A Markov model acoustic phonetic component for automatic speech recognition. *Int. J. Man-Machine Studies*, 9:363–373, 1977.

[314] Y. Tohkura. Features for speech recognition. In *Proc. IEEE Int. Conf. on Acoustics, Speech, and Signal Processing*, New York, NY, May 1983.

[315] Michael Tomasello. *The Cultural Origins of Human Cognition*. Harvard University Press, Cambridge, MA, 1999.

[316] M. Tomita. *Efficient Parsing for Natural Language*. Kluwer Academic, Boston, 1985.

[317] J. T. Tou and R. C. Gonzalez. *Pattern Recognition Principles*. Addison-Wesley, Reading, MA, 1974.

[318] J. M. Tribolet, L. R. Rabiner, and M. M. Sondhi. Statistical properties of an LPC distance measure. *IEEE Trans. Acoust. Speech Signal Process.*, ASSP-27:550–558, 1979.

[319] A. M. Turing. On computable numbers with an application to the Entscheidungsproblem. *Proc. London Math. Soc.*, 42: 230–265, 1937.

[320] A. M. Turing. Computing machinery and intelligence. *Mind*, pages 433–460. 1950.

[321] V. M. Velichko and N. G. Zagoruyko. Automatic recognition of 200 words. *Int. J. Man-Machine Studies*, 2:223–234, 1969.

[322] T. K. Vintsyuk. Recognition of words of oral speech by dynamic programming methods. *Kibernetika*, 81(8), 1968.

[323] A. J. Viterbi. Error bounds for convolutional codes and an asymptotically optimal algorithm. *IEEE Trans. Inform. Theory*, IT-13, March 1967.

[324] J. Von Neumann and O. Morgenstern. *The Theory of Games*. Princeton University Press, Princeton, NJ, 1950.

[325] H. Wakita. Direct estimation of the vocal tract shape by inverse filtering of acoustic speech waveforms. *IEEE Trans. Audio Electroacoust.*, AU-21:417–427, 1973.

[326] D. E. Walker. The SRI speech understanding system. *IEEE Trans. Acoust. Speech Signal Process.*, ASSP-23:397–416, 1975.

[327] A. G. Webster. Acoustical impedance and the theory of horns. In *Proc. Nat. Acad. Sci.*, volume 5, pages 275–282, 1919.

[328] E. P. Wigner. On the unreasonable success of mathematics in physics. *Comm. on Pure and Applic. Math.*, 13(1), 1960.

[329] R. Weinstock. *Calculus of Variations with Applications to Physics and Engineering*. Dover, New York, 1974.

[330] N. Wiener. *Cybernetics or Control and Communication in the Animal and Machine*. MIT Press, Cambridge, MA, 2nd edition, 1948.

[331] E. O. Wilson. *Consilience: The Unity of Knowledge*. Knopf, New York, 1998.

[332] J. J. Wolf and W. A. Woods. The HWIM speech understanding system. In *IEEE Int. Conf. on Acoust., Speech, Signal Processing*, pages 784–787, Hartford, CT, 1977.

[333] W. A. Woods. Transition network grammar for natural language analysis. *Commun. ACM*, 13:591–602, 1970.

[334] W. A. Woods. Motivation and overview of SPEECHLIS: An experimental prototype for speech understanding research. *IEEE Trans. Acoust. Speech Signal Process.*, ASSP-23:2–10, 1975.

[335] W. A. Woods. Syntax semantics and speech. In D. R. Reddy, editor, *Speech Recognition*. Academic Press, New York, 1975.

[336] W. A. Woods. Optimal search strategies for speech understanding and control. *Artificial Intelligence*, 18:295–326, 1982.

[337] W. A. Woods, M. A. Bates, B. C. Bruce, J. J. Colarusso, C. C. Cook, L. Gould, J. A. Makhoul, B. L. Nash-Webber, R. M. Schwartz, and J. J. Wolf. Speech understanding research at BBN. Technical Report 2976, Bolt, Beranek and Newman, 1974. Unpublished.

[338] D. M. Younger. Recognition and parsing of context free languages in time n^3. *Inform. Control*, 10, 1967.

[339] P. L. Zador. Topics in the asymptotic quantization of continuous random variables. Technical report, Bell Labs, 1966.

[340] P. L. Zador. Asymptotic quantization error of continuous signals and the quantization dimension. *IEEE Trans. Inform. Theory*, IT-28:139–148, March 1982.

[341] W. Zhu and S. E. Levinson. Edge orientation-based multiview object recognition. In *Proc. IEEE Int. Conf. on Pattern Recognition*, Vol. 1, pages 936–939, Barcelona, Spain, 2000.

[342] W. Zhu and S. E. Levinson. PQ-learning: an efficient robot learning method for intelligent behavior acquisition. In *Proc. 7th Int. Conf. on Intelligent Autonomous Systems*, pages 404–411, Marina del Rey, CA, March 2002.

[343] W. Zhu, S. Wang, R. Lin, and S. E. Levinson. Tracking of object with SVM regression. In *Proc. IEEE Int. Conf. on Computer Vision and Pattern Recognition*, pages 240–245, Hawaii, USA, 2001.

[344] V. W. Zue. The use of phonetic rules in automatic speech recognition. *Speech Commun.*, 2:181–186, 1983.

Index

accepting state, 124
acoustic pattern, 27
acoustic tube, 10
acoustic-phonetic model, 162
active constraint, 70
adiabatic constant, 13
allophone, 158
allophonic variation, 157
ambiguity function, 25
area function, 22
ARPA, 2
ARPAbet, 53
articulation
  manner of, 52, 54
  place of, 52, 54
articulator, 52
articulatory configuration, 58
articulatory mechanism, 99
articulatory synthesis, 169
Artificial Intelligence, 213
associative memory, 230, 232, 238
asynchronous parsing methods, 130
autocorrelation function, 22
autocorrelation matrix, 87
autonomous robot, 235
autoregressive coefficients, 106
autrogressive process, 87
auxiliary function, 83, 104
average sentence length, 148

backward likelihood, 91
backward probability, 59
Baker's algorithm, 140
Baum algorithm, 122, 137, 166
  geometry of, 67
Baum, L., 3, 63
Baum-Welch algorithm, 62
Baum-Welch reestimation, 65
Bayes law, 32
Bayesian decision, 50
beam search, 160
behaviorism, 233
best-first algorithm, 160
binary tree, 112
binomial distribution, 37
boundary conditions, 14
broad phonetic categories, 52, 109

Cantor, G., 205, 206
categorical perception, 27
Cave–Neuwirth experiment, 107, 109
center embedding, 118
central limit theorem, 35
cepstral coefficients, 23
characteristic grammar, 114
Cholesky factorization, 107
Chomsky hierarchy, 110
Chomsky normal form, 111
Chomsky, N., 2, 110, 234
Church-Turing hypothesis, 205, 212
class membership, 30
class-conditional probability, 30
classification error, 33
clustering, 42
coarticulation, 54
Cocke-Kasami-Younger algorithm, 12

Mathematical Models for Speech Technology. Stephen Levinson
 ISBN: 0-470-84407-8

code, 143
combinatorial optimization, 50
communication process, 175
connectionism, 43
connectivity matrix, 145, 150
consciousness, 229
consistent estimator, 37
consistent grammar, 114
constrained optimization, 64
constructive theory of mind, 212, 228
context sensitive language, 110
context-free language, 110
continuous observations, 80
continuum hypothesis, 207, 210
convective air flow, 16
correlated observations, 88
correlation matrix, 22
cost function, 33
covariance matrix, 34
critical point, 66, 69
cross entropy, 83
cybernetic paradigm, 202
cybernetics, 6, 199
CYC project, 216

decideability problem, 200
decision rule, 30
decoding error, 144, 154
density
  beta, 35, 37
  binomial, 35
  Cauchy, 35
  chi-squared, 35
  Laplace, 35
  Poisson, 35
derivation operator, 109
determinism, 225
diagonalization, 206
dialog, 56
digamma function, 96
directed graph, 112
discrete acoustic tube model, 12
discrete Fourier transform, 19
distribution
  gamma, 89
duality, 225
Dudley, H., 5
durational parameter, 95
dynamic programming, 47, 59, 120, 123, 126, 163
dynamic time warping, 49

effective procedure, 214
effective vocabulary, 155
efficiency, 155
eigenvalues, 41, 151
eigenvectors, 41, 151
Einstein, A., 222, 226, 227
elliptical symmetry, 82
EM algorithm, 107, 138
embodied mind, 231
emergence, 220
ENIAC, 216
entropy, 143
epsilon representation, 153
equivocation, 154
Euler's constant, 96
  theorem of, 68
Euler-Lagrange equation, 46
event counting, 139
expectation operator, 34, 107
expected value, 34

Fano bound, 154
finite language, 111
finite regular language, 126
finite state automaton, 115
finite training set, 75
first moment matrix, 152
first-order predicate calculus, 190, 192
fixed point, 69
Flanagan, J. L., 9, 14
Fletcher, H., 5
fluent speech, 166
fluid dynamics, 16
formal grammar, 5, 109
formant bandwidths, 13, 23
formant frequencies, 11, 22, 54
forward likelihood, 91
forward probability, 59, 102
forward-backward algorithm, 59
foundations of mathematics, 190
free-will, 225

fricatives, 52
functional equivalence, 213

Gaussian density function, 34
Gaussian mixture, 35, 81, 85
generating function, 111, 152
Godel numbering, 208, 209
Godel, K., 205, 207
Gold's theorem, 137
grammar, 52
grammatical constraint, 144, 148, 150
grammatical inference, 137
grammatical rules, 109
growth transformation, 66

heat capacity, 13
Hessian, 70
heuristic function, 160
hidden Markov model, 2, 57
Hilbert, D., 44, 200, 205
Hodgkin Huxley model, 43
homogeneous polynomial, 59, 66
human machine communication
  error recovery in, 183
human-machine communication system, 188

impredicative set, 205
incompleteness theorem, 205, 207, 208
inequality constraint, 76
inflected forms, 167
inside probability, 140
instinctual behavior, 238
introspection, 215
Ionian Enchantment, 221
Itakura method, 49

Japanese phonotactics, 132
Jelinek, F., 2

k-means algorithm, 42
Kolmogorov representation theorem, 43
Kuhn–Tucker theorem, 70
Kullback–Leibler statistic, 83, 163

Lagrange multiplier, 41, 64, 65, 149
lambda calculus, 195, 205
language engine
  integrated, 157
  modular, 157
language identification, 157, 170
language translation, 157, 171
left-to-right HMM, 77
letter-to-sound rules, 167
lexical access, 162
lexical assignment rules, 134, 140
lexical semantics, 193
lexicon, 157
liar's paradox, 208
likelihood function, 83, 84
linear prediction, 21
linear prediction coefficients, 21
linear prediction transfer function, 21
linguistic structure, 51
logical inference, 193
logical operators, 191
logical quantifier, 192
loss function, 33
LPC power spectrum, 23

Mahalanobis distance, 36
majority vote rule, 31
manifold, 65
Markov process
  entropy of, 147
  stationary probabilities of, 151
mathematical realism, 223
mathematical truth, 190
maximum a posteriority probability, 33
mean vector, 34
mel-scale cepstrum, 23
membership question, 116
metric, 30
  Chebychev, 31
  Euclidean, 31
  Hamming, 31
Miller, G., 143, 156
mind-body dichotomy, 225
minimax criterion, 43
minimum risk rule, 33
Minkowski *p*-metric, 31
modular architecture, 161
morphology, 55

multi-layer perceptron, 43
mutual information, 189

n-gram statistics, 133
nasals, 52
Navier-Stokes equations, 17
nearest neighbor distance, 31
nearest neighbor rule, 39
neural network, 43
Newton technique, 70
Newton's method, 96
non-ergodic HMM, 77, 142
non-parametric decision rule, 35
non-parametric estimator, 50
non-stationarity, 16, 25, 57, 99
non-terminal symbol, 109
null symbol, 109
Nyquist rate, 20

observation sequence, 59
ontology, 193, 197
operator precedence, 191
optimal decision rule, 34
optimal decoder, 144
orthogonal polynomial, 35
orthonormality constraint, 41
outside probability, 141

paradigm shift, 241
parameter estimation, 58
parametric decision rule, 35
PARCORs, 22
parse tree, 139
parsing, 119
parts-of-speech, 117
Parzen estimator, 35
pattern recognition, 28
perceptron, 43
philosophy of mind, 201
phonetic inventory, 53
phonology, 52, 116
phonotactics, 52, 55, 109
phrase structure language, 110
Pierce, J., 2
Platonic theory of forms, 27, 202
plosives, 52
point risk, 34
point-set distance, 29
point-spread function, 25
polygamma function, 96
Poritz experiment, 107, 109
Poritz, A., 87
postmodernism, 226
potential function, 35
power spectrum, 15, 19
pragmatic information, 148
pragmatics, 56
predicate
  logical, 191
predicate argument structure, 195
prior probability, 33
priority queue, 160
probability density function, 33
production rules, 109
pronouncing dictionary, 157
propositional logic, 190
prosody, 55
prototype, 30
psychological truth, 190
push-down automaton, 115

quantization, 42, 58
quasi-stationarity, 99, 159

radiation impedance, 14
reasoning, 193
recursion equation for Webster equation, 15
reductionism, 220
reestimation formula, 62, 66, 93, 104–106
reflection coefficients, 22
regression coefficient, 99
regular grammar, 111, 112
regular language, 110
  entropy of, 150
reinforcement learning, 230, 232
relative frequency, 145
relative redundancy, 144, 150
residual error, 22
resonance frequencies, 15
rewriting rules, 109
Richard paradox, 208

Riemann zeta function, 96
right-linear language, 119

scaling, 73, 78, 97
Schorske, C., 214
scientific method, 222
self-punctuating code, 148
semantic analysis, 174
semantic category, 180
semantic information, 148
semantic net, 190
semantics, 55, 234
  natural language, 189
semi-Markov process, 88
sensorimotor function, 230
separatrix, 84
Shannon, C., 143, 203
short duration amplitude spectrum, 17
short time Fourier transform, 19
singular matrix, 86
sociobiology, 221
Sokal hoax, 227
sound spectrograph, 1
source-filter model, 1, 17
source-filter model diagram, 19
spectrogram, 19, 20, 24
speech coding, 170
speech understanding, 3, 173
start symbol, 109
state duration, 88
state sequence, 61
state transition diagram, 124, 144
state transition matrix, 58
stationarity, 44
statistical decision theory, 28
stochastic grammar, 113
string derivation, 109
strong theory of AI, 213, 228
structure building rules, 134, 140
sub-word model, 157
sufficient statistic, 107
syntactico-semantic relationship, 189, 194
syntax, 55, 116

terminal symbol, 109
test set, 30
Text analysis, 167
text-to-speech synthesis, 167, 168
thermal conduction, 12
thermal losses, 11
thermos bottle parable, 219
time scale normalization, 108
time-frequency resolution, 25
Toeplitz matrix, 22, 87
tolerance region, 37
training set, 29
transfinite numbers, 206
transformation
  information preserving, 39
transitive closure, 109
trellis, 160
triphone, 157, 158
Turing machine, 210
Turing test, 228
Turing, A., 6, 200, 205, 217
Turing-equivalence, 119
two-dimensional Fourier transform, 24

unambiguous grammar, 114, 144
undecideability theorem, 211
universal Turing machine, 213, 235

variable duration model, 91
variational problem, 46
viscous friction, 12
viscous losses, 11
Viterbi algorithm, 74, 120, 127
vocabulary, 59
vocal tract constriction, 52
vocal tract model, 9
vocal tract transfer function, 15
vocoder, 1
voice-operated typewriter, 173
voicing, 53
vowels, 52

Webster equation, 10
Wiener, N., 6, 199
Wigner transform, 24
Wigner, E., 222
word lattice, 131

RECEIVED
MAY 26 2005